TRANSOCEANIC STUDIES
Ileana Rodríguez, Series Editor

IN SEARCH OF AN ALTERNATIVE BIOPOLITICS

ANTI-BULLFIGHTING, ANIMALITY,
AND THE ENVIRONMENT
IN CONTEMPORARY SPAIN

KATARZYNA OLGA BEILIN

THE OHIO STATE UNIVERSITY PRESS | COLUMBUS

Copyright © 2015 by The Ohio State University.
All rights reserved.

Library of Congress Cataloging-in-Publication Data
Beilin, Katarzyna Olga, 1966– author.
  In search of an alternative biopolitics : anti-bullfighting, animality, and the environment in contemporary Spain / Katarzyna Olga Beilin.
    pages cm — (Transoceanic studies)
  Includes bibliographical references and index.
  ISBN 978-0-8142-1290-5 (cloth : alk. paper)
  1. Bullfights—Political aspects—Spain—21st century. 2. Animal rights—Spain—21st century. 3. Biopolitics—Spain—21st century. I. Title. II. Series: Transoceanic studies.
  GV1108.7.M67B45 2015
  791.8'2—dc23
                    2015025468

Cover design by James A. Baumann
Text design by Juliet Williams
Type set in Adobe Minion Pro
Printed by Thomson-Shore, Inc.

♾ The paper used in this publication meets the minimum requirements of the American National Standard for Information Sciences—Permanence of Paper for Printed Library Materials. ANSI Z39.48-1992.

9 8 7 6 5 4 3 2 1

For Sai

CONTENTS

List of Illustrations • ix

Acknowledgments • xi

Preface • xv

INTRODUCTION
Bulls, Apes, Genes, and Clouds • 1

CHAPTER 1
Bullfighting Biopolitics and Debates on National Culture • 49

CHAPTER 2
Anti-Bullfighting Activists as Marginal Intellectuals • 78

CHAPTER 3
Science, Politics, and Animals:
Tiempo de Silencio as an Anti-Bullfighting Novel • 106

CHAPTER 4
Meanings of Animalization and Humanity as Anthropomorphism • 129

CHAPTER 5
Animal Rights Movement for an Alternative Biopolitics • 159

CHAPTER 6
Bullfighting and the War on Terror:
Debates on Culture and Torture in Spain, 2004–11 • 187

CHAPTER 7
Die or Laugh: Biopolitical Crisis in *Biutiful* and *Nocilla experience* • 211

CHAPTER 8
Debates on GMOs in Spain and Rosa Montero's *Lágrimas en la lluvia* (with Sainath Suryanarayanan) • 235

CONCLUSION
In Search of Alternative Biopolitics • 262

Works Cited • 273

Index • 297

ILLUSTRATIONS

FIGURE I.1
Activists of Fundacion Ecuanimal and WSPA-International in Bilbao • 2

FIGURE 1.1
Cover of Eugenio Noel's *La novela de un pueblo en capea* • 56

FIGURE 5.1
Cartoon from *El País*, by Forges: "Señas/Sañas de identidad" • 161

FIGURE 5.2
Activists at the International Day of Animal Rights, Plaza Sol, Madrid • 172

FIGURE 5.3
An activist crying over a dead piglet during the International Day of Animal Rights, Plaza Sol, Madrid • 173

FIGURE 5.4
Alaska poses in front of the poster designed by Juan Gatti, "La verdad al desnudo; Tauromaquia es cruel" (The Naked Truth: Bullfighting Is Cruel) • 176

FIGURE 5.5
"Eau de toroture" by Albert Riera • 180

FIGURE 7.1
Wet stain growing on the ceiling of Uxbal's room • 217

FIGURE 7.2
Screenshot from *Biutiful* showing smokestakes of Sant Adrià de Besòs • 217

FIGURES 7.3A AND B
Comparison between Uxbal's body and the city • 221

FIGURE 7.4
A shark made of hundred-euro bills • 222

FIGURE C.1
Ramón Rodríguez's vision of the king's abdication • 264

ACKNOWLEDGMENTS

THIS BOOK owes its existence to many humans and nonhumans, but it would not have been conceived without Sai, who gave me ideas, convictions, and strength. He is also a coauthor of the last chapter of this book, whose somewhat different version had been written by both of us as an article for *Hispanic Issues*, but also helped me think through various other sections. Nothing would have been possible without children: Felix, whose moral strength has kept our family vegetarian, and Mundo, whose emotional strength keeps us in peace. I am grateful to my parents, Barbara and Wiesław Szymczyk, for continuous support, for taking care of the boys while I wrote, and for being great friends.

Additionally, I could not have finished the manuscript without the generous, friendly, and instructive editorial help of Nancy Soth, from whom I learned a lot. Jorge Riechmann, Paula Casal, Joaquín Araujo, Pablo de Lora, Oscar Horta, Fernando Turró, Carmen y Manel Cases, Albert Riera, and the activists of Ecuanimal, Igualdad Animal, and ADDA taught me how to think about their struggle for an alternative biopolitics and how to understand it in a Spanish context. I feel an intellectual debt and gratitude for their time and thoughts. I need to say "thank you" to Andrea Gutiérrez, Alberto Martínez Rivas, and Ángela Montenegro for helping me to establish contacts in Barcelona. I am very grateful to Juan Egea, Luis Martín-Estudillo, John Beusterien,

Sebastiaan Faber, Jorge Riechmann, Paula Casal, Oscar Horta, Glen Close, and Mario Santana for the invaluable feedback they gave me on the manuscript or its parts. Thanks to Robert Nixon, Steven Hutchinson, Susan Friedman, and Michael Ugarte for supporting this project and helping me find grants to pursue it.

I also need to express my gratitude to the Institute for Research in the Humanities, the Center for European Studies, and the Graduate School at the University of Wisconsin, Madison, as well as the Spanish Ministry of Culture for honoring me with the grants and stipends that made my research and writing possible. Thanks to Ecuanimal for permission to use the photograph of the performance in front of Guggenheim Museum in Bilbao, to Albert Riera for his ingenious image of "Eau de Toroture," and to Antonio Fraguas Forges for permission to use his drawing of "Se/añas de identidad." Thanks to John Beusterien for the antic edition of Eugenio Noel's novel, whose front cover became one of my illustrations. Thanks to Hanwen Dong for his miracles in editing images. Thanks to the *International Journal of Iberian Studies*, to *Comparative Drama*, and to *Hispanic Issues* for permission to republish parts of the articles formerly printed in each. I also want to say "thank you" to the graduate students from the University of Wisconsin, Madison, whose contributions to the seminar discussions helped me think through this book. In particular, I am indebted to Daniel Ares López, Marilén Loyola, Joseph Patteson, and Jenny Jeong from whom I learned things that made it into this book. I began to think about Anthropocene during the discussions of the Faculty Seminar on Environmental Humanities led by Rob Nixon, who became my role model as an intellectual activist who fights for a better world by telling stories of other fights. To an extent, my book attempts to follow this path.

I have benefited intellectually from the exchanges during an Animal Studies Mellon Workshop led by Mario Ortiz in 2012. The writing collaborations that I have developed while working on this book with William Viestenz, Eugenia Afinoguenova, Luis Prádanos-García (Iñaki), Carmen Flys, Susan Larson, Matthew Feinberg, John Threvatan, Paul Begin, Sebastiaan Faber, Daniel Ares, Pablo de Lora, Tonia Raquejo, John Beusterien, Luis Martín-Estudillo, and Nicolas Spadaccini have been enormously stimulating and made work a joy. Finally, I want to say "thanks" to my department's consecutive chairs Rubén Medina and Luis Madureira for creating a wonderful intellectual atmosphere for research.

Among the nonhumans, my thanks go to the squirrels that chase each other through the trunks of the trees in my garden and across the electric wires as if they were still on tree branches, to the cardinals that sit at the railing of my terrace looking at me with one eye, and to a humming bird that appears like a hallucination over some flowers to disappear before anyone else sees him. I need to acknowledge bunnies that live under the terrace and eat our tulips and move their noses in a very funny way, the neighbor's cat, Krystynka, who would live with us if she could, the wasps in the window frame that never stung, and the generous maples that give me millions of leaves to rake and a hiding place for my little boy in the fall. Finally, thanks to the prairie in Hoyt Park were best ideas always come to my mind.

PREFACE

"Pensamiento renovador es siempre minoritario."
—Joaquín Araujo

EACH TIME I am carried away with enthusiasm, thinking that I am writing a book on a significant change that is taking place in Spanish contemporary culture, I remember my conversation with Jorge Riechmann, one of the most important intellectual leaders of this change. When I asked him about the cultural significance of the transformation, he looked at me seriously and said that change is affecting less than one percent of society, though that one percent think that they are the ninety-nine percent. In his opinion, a new understanding of the world that comes together with the awareness of the need to change happens only in microworlds of new social movements narcissistically constructed around their own websites, whose links to other microworlds are not frequently used. Riechmann's assessment may be connected to the result of the Spanish parliamentary elections in the fall of 2011, in which none of the parties related to the new social movements managed to collect above one percent of the votes ("Elecciones generales" 2012).

Numbers, however, are far too exact to measure changes in culture, society, and politics, which are often governed by surprising dynamics and where what seems impossible sometimes happens. Joaquín Araujo, another highly important figure in the Spanish Ecologist movement, told me in his apartment in Madrid that a thought that brings changes is always marginal, at first. He added that conditions most likely will have to get much worse in order for them to get better, but he implied that the change started by few today will have to spread and become a new norm sometime in the future, although no one knows how soon. For Araujo, ecological culture is reunifying and truly global in terms of its resistance to corporate domination. Although the fundamental rule of the ecological culture is respect for all sorts of life, its ramifications are diverse. They range from the desire to eat organic food to the belief that one should treat animals well, look for new sources of energy, decrease contamination, and change institutional relations. Araujo sees things connected, as he tells me, "El desprecio del animal es un modelo de pensamiento que lleva a la exarbación del consumiso, que lleva al cambio climático y que lleva a la destrucción en general" (The lack of respect for animals is a model of thought that leads to excessive consumerism, to climate change, and to destruction in general).[1] He believes that social thought should look for a big picture and that excessive specialization destroys life because it leads to seeing things separated from their context.

In another apartment in Barcelona, political philosopher and ICREA professor Paula Casal explained to me in a low voice, so as not to wake up her children, that we are all used to the idea of everything coming late to Spain. Regarding various issues, however, Spain has in recent years moved to the vanguard in many matters of life, including those in which it had been notoriously late, such as animal rights. For example, internationally, the country has always been associated with the traditional Iberian machismo, and yet it now has one of the most progressive gay marriage laws in the world. Spain used to be associated with a bureaucracy that would not move faster even if people's lives depended on it, and yet it has one of the most flexible and dynamic systems of organ donation in the world, now taken as a model for other countries to follow. And, of course, Spain is well known not only for bullfighting but for all sorts of animal abuse, against which the British, for instance, have been campaigning for many decades. Now, however, there is a strong opposition to bullfighting by many active animal rights groups, and Spain was the first European country to discuss the Great Ape Project in its Parliament. In all these cases, she noted, we have stopped following a morality focused on supernatural entities and the "other" world to focus on living creatures and on how to make real lives go well. "It all has to do with a determination to minimize suffering," argued Pablo de Lora, professor of Law and Bioethics at the Universidad Autónoma in Madrid, during a lunch in Madison.

Jorge Riechmann, Joaquín Araujo, Paula Casal, and Pablo de Lora have been my guides and my inspiration in exploring this change, which germi-

---

1. Personal communication, Madrid, 13 Dec. 2011.

nates still at the margin but promises to have a significant influence on Iberian cultures in the not-too-distant future. My conviction that the violence among humans is deeply related to the human attitude toward the nonhuman world has been inspired also by the U.S.-based intellectuals. I became aware of this importance of nonhuman life in 2009, during Cary Wolfe's lecture at the University of Wisconsin, Madison. The ground was already prepared by conversations with Sai Suryan, who looked at me with diminished friendliness when during our never-ending debates I insisted that human issues should have priority and that devoting excessive attention to animals was a sort of "a waste" of energy and resources that could be otherwise devoted to help the poor and the persecuted, stop wars, and so on. Wolfe's lecture brought about a great change in the way I understand the world. Today, I still remember the two main turns of his argument that convinced me.

Wolfe compared the animal rights movement to two of the most important civil rights movements—that for the rights of women and for the rights of blacks—and gave us a sense of how scandalized white men were that beings of another gender and/or race than theirs would want to be free of discrimination and have rights. Today, many people—and I was among them—react similarly to the animal rights movements' claims and demands, believing that it is outrageous to demand any sort of equality between humans and animals. The way Wolfe talked about the historical process of inclusion into the ethical realm of different kinds of "others" made it seem only logical and right that animals become the next subjects of our moral questioning—especially if we do not consider that one's rights are directly related to one's cognitive abilities but rather that the capacity for suffering, vulnerability, fragility, and mortality are indeed at the very heart of morality, as Bentham suggested in his widely quoted text on slavery in 1789.

Just as Araujo, Casal, and Riechmann, I believe that the problem of mistreatment of animals, while important in itself, is a model of other kinds of injustice affecting the world. Because of this, I have connected an animal studies perspective with an environmental one, and I consider them as fundamental fields of biopolitical transformation. The relations between the question of the rights of animals and other issues are traced obsessively in this book. Among them are unjust social hierarchies, political torture, war and peace, genetic modification of species, destruction of the earth's climate, as well as disregard for the poor, old, handicapped, nonwhites, and the "illegal." I believe that changes are taking place through a constant buildup of the connections between the areas where things go wrong. In all these contexts, Spain is a unique case because of pre-existing cultural conditions. Since bullfighting has been considered for a long time as a National Feast, the question of animal rights is inseparable from the debate on national identity, tradition, and other political disputes of utter significance. As the result of synergies among these debates, the postulated changes have been more radical and all-embracing.

The real object of research in this book is language and the conceptual frameworks that structure both the state biopolitics and the discourses of the resistance. Jorge Riechmann, whose pessimistic remarks began this preface, is not a pessimist without faith. When I asked him what, in spite of his conviction of doom, gives him the strength to write so many books, attend so many conferences, and even talk to strangers like myself, he answered that it was love and poetry. He believes that language is the cause of "human sickness" (*Poemas lisiados*), but it can also be the means for transformation. This would be a movement of empathy, turning language into a bridge toward other forms of being. A language tenderly faithful to the world can bring new knowledge that would take us forward, to the otherness of animals and on to the stars.

INTRODUCTION

BULLS, APES, GENES, AND CLOUDS

BETWEEN 2008 AND 2011, members of the Spanish movement against cruelty toward animals gathered in several towns across the Iberian Peninsula to physically represent a wounded bull, filling its outline with their own bodies, painted in red and black (see fig. I1).

This image of the huge bull formed by human bodies, which only a bird's eye can fully appreciate (analyzed in detail in chapter 5), encapsulates a number of arguments and debates addressed in this book. It represents the relationship between concepts structuring national culture: the bull and the man. But contrary to the traditional opposition that a man establishes against the bull in the arena, this figure visualizes animality as an all-embracing reality of flesh in which both human and nonhuman animals are immersed. The red, bloody stain on the bull's back, formed by various humans, attracts attention to the vulnerability and pain that are sentient creatures' common predicament. Animals have been constructed culturally as "the other," and yet animality is a shared condition of humans and nonhumans. This performance communicates an attempt to modify Spanish culture by transforming human attitudes toward animals, in particular the cultural use of the bull, which is one of the main themes of this book and one of the most popular debates in the Spanish part of the Iberian Peninsula. This performance can be also read as a figure of an "affirmative biopolitics" that, according to Timothy Camp-

FIGURE 1.1
Activists of Fundacion Ecuanimal and WSPA-International in Bilbao

bell (2008, xxxviii), Roberto Esposito searches for in *Bíos* (2008). Affirmative biopolitics restrains the system of self-defense for the sake of preservation of as many forms of life as possible, without sacrificing any.

In the opening pages of his *Bíos,* Esposito sums up a number of thrilling stories from our present-day political stage, where some lives are destined to death so as to protect others, such as in the "humanitarian" bombardments of Afghanistan or Russian police killing 128 hostages in the process of freeing them in Dubrovska Theater in 2002. He also tells a terrifying story of the Chinese village of Donghu in the province of Henan, whose inhabitants sold the plasma of their own blood out of poverty and were in turn injected with blood infected with HIV. In this way, they were en masse condemned to death from AIDS, while rich buyers benefitted from their plasma, which was well tested in rich laboratories. Through these examples, Esposito defines negative biopolitics as verging on thanatopolitics, a politics of life nourished by death through a process of discrimination between the chosen and condemned. He writes, "Biopolitics has to do with that complex of mediations, oppositions, and dialectical operations that in an extended phase made possible the modern political order" (15). These "mediations, oppositions, and dialectical operations" involve language and are sensitive to historical circumstances that are different in each cultural setting. If the term "biopolitics" reflects the understanding that politics penetrates life, making it different from itself, culture penetrates both politics and life because concepts designating life and politics are heavily driven by cultural meanings. For this reason, it is important to investigate what particular shapes these "mediations, oppositions and dialectical operations" have taken in the given cultural context, and how they can vary to bring different results. This book retraces interactions between the hegemonic cultural discourses and biopolitics as well as heterodoxies in search of an alternative, less deadly, and more sustainable administration of life within the same culture.

Spanish deadly biopolitics has been represented symbolically in the spectacle of bullfighting, where subjectivity and humanity are defined against the animality of the bull and by the right to kill the animal. Matador, the national hero, is the one who kills. The performance in front of the Guggenheim Museum in Bilbao has a radically different symbolic meaning, coinciding with Esposito's idea of an "affirmative biopolitics." The idea of life expressed by the shape of bull filled by human bodies seems to engage in a dialogue with his words: "Anything that lives needs to be thought in the unity of life—[which] means that no part of it can be destroyed in favor of another: every life is a form of life and every form refers to life" (194). In the huge figure of the bull in Bilbao, instead of two confronting individuals, there is a community where individuals cooperate rather than compete, nurturing common life.

In Spanish literature, film, and theater of the last two centuries, there have always existed various alternative ways of portraying and valuing life, as opposed to those where *superior* humans were viewed as authorized to feed off and destroy "inferior" animals and humans. Writers, such as José Mariano de Larra, Leopoldo Alas (Clarín), Eguenio Noel, Luis Martín-Santos, as well as contemporaries, Juan Mayorga, Jorge Riechmann, Rosa Montero, and various others, who are dissatisfied with the existing relationships between human civilization and other forms of life, reveal in their works that the current treatment of nonhuman life rests on a great deal of ethical inconsistencies. In particular, they notice that the discrimination of "inferior" forms of life is a result of drawing ethical and political consequences from differences that are not ethically relevant, such as different capacities, for example. The ethical principle of respect for all life is empty since it is constantly compromised when it does not match with the interests of the powerful. Various political concepts, such as citizenship and sacrifice, as well as discursive strategies, such as pragmatism and the argument of economic gain, seem to justify the destruction of life. Mass media debates and activists' campaigns in Spain since 1975, and most intensely since 2004, show the relevance of blurring the human-animal divides for the discourses on war and peace, torture, gender, economy, environment, climate change, and even for the future forms of life that may appear as a result of genetic engineering and synthetic biology.

The first and the basic opposition that modern biopolitics rests upon so as to establish the right to kill is the one between human and animal, and in Spain, that emblematic animal is the bull.[1] In the first chapter of this book, I analyze the modern debates about the use of the bull in the national culture, where the meaning of the human fight against the animal, thought of as a mark of human and particularly of Spanish superiority, dangerously slides to the other side, ending up by animalizing the human in Luis Buñuel's and Pablo Berger's visions. In terms of biopolitics, this chapter represents the transformation of the immune paradigm into the autoimmune one, whereas an obsessive rejection of otherness perceived in the animal leads to the destruction of oneself. The next two chapters focus on alternative and critical visions of the right to kill in both human/animal relations and in a strictly human domain in Spain, promoted by marginal intellectuals such as

---

1. See Stanley Brandes (2009) for a discussion of the emblematic animals of autonomic provinces of Spain.

Larra, Noel, and Martín-Santos, who often pay for the lack of accommodation in the national culture with personal failures and sacrifices. Yet these marginal intellectuals-activists manage to subvert the national culture discourses and provide alternative metaphors that have slowly led to significant cultural transformations. Chapter 4 analyzes the implications of varying figures of animalization and anthropomorphism for the construction of biopolitical discourses and for political action. It shows that both concepts of "humanity" and "animality" are arbitrary constructions that need to be revised before any biopolitical changes can be implemented. Further, I argue that paradoxically, in order to regain lost humanity, we may need to look at animals for inspiration. Chapter 5 returns to the cultural analysis of the performance at the Bilbao Guggenheim Museum and other performances, writing their meaning into the recent history of the Spanish Movement for Protection of Animals. Performances in public spaces by anticruelty organizations are analyzed as "ruptural performances" (Perucci, 2009, 1) that seek to challenge the prevalent values of society by creating a shock and suspending "automatism of perception" (Shklovsky, 1965, cit. by Perucci, 5) through defamiliarization. In connection to a shock, the vagueness of most successful ruptural performances allows them to establish a multiplicity of metaphorical associations that appeal to different viewers and build bridges to other new social movements. In Spain, most of the performances for the defense of animals establish connections to bullfighting as symbolic of the national identity, but they enact subversive versions of the inherited cultural scenarios, where the difference with which the archive (Taylor, 2003) is reenacted corresponds to the intended cultural transformation.

Chapter 6 focuses on one particular case of frame-bridging between the movement for the defense of animals and the antiwar movement during the times of the War on Terror. It analyzes the commentaries on torture published in *El País* and *La Vanguardia* during the years 2004–11 in two parallel debates: first, on bullfighting and human/animal relations, as Barcelona was preparing to manifest itself as an "anti-bullfighting" city (2004) and later on while debating the final ban on bullfighting (2010–11); second, on torture in the War on Terror when photographs from Abu Ghraib were first revealed (2003–4). It shows that as a result of the synergy between these two debates new meanings were coined that connected war and other forms of violence in the human domain to the violence exercised by humans on animals. While the first section of this chapter deals with articles in the press, in the second section I discuss Fernando Savater's book, *Tauroética* (2011), against the animal rights movement; Jesús Mosterín's pro-animal-rights text, *A favor de los toros* (2010), which dialogues with Savater's[2]; and Juan Mayorga's play *La paz perpetua,* (2007), which is inspired by the photographs from Abu Ghraib.

The activists who painted their bodies in black to *be* the bull in front of Bilbao's Guggenheim Museum transform the modern use of this animal as human antagonist to that of a human living environment. At this point, the search for alternative biopolitics coincides with environmentalism. In this way, the meaning of environment is also transformed; it is formed by interconnected human and nonhuman bodies. This is the main theoretical framework of chapter 7. The connections between all the breathing bodies become the great matrix of life, whose predicament is "becoming with" each other, as in writings of Donna Haraway (2008, 4) and whose survival is conditioned by coexistence, as in theories by Timothy Morton (2007, 2010a, and 2010b). The activists in front of the Bilbao museum imagine the bull as home, an extension of their individual lives. In the environmental framework, the modern distinction between human and nonhuman (animals and other life in the environment) is blurred, implying fluid connections, overlapping, and mutual dependency and as a result a vision of culture that not only includes but is formed (this verb implies *agency*) by a number of nonhuman elements. The conceptual change leads to a transformation of attitude toward animals and toward human animality and also toward the surrounding environment. This is the vision of reality that emerges in Alejandro González Iñárritu's *Biutiful* (2010) and in Agustín Fernández Mallo's *Nocilla experience,* (2008) analyzed in the seventh chapter. In both the film and novel, balance of life is destructively mutated by human greed. *Biutiful* focuses on the corrupting influence of money, which has transformed the functioning of the universe, becoming its food and fiber, an addictive poison that needs to be dosed continuously to maintain the functioning of a world. The "narrativa transpoética" that *Nocilla experience* exemplifies is filled with hybrid artifacts: mixtures of human and nonhuman, present and past, real and imagined, all of which merge together as if through morphing. The novel challenges concepts that are arguably responsible for environmental disasters, such as the idea that humans are distinct from their environments.

The last chapter analyzes debates on the cutting edge biopolitcal technologies such as genetic engineering and synthetic biology in the historical context of Spanish cultural discourses on science and in dialogue with a science fiction novel by Rosa Montero, *Lágrimas en la lluvia* (2011). Sainath Suryanarayanan, a co-author of this chapter, and I search for answers to the question

---

2. Large parts of the texts of Mosterín's and Savater's books were previously printed in the form of articles in newspapers, which is where both philosophers established a dialogue.

of why Spain appears systematically in various polls as the most enthusiastic supporter of biotechnology in Europe. We tentatively connect this fact with Spain's desire today to distance itself from its anti-Enlightenment discourses that separated the country from Europe by a self-proclaimed "difference." We notice, however, that in expressing enthusiasm towards biotechnology, Spain may be in fact repeating some of the old patterns that it wants to leave behind, namely subservience to the regimes of authority and power, nowadays represented by multinationals and scientific regulations. Our analyses of the debates on biotechnologies in the context of the new form of capitalism, called bioeconomy by Pavone (2012), lead us to distinguish between precautionary approaches to the cutting-edge technologies of life and emancipatory ones, recognizing their complex historical roots and conditionings.

The book opens with the figure of a bull, evoking its crucial place in passionate debates on Spanish national identity, masculinity, love, and death. In a brief history of anti-bullfighting thought, it counterposes bullfighting and anti-bullfighting worldviews in Spain and compares the cultural positioning of the intellectuals who supported each. But, apart from the bull, there are other key life figures, as well. Apes, such as Copito, the legendary white ape of the Barcelona zoo and a protagonist in Mayorga's theater, appear as the caricaturesque imitator of the human, mocking human superiority and bringing to the surface more similarities than differences. The debates analyzed in the book are prompted by the reconsideration of the right to kill and master life, which is constructed over the boundaries between human and nonhuman life. This questioning also occurs beyond the bullfighting contest, reflecting on hunting, farming, pets, zoos, and experiments performed on animals in scientific facilities, but also on torture performed on animalized humans and on the genetic modification of organisms whose life is turned into a form of capital.

The next part of this introduction aims to familiarize readers with theoretical frameworks employed in this book that transform discourses of life, blurring modern distinctions between the human and the nonhuman domains. The following part is divided into four sections (bulls, apes, genes, and clouds), and it sets the stage for an in-depth analysis of conceptual transformations presented in the subsequent chapters by summarizing recent debates in Spanish media on bullfighting, animal rights, genetic modifications of organisms, and the environment. The last section of the introduction constitutes a brief insight into Hispanism's complicity with the bullfighting culture. It characterizes discourses that dominate the field on human-animal relations and retraces their origins to the bullfighting worldview inherited from the Generation 27—which included writers Frederico García Lorca,

Rafael Alberti, Luis Cernuda, all of whom were fascinated with bullfighting—but it also discusses works of the cultural critics whose voices aligned themselves alternatively.

There have been thousands of books written about bullfighting in Spain by its fans, but few by its critics. While a number of publications by U.S. Hispanists are devoted to the symbolism of bullfighting in literature, very few of these reflect on the significance of anti-bullfighting thought that during the last two centuries presented an alternative vision of a relationship between human and nonhuman life, as well as an alternative vision of life and bioethics in general. This book attempts to fill this lacuna, stressing the importance of these alternative visions for the transformation of ethics and politics of life on Earth in order to let it last a bit longer. It argues that the revision of the relations between human and nonhuman (or not-quite-human) lives is the first step in search for an alternative biopolitics.

LIFE

Life has always been divided not only into hierarchies of *taxa*: domains, kingdoms, families, orders, species, and more, where *homo sapiens* is only one of thousands of possibilities. It has been thought of in terms of physical processes, examined by biology, and also the so-called "meaning of life," which has been mostly a human domain, analyzed by religions and humanities. It is hard to understand how the processes and their meanings could be considered separately, how they could be the subject matter of different disciplines. In this way, the human being has also been split in half—an animal body, tested in labs, but with a superior soul, analyzed by soul gurus. This division persisted even after the religious worldview was largely displaced by a cultural turn. Literary studies, art criticism, and philosophy displayed limited interest in overcoming the divide. The cultural, until recently, had been invariably understood as purely human, consisting of built environments and social relations. Humanities dealing with human cultural production were alienated from the material realities of the world surrounding and sustaining them. The "biopolitical turn" in social sciences and humanities (Campbell and Sitze, 2013, 4) has brought politics (and culture) together with biology, showing that they are intermingled and, in fact, inseparable because there is hardly any nonhuman life unaffected by humans, existing independently and in alienation from the global system of power fluxes. Biopolitics is a concept of the Anthropocene as it constitutes a realization that planet life is driven by human activities.

Recent discoveries of the rapid change in planet temperature announce the possibility of a dramatic global deterioration of planetary conditions that may bring catastrophic consequences. The realization that this change is man-made, that humans have become a geological force on the planet, gave rise to an assumption that we are now living in a new geological epoch called Anthropocene.[3] This new geological placement requires an adjustment in the understanding of humanity and its relations with the nonhuman realm. We have become a decisive factor of the climate, comparable to oxygen or nitrogen, but decisively harmful, pushing the planet out of the Holocene equilibrium, which favored life. According to Nigel Clark (2011), as we acquire new knowledge of our relationship with what used to be considered beyond our influence but is no longer so (weather, natural catastrophes), humanities and social sciences should "return to earth" (iii). Clark urges humanists to become engaged in physical sciences research in environmental matters to "come to terms with the planet" and explore "how better we can live with other things and with each other—in the context of a deep elemental underpinning that is at once a source of profound insecurity" (v). In light of Clark's writing, it is the humanities' lack of engagement with the processes considered to be the domain of the sciences that may have contributed to the political and environmental crisis that we are facing today. (Science is too important to be left to scientists). On the other hand, it may be the humanists' lack of engagement with the materiality of Earth and its life that makes the humanities seem irrelevant to those who work on the life's hardware. Hence, it is through an engagement with the scientific domain of environmental processes that the humanities might get back on track in fulfilling its role in the contemporary world.

Theories by the late Michel Foucault (2003), Bruno Latour (1993), Donna Haraway (1991, 2008), Timothy Morton (2007, 2010a, 2010b, 2013), Martha Nussbaum (2009), Giorgio Agamben (2004), and Roberto Esposito (2008, 2013a, 2013b) reformulate the relationship between humans, animals, and other forms of life, and Jane Bennett (2010) questions even the habitual distinction between living and nonliving matter. For Foucault, in his famous essay "Right of Death and Power over Life" (1976), the power's concern with life as biological existence is defining of modernity. It is when the politics begin to consciously regulate and master life for the purpose of the national well-being and prosperity (the stress on each of these two goals varies strongly in different moments of history) that the division between culture

---

3. The term has been coined by the Nobel Prize winner Paul Crutzen and has since been widely discussed by sciences, social sciences, and environmental humanities. See chapter 7 for more detail.

and nature, established vigorously by sciences, effectively decreases in reality. In this sense, biopolitics is in fact about life (human and nonhuman) growingly transformed and mastered by humans. As in Foucault's (2003) lectures gathered in *Society Must be Defended*, it is about the processes of the transformation of life into more than just life, crowned with the genetic modifications by biotechnology. This book tells the story of how the biopolitics in Spain evolves from the right to kill (as in bullfighting) to the right to transform life into something else, to use it for the purposes of economic growth by blowing it into new proportions through genetic manipulation. In parallel, it tells the story of the search for alternatives.

Latour notices the modern discourses' bluff in classifying nature and life as subjects of science and separate from the society that is the subject of politics, while politics, science, nature, and society are progressively more and more intertwined. In his widely quoted *We Have Never Been Modern* (1993), Latour criticizes modern social sciences for building false divides between nature and society, object and subject, as well as for creating various other divides across disciplines that blind us to the way the world really is. He postulates rethinking disciplinary discourses and divisions, and building new concepts to avoid those misleading separations. Latour's actor-network-theory (ANT) describes reality as a series of relations between humans, animals, plants, and objects that are viewed as equipped with agency (although not intentionality) and where processes occur as a result of accumulation of interactions. Actors, humans, and what Latour (1996, 2005) calls the nonhuman *actants* are equalized in his theory: "Actors are not conceived as fixed entities but as flows, as circulating objects, undergoing trials, and their stability, continuity, isotopy has to be obtained by other actions and other trials" (1996, 377). Latour comments on his theory as a poststructural inheritance extended to the real world, "extending the semiotic turn to this famous nature and this famous context it had bracketed out in the first place" (378).

Nature-culture is one of those concepts emerging from Latour's work that allows seeing humans and nonhumans linked functionally and materially as they have always been in life. Natureculture acquires new meanings and political applications in Donna Haraway's *When Species Meet* (2008) and is also significant for this book, which, inspired by Latour and Haraway, argues that seeing connections instead of separations between different forms of life is a fundamental step in the search for an alternative biopolitics, that can decrease suffering. In the feminist framework of Haraway, the concept of naturecultures challenges the human treatment of animals in scientific labs, serves to criticize science's attitude toward life in general, and works to develop a political vision. In Haraway's politics a "situated knowledge" (1991, 183–202), conscious of its limitations and communicable through alliances, is preferable to an objective point of view from nowhere in a nonexistent God-point that is responsible for hierarchies and the subjugation of life. For Haraway, vulnerability of the knowing subject is a criterion of the validity of her knowledge. She proposes a transformative criticism that would refuse to disregard suffering, human and animal, and that in political terms would amount to substituting the capacity to control with the capacity to produce change, to nurture and empower others. For her, the production of innovating knowledge can be compared to—as indicated in the title of her article by the same name—"A Game of Cat's Cradle" (1994), where discourses are restructured by teams of thinkers, taking better and better shapes, like the threads tangled around the fingers of a group of girls who pass the game to each other in a courtyard. Haraway's belief that "knowledge is better from below" (1991, 190) connects to the intellectual agenda of the movement against cruelty for animals, which aims at the deconstruction of discourses justifying harm and the destruction of life.

As Latour and Haraway, Morton (2010a) criticizes the concept of "Nature" as distinct from culture and suggests substituting the vision of separate domains for the concept of "a mesh" (28, 33), which expresses the idea of ecological interdependence. The interdependence begins on the conceptual level where anything that exists acquires its identity as different from something else, as in Jacques Derrida's *Of Grammatology* (1967), and also etymologically because everything derives from other entities preceding it. Interdependence is also a result of the fact that all that exists consists of the same physical particles on some basic level. Similarly to Latour, Morton (2010a) compares the structure of life to that of language as imagined by Derrida, and he shows that both the language and the system of life-forms are characterized by an infinite and illogical multiplication of differences. Because they contain otherness, Morton calls all life-forms in the mesh "strange strangers," a name that is ethically and existentially consequential. "Strange strangers" (38) cause curiosity, bring respect, and demand hospitality. In spite of its strangeness, however, every nonhuman life-form is perceptibly familiar to humans because we have descended from it. This mixture of familiarity and difference that all life presents us creates, in Morton's view, the disquieting sensation of "uncanny" (2007a, 52).

The disquiet is not only caused by the ultimate impossibility to know "strange strangers," but it is also due to the invisibility of destructive processes taking place in the human environment that Morton (2013) calls "hyperobjects" and that are all-embracing and as unstoppable as climate change, sixth great species extinction,[4] and nuclear radiation that will go on for hundreds or even thousands of years after we are gone. Hyperobjects are interobjective because all life-forms are affected by them, but they are invisible. The planetary scale in which they unfold makes them, in spite of their imperceptibility, more real than what we are used to conceiving as real. They revert the order of the real, contributing further to postmodern art's quasi-fantastic aura. It may be due to the silent work of the hyperobjects that in various literary works and films in the twenty-first century, protagonists question the matrix of reality and of identity, discovering that what they had thought was obvious and commonsensical is a construction that often needs to be undone. These questionings often involve a reevaluation of both human and nonhuman life, as in the theater of Juan Mayorga, *Nocilla experience* by Agustín Fernández Mallo, *Lágrimas en la lluvia* by Rosa Montero, *Biutiful* by Alejandro González Iñárritu, all featured in this book but also in *Un lugar sin culpa* by José María Merino, *El año de Gracia* by Cristina Fernández Cubas, and others.

The meaning of "human" is altered by the new environmental consciousness, by the emerging biotechnologies of life that aspire to overcome the natural limits and also by the transformed notions of animal life. As a result of progressing destruction of the environment, apocalyptic scenarios announcing the end of our civilization appear in film and fiction with growing frequency. Different visions of the end provide for variances in the meaning for humanity. In his recent bestseller, *The World without Us* (2007), Alan Weisman imagines an Earth healing after humanity's "brief" intrusion on its surface. Humanity appears in this documentary-style work by imaginative reporting from the future as devoid of its spiritual glory due to all the damage it has done to other life forms. It is an object of current debate whether the ecological crisis that we are facing is just another opportunity for humans to overcome its difficulties or rather an announcement of the end of human progress and a limit to human freedom. Since the detrimental change of climactic conditions on Earth is the result of human production and consumption patterns, these patterns need to be transformed to insure that our planet remains habitable. Apart from the consumption of mineral fossils, an important change should take place in patterns of human consumption of animals. This is not only an important ethical issue but also an environmental one. For example, given that one fifth of methane in the atmosphere, one of the three gases contributing to the global warming, proceeds from animal hus-

---

4. See, for example, http://www.theguardian.com/environment/2014/dec/14/earth-faces-sixth-great-extinction-with-41-of-amphibians-set-to-go-the-way-of-the-dodo.

bandry, eating meat may need to be abandoned or strongly limited. Other options for the future include production of in vitro meat and further genetic modifications of cattle (which has already been considerably transformed).[5]

HUMAN/ANIMAL LIFE AND LANGUAGE

Frederico García Lorca was among those who condemned the modern city economy based on the massive slaughter of animals. An apocalyptic vision of a polis whose financial success is nourished by the blood of millions of animals and the poor that moves the grids of the machines appears in *Poeta en Nueva York*, written between 1929 and 1930 and published for the first time in 1940, after the poet's death. The city's life rendered by the poems is underscored by the pain of all those invisible victims with whom the poet empathizes. The unfamiliarity with the highly industrialized American civilization causes Lorca, who loves bullfights in Andalucía, to become suddenly acutely aware of animal suffering and death in New York. In "Oficina y denuncia," (Office and Condemnation), Wall Street's calculus of gain reminds us of a genocide report, whose laconic form reflects on the indifference of a systemic killing, many years later called "Eternal Treblinka" by Charles Patterson (2002) and "Holocaust on Your Plate" by Nathan Sanza (2004):

Todos los días se matan en Nueva York
cuatro millones de patos,
cinco millones de cerdos,
dos mil palomas para el gusto de los agonizantes,
un millón de vacas,
un millón de corderos
y dos millones de gallos
que dejan los cielos hechos añicos.
(Lorca, 2009, "New York," *Poeta en Nueva York*, lines 16–23)

Every day New York slaughters
four million ducks,
five million pigs,
two thousand doves for the dying,

---

5. While for some, producing in vitro meat is a way to prevent the unnecessary death of animals, others argue that it is a waste of money and resources and that different kind of changes in redesigning food systems are needed.

> one million cows,
> one million lambs
> and two million roosters
> that tear the sky to pieces.⁶

While for the Wall Street offices, the meaning of the numbers is that of an economic growth, for the poet it is the meaning of a massacre. The screams of killed animals pierce the heavens, but they are not heard in the modern city, which keeps all its killing hidden and its victims silenced in slaughterhouses whose walls are not transparent, but soundproof. Antonio Lafora (2004) writes that after Lorca returned from the trip where he visited a slaughterhouse, he began to reconsider his attitude toward bullfighting (256). Paul McCartney famously stated that "if slaughterhouses had glass walls, everybody would be vegetarian," and afterwards PETA and various other organizations against cruelty toward animals placed videos on the Internet representing the horrific treatment of animals in farmhouses and of their slaughter, which had a great impact on many of us.

Descartes, who was deeply convinced of the superiority of the immaterial and spiritual soul over any manifestations of materiality, judged that since animals lacked consciousness, they could not suffer. For that reason it was perfectly fine to use animals "as any natural resource without moral scruple" (Steiner, 2005, 135). It only was an appreciation of purely physical pain as a voice of the body by the Enlightenment laical worldview, which became dominant during the following two centuries and made it possible to imagine that both humans and animals shared a common predicament. At the end of the eighteenth century, Jeremy Bentham (1789) pointed out that if animals suffered, ethics should also consider them.⁷ But it was only recently

---

6. My translation.

7. "The day has been—and I am sad to say in many places it is not yet past—in which the greater part of the species, under the denomination of slaves, have been treated by the law exactly upon the same footing, as, in England for example, the inferior races of animals are still. The day *may* come when the rest of the animal creation may acquire those rights which never could have been withheld from them but by the hand of tyranny. The French have already discovered that the blackness of the skin is no reason a human being should be abandoned without redress to the caprice of a tormentor. It may one day come to be recognized that the number of the legs, the villosity of the skin, or the termination of the *os sacrum* are reasons equally insufficient for abandoning a sensitive being to the same fate. What else is it that should trace the insuperable line? Is it the faculty of reason or perhaps the faculty of discourse? But a full-grown horse or dog is beyond comparison a more rational, as well as a more conversable animal, than an infant of a day or a week or even a month old. But suppose the case were otherwise, what would it avail? The question is not, Can they *reason*? nor, Can they *talk*? but, Can they *suffer*?" (Bentham, 1789, chapter 17, section 1, footnote 2; http://www.laits.utexas.edu/poltheory/bentham/ipml/ipml.c16.s05.html).

that scientific proof has been obtained that sentient animals (vertebrates and some nonvertebrates) feel degrees of pain comparable to humans. It was in the 1980s when research on animal pain was published that contributed important arguments to animal rights movement allegations. For example, in their article, "Pain, Suffering and Anxiety in Animals and Humans," David DeGracia and Andrew Rowan present scientific evidence that only confirms what was previously known as a matter of common sense, that vertebrates not only experience pain but also anxiety. In 1987, the journal *Florida Entomologist* published an article by Jeffrey Lockwood in which the author argues for existing evidence that insects should also be qualified as sentient and that their lives should be included in moral deliberations.[8] The development of ideas on animal pain bring the realization that, as Bentham once suspected, in pain they are indeed like us, which in a world governed by empathy and consideration for suffering should lead, and to certain degree have led, to a change of practices. For example, after it was scientifically acknowledged that animals indeed feel pain, veterinarians began to use anesthesia while operating on animals, while this was not a common practice before the 1980s.[9]

The Spanish philosopher Fernando Savater states that if humans do not stop using animals as food and entertainment, these animals will disappear from our lives (2011, 46). The argument proceeds from Ted and Shemane Nugent's *Kill It and Grill It* (2002), which suggests that the best way to save a species from extinction is to start to eat it: "Then it will be managed—like chickens, like turkeys, like deer, like Canadian geese" (6) and will be blown up to reach record numbers. Savater similarly argues that various animal species would become extinct if humans had no use for them. In "Death by Birth," Alastair Hunt (2013) shows, however, that contemporary technologies of breeding turn animals into something else, keeping them alive only for killing purposes but then taking life away from them. Hunt claims that in factory farms, animals are born not to live but only to be killed. While a common sense point of view is that the meaning of animal agriculture is that of animal life, in fact, these animals have no life. They are already born dead because their birth does not just precede their death but constitutes a technology producing it. They are just condemned to "death *by* birth" in which birth itself kills or is a technique for killing. Hunt points out that the real

---

8. See also Lockwood's "Not to Harm a Fly" (1988).

9. Another reason animals were thought to lack consciousness and higher forms of sensibility was "Morgan's canon," which was very influential in biological sciences and stated: "In no case may we interpret an action as the outcome of the exercise of a higher mental faculty, if it can be interpreted as the exercise of one which stands lower in the psychological scale" (Tokarczuk, 2012, 35).

predicament of modern farm animals is indeed not being killed, but being born. Most of the animals bred in this way are not even able to copulate and multiply in a natural way, some cannot walk. To conclude, Hunt brings up Derrida's reflection on species extinction as paired with "extermination of life proper to animals" (2008, 26) by their forced birth in farm factories. Thus, he suggests that this species will be saved from extinction but will be also saved from life, being born "dead" as a "living corpses" awaiting consumption. Hunt hints at the existence of parallelisms within the biopolitics of human and animal lives. He evokes Hannah Arendt's reflection on totalitarianisms as "an insane mass manufacture of corpses" preceded by "the historically and politically intelligible preparation of living corpses" (2013, 447). (Are the futuristic dystopias such as *Matrix* a pure science fiction fantasy?)

Humanity and animality are two discursively opposed forms of life, which transcend each other in reality and in various aspects are identical. Agamben (2004) argues that the discursive attempts to separate humanity from animality, which he conceptualizes as the "anthropological machine," led to the construction of societies where those who are not sufficiently human are rejected or abandoned like animals (33–38). This happens because the division between humanity and animality runs not only outside but also within a human being and human society. In literature and film, this division is reflected by the rhetorical figure of animalization, which may be interpreted not only as the representation of humans as animals but rather more broadly as a construction of inferiority, which justifies the rejection of certain forms of life. The same idea appears in Haraway's *When Species Meet* (2008): "The discursive ties between the colonized, the enslaved, the non-citizen, and the animal—all reduced to type, all others to rational man, and all essential to his bright constitution—is the heart of racism and flourishes, lethally, in the entrails of humanism" (18).

To transform this state of things, according to Cary Wolfe (2010) our anthropocentric perspective needs to be overcome, and simultaneously, our categorizations of nonhumans and animals may need to be reconsidered. This task amounts to a revision of the concepts, metaphors, and discursive frames that order the world and justify the uses of power against the vulnerable life. As, following Wolfe, Georgina Dopico Black (2010) states in her article, "The Ban and the Bull: Cultural Studies, Animal Studies and Spain," that if we do not question the humanistic worldview that reproduces "mode of subjectivity" (236) within which the human is posed as superior, we will not be able to dismantle hierarchies of power and subjugation. The task of overcoming anthropocentrism is challenging because the language where our thinking is embedded reflects this anthropocentricity. Radically overcoming anthropocentrism can be accomplished as a resignation of meaning and of language as the focus of subjectivity altogether as in Jorge Luis Borges's "El Inmortal" (The Immortal, 2011) or in *Planet of Apes* (1968) where in the imagined future humans indeed regain their animality and lose the capacity to speak. More moderately, it has to be limited to a revision of the ways of seeing, structured by concepts and metaphors that order the world. These efforts of reworking and revising our conceptual framework from within will not liberate us fully from an anthropocentric perspective but may rid us partially of speciesism, which works as an animalization that kills. Unless we stop thinking altogether we will never stop thinking like humans, but we may strive for a humanity that does not harm nonhuman life in the name of our alleged superiority but rather rejoices transcending toward the nonhuman perspectives with care, leading to an alternative biopolitics. This transformation of naturecultures through new metaphors occurs in highbrow philosophical essays, in literary works, in theater, in rock concerts, blogs, street happenings, publicity, press, and new laws and slowly results in new patterns of behavior. The struggles about concepts are in fact about reality.

IBERIAN DEBATES

Bulls

Achille Mbembe's essay, "Necropolitics," a particularly violent form of biopolitics, reflects on the defiance of death as a basis for various kinds of modern spirituality and politics. Mbembe argues that in certain modern discourses, originating in Hegel, "the human being truly becomes a subject—that is, separated from the animal—in the struggle and the work through which he or she confronts death (understood as violence or negativity)" (2013, 163–64). These discourses, where the sovereignty and power are achieved by living "as if death were not" (165) are intensely present in bullfighting culture and politics, characterized in the words of Lorca by "intimacy with death" (164). On the other hand, however, the bullfighting spectacle that symbolically evokes war and struggles for sovereignty is also painfully real. In the process, a live animal is being teased, poked, and killed so as to prove human superiority and, as a result, establish symbolically the human right to kill.

Even if the attitude toward bullfighting has significant political implications, the divide between bullfighting and anti-bullfighting Spain does not correspond to the "two Spains" from the famous poem by Antonio Machado; neither does it correspond to the divide between left and right. Almodóvar was not the only leftist artist or intellectual who attempted to fit bullfighting into the new democratic status quo of post-Franco Spain. It can probably be argued that most publications on bullfighting during the first fifteen years after Franco's death were deeply involved in this undertaking. Timothy Mitchell's *Blood Sport* (1991) contains an exhaustive analysis of the most striking arguments for justifying bullfighting as part of the new status quo in democratic Spain. Enrique Gil Calvo (1989), Spanish sociologist and a journalist publishing regularly in *El País,* wrote a book to prove that bullfighting contributed to Spanish democracy as it played an important role in introducing Spain to the market economy. Mitchell sarcastically remarks that Gil Calvo went as far as to argue that the experience of the agony of a bull is a class-liberating event, and the rivalry among bullfighters represents a competition present in the capitalist market. From another perspective, José Bergamín in *La música callada del toreo* (The silent music of bullfighting, 1981), insisted on Lorca's idea that the sublime beauty of bullfighting is only accessible for true Spaniards, who have a uniquely Spanish taste for it. On a different note, Fernando Savater, who regularly publishes on bullfighting in *El País,* argued that bullfighting celebrates Western men's relationship with nature, which is the basis of our civilization and which he contrasted with "una especie de budismo que ha permeado la tradición ética del occidente" (a kind of Buddhism that recently transcended the Western tradition; cit. by Petit, 2010, n. pag.). Savater considers the animal rights movement as a threat to Western civilization because it challenges human superiority over animals, which is the foundation of this civilization (2011). Enrique Tierno Galván, City Major of Madrid (1979–86), important for his reconstruction of the city social space, and known for his support for the cultural movement *La Movida,* argued that bullfighting educated Spanish people socially and politically, developing "a collective act of faith . . . in the male of the species. . . . The bullfighter presents himself as the standard-bearer of manliness, and ratifies in each moment of the bullfight, that the faith in the certain kind of man, in which the public believes, makes a complete and continuing sense" (1988, 74–75). Continuing cultivation of this symbolic masculinity by a democratic socialist city mayor may have been connected to the fact that the institution of a democratic system had not immediately transformed the idea of citizenship and that, as some argue, the system remained in part, authoritarian. Perhaps the remaining authoritarianism prevented deeper changes in thinking and behavior so that various socialist politicians during the Transition were as devoted to bullfighting as their predecessors. The state continued sponsoring bullfighting because it was as useful for the Spanish Socialist Workers' Party's (PSOE) centralist and corrupted mode of government as it had been for Franco's regime. Mosterín (2010) reminds us that the PSOE's vice-minister, Alfonso Guerra, brought his child to bullfights when it was illegal to do so and that in 1992 the minister of Interior Affairs, José Luis Corcuera, introduced a new *Reglamento de Espectáculos Taurinos* (Regulation of Bullfighting Spectacles), which allowed entrance of minors to the bullfighting ring, annulling the progressive law from 1929 by the dictator Primo de Rivera that had forbidden it. Similarly a probullfighting attitude was shown by PSOE's *Junta de Andalucía*, which promoted the popularity of bullfighting among secondary school students, implementing a program called *Cancelas abiertas* (Mosterín, 46). Manuel Vicent writes in his *Antitauromaquia* (2001) that these attitudes betrayed the original spirit of the foundational Revolutionary Manifesto of PSOE which, in 1917, demanded the abolishment of bullfighting. There is a debate whether bullfighting is a lower-class entertainment today or whether it is being manipulated to be viewed as such. This would serve as a way to justify state subventions that contribute to the financial growth of the bullfighting businesses. It would also help to proclaim it as a national heritage, which would protect it from autonomous governments' attempts to ban it.

Bullfighting, as do many other violent traditions, puts on stage the national politics of life in order to naturalize it and stimulate community bonding. The purpose of tradition is then to make the working of hierarchies and their tolls seem natural. Before the Catalan Parliament banned bullfighting in 2010, as well as during the three years afterward when the Popular Party arranged for bringing it back, among all the arguments in defense of bullfighting, citing it as a cultural tradition has been particularly prominent. Tradition has also been an argument for the defense of hunting and kosher slaughter, which results in a slow torturous death of animals. In the times of multiculturalism, tradition has become a political issue, and the European Union (EU), after various debates, did not force Spain to abolish bullfighting.[10] There are, however, various questionable traditions that put in danger human and nonhuman life, many of which have been banned in modern times. Perhaps one of the most misunderstood interventions during the debates of the Catalan Parliament on 3 March 2010 was Mosterín's claim that the status of a Spanish cultural tradition should not necessarily protect bullfighting from ethical questioning as it does not protect from questioning customs belonging to other cultures such as clitoral mutilation (widely discussed in Spanish media due to growing immigration from Sub-Saharan Africa). The

---

10. See an analysis of those debates in chapter 5. The details can be also found in Pablo de Lora's *Justicia para los animals* (2003, 298–303).

anger of various media commentators caused by this comparison showed that even if different violent traditions are comparable for a philosopher, they are not necessarily considered to be so to interested parties.[11] Mosterín believes that ethical improvement involves constant rational questioning of those cultural practices that are harmful. This belief, however, may stand in opposition to the spirit of multiculturalism, in which tolerance or even support for differences in cultural practices is the ideal. For multiculturalists, a forceful integration that does not grant legal exceptions to traditional practices of different ethnic groups is a "destructive integration" (Shweder, 2003). Alert to the hazards of ethnocentrism, multiculturalists argue for abstaining from a moral judgment while observing and analyzing other cultures' practices, even if these practices violate the moral rules of the observers' worldview. This romantic respect for traditions, however, often ends up favoring the agendas of nationalistic regimes and overlooking individual suffering.

The critics of multiculturalism ask whether stoning of adulterous women, female genital mutilation, preventing children from attending schools, or torturing animals should indeed be tolerated because they are traditions. It is not always clear how conflicts between group interests and individual interests should be handled. While multiculturalists tend to give priority to group rights over individual rights, they define groups in terms of culture rather than gender or age. A famous article and then book by Susan Okin, *Is Multiculturalism Bad for Women?* (1999), states that multicultural rights that grant exceptions to minority groups due to their alleged cultural differences are often detrimental for the women belonging to those minority groups. In dialogue with Okin, Spanish philosopher Paula Casal argues in "Is Multiculturalism Bad for Animals?" (2003) that cultures should not possess special rights to practice rituals that are harmful to animals. While in debates on multiculturalism "the desire to side with the underdog" (9) is often an argument for minority rights, in the context of ritual sacrifice, the underdog may literally be a dog.

In Spanish media, bullfighting has been defended as an integral and historically rooted part of Spanish culture and as a form of art (Grandes, 2010; Prada, 2009; Wolff, 2011; Savater, 2011). The interventions of those intellectuals show how successful the process of cultivating bullfighting as tradition for internalization of necropolitics has been. Prada, for example, claims that only

---

11. The idea that torture and monstrosities are in fact integral parts of culture and that their cultural character does not make them less abominable had appeared already in Mosterín's *La cultura de la libertad* (2005) and was reiterated afterwards in *A favor de los toros* (2010), which contains a long list of torturous cultural practices: cranial deformations, body mutilations, addictions, wars, and terrorisms.

Catholics have access to the mystery of the bullfighting celebration because this religion gives them the sensibility to move naturally between life and death. For this reason, Prada believes that attacks against bullfighting are in fact against the Catholic religion. Similarly, Grandes (2011, n. pag.) calls bullfighting "a miracle" that only the chosen can see.

There is an obvious split among Spanish intellectuals and politicians. Jorge Riechmann, Jesús Mosterín, Paula Casal, Pablo de Lora, Rosa Montero, Rosa Regàs, Elvira Lindo, Manuel Vicent, Oscar Horta, Marta Tafalla, Juan José Millás, Antonio Muñoz Molina, and politicians such as Cristina Narbona, Joan Herrera, Francisco Garrido, and Pilar Rahola side with the bull. In addition, many of them associate *tauromaquia* since Ferdinand VII with tyranny, violent masculinity, and a primitive way of understanding our relation with nature, where human survival can only be guaranteed by the destruction of the nonhuman. The symbolic meaning of this traditional ritual is in their view incompatible with that of a democratic transformation that focuses on the elaboration of new models of masculinity, equitable participation, and more sustainable attitudes toward nature.

Advocates of bullfighting, however, also appeal to democracy. For example, in an article published in *El País*, Javier Marías (2010) compared the debated ban on bullfighting to the prohibition of smoking in public places or prohibition of gay bars, and he condemned both laws as limiting liberties and antidemocratic. Similarly, Savater claimed that the banning zeal of Catalans brings to mind the Inquisition and Franco's censorship on forms of public life in his "Rebelión en la granja" (2010), and in *Tauroética* (2011), he compared it to a ban on abortion. Both these intellectuals denounced attempts to regulate citizens' behavior through governmental prohibitions. Felix Ovejero, Pablo de Lora, and José Luis Martí (2010), engage with the argument that the state should not intervene in its citizens' way of life and that, in other words, it should be forbidden to forbid. They remind the readers that every civil law is full of restrictions on individual freedom that are imposed for the sake of the protection of other's freedom or for the sake of another greater good. If the life and suffering of an animal is not considered as an ethically relevant good, if in other words, animals' lives do not matter or matter less than a human's freedom to poke them, indeed such a prohibition does not make sense. In the case of granting ethical importance to animals' lives, it is necessary to protect them by limiting human freedom to inflict pain on them.

Francis Wolff, French philosopher at the Sorbonne, in *Cincuenta razones para la defensa de los toros* (Fifty Reasons to Defend Bullfighting, 2011) mentions that bulls as a species would be extinct without bullfighting. According to Wolff, Gómez Pin (2009), and others, due to all the efforts to create good conditions for bulls bred for bullfighting, this blood sport should be considered in line with the worldview concerned with ecological crisis. The terrains destined for bulls' breeding have turned to ecological reserves, rich in biodiversity. If bullfighting were banned, it would put an end to the ecological reserves and contribute to the extinction of these two hundred thousand bulls that are now alive in the reserve. According to this argument, bullfighting is in fact saving the lives of more bulls that it takes. Wolff (2011) and Savater (2011) attempt to show that cruel entertainments, such as bullfighting or hunting, are essential for the maintenance of life-filled enclaves that have not yet been taken over for agricultural or industrial purposes. The argument that bullfighting brings revenue was brought to the debates in the Catalan Parliament by Salvador Boix, the legal representative of the bullfighter, José Tomás, who testified that when Tomás appears in Barcelona, everyone earns more: taxi drivers, hotels, restaurants ("Violentos, torturadores, inmorales," n. pag.). All these arguments suggest that for animal life, their natural environments are only possible as long as they are rentable. It follows that nonhuman life has the right to exist as long as it is transformable into food, entertainment, or some other form of material resource. The visions of these two philosophers and of Boix are obviously reflecting on the dynamic of life under a neoliberal economy. It is surprising, however, that the philosophers find this state of things ethically satisfying.

When bullfighting is defended not as a tradition connected to Spanish spirituality but rather as a way to make money, Mbembe's vision of biopolitics, where spirituality and politics grow out of death, transcends to Warren Montag's "necro-economics." Based on the rereading of Adam Smith's description of the functioning of the market, Montag argues that Smith made a virtue out of greed by convincing his followers that private vices are public virtues and that the rich "in spite of their natural selfishness and rapacity . . . are led by an invisible hand to make nearly the same distribution of the necessaries of life, which would have been made had the earth been divided into equal portions among all its inhabitants" (qtd. in Montag, 2013, 197). "The invisible hand," according to Montag, functions as a Providence controlled by God that justifies the worldly inequalities making us believe that the poor die happy and that acting in self-interest constitutes "the only true way to reason and justice" (203). This vision of the world where the market guarantees the universal good is partially responsible for the ease with which the biopolitical distinction between life worth sustaining and life left to let die or be killed is made.

In response to the argument that bullfighting brings material gain to Spaniards, de Lora (2003), Ovejero et al. (2010), Mosterín (2010b), and Vicent

(2001) state that it is not right to earn money at the expense of the suffering. Ovejero et al. counters the argument of species extinction stating that if we breed species just to torture and kill their members for entertainment purposes, it seems better that they become extinct. No individual in particular will suffer if the species becomes extinct because nonexistent life does not suffer. He claims that alternatively species could be preserved in existing reserves turned into national parks for their conservation. Although the main argument of the anti-bullfighting movement is that ethics are more important than financial gain, it is also true that that bullfighting has been heavily subsidized by the state, and possibly without these subsidies, it would not be profitable. The money used to subsidize bullfighting could be used to maintain ecological reserves in which bulls and other animals could live and die freely. According to Mosterín (2010b), the bullfighting industry received 600 million euros in subsides from the Public Administration. Since both the 2000 and 2006 Instituto Gallup polls find that only 10 percent or less of all Spaniards consider themselves bullfighting fans, continuing to subsidize it seems unjustified (Gallup, 2002, "For a Bullfighting Free Europe," 2008; Lafora, 2004; and others).[12]

Bullfighting fans argue further that even if we cannot know for sure if the bull really likes bullfighting, the death in the ring after a happy, free life on the meadows must be preferable for him to a slaughterhouse that ends the lives of so many other animals. Various authors state that animal activists should be first concerned with farm factories, slaughterhouses, and even with fishing (Ansón, 2004; Wolff, 2011), not to mention the terrible conditions surround-

---

12. The 2006 Instituto Gallup poll of Spanish opinions on bullfighting has shown that "72.10% of Spaniards are not interested at all in bullfighting and just 7.40% are very interested; in Catalonia over 80% show no interest at all." (n. pag.). A very exact analysis of this and other polls on bullfighting can be found in Lafora (2004): 220–30. Even higher dislike of bullfighting emerges from the polls commissioned by International Humane Society; according to their website, "seventy-six percent oppose use of public funds to support the bullfighting industry." Further:

- Only 29 percent of the population support bullfighting (just 13 percent support it "strongly")
- Seventy-five percent of respondents said they hadn't attended a bullfight in the last five years
- Seven percent of respondents said they attended a bullfight "about once a year," compared with 20 percent who said they visited a museum/art exhibition; 19 percent who made theatre visits; and 12 percent who attended football matches
- Sixty-seven percent agree that children under 16 should not be allowed to attend bullfights. ("Bullfight Opinion Polls," 2013, n. pag.)

Even the polls commissioned by *El País* find that 60 percent of Spaniards do not like bullfighting ("Polls, most Spaniards do not like bullfighting," 2010).

ing the lives and deaths of millions of people all over the world due to poverty and wars (Cortina, 2009). Wolff pointedly notes that it is not the fact of the torture and killing of the bull that is questioned by the anti-bullfighting movement but rather its visibility. He suggests that hidden death would be even crueler for the animal (15).

Most Spanish animal rights activists agree that the systemic abuses of animals in farm factories and research facilities as well as their massive killings in slaughterhouses constitute a more important problem than bullfighting, but due to the invisibility of these practices, as well as the widespread belief throughout society in the need for eating meat, they constitute a far more difficult target than bullfighting, which occurs in front of everyone's eyes as entertainment. During the debates in the Catalan Parliament, Boix discussed the possibility that a bullfighting ban would lead to further demands made by the Animal Rights Movement as an argument against any concession: "¿Cerrarán luego las granjas de cerdos o pollos o prohibirán también la caza?"("El debate," 2010, n. pag.) (Soon they will close the farm factories of chickens and pigs and forbid hunting). Most true animal rights activists indeed consider the ban of bullfighting as a first step on the way to questioning the practices of factory farming and slaughterhouses. For strategic reasons, however, this could not be stated during the debates on bullfighting. Bullfighting fans argued that slaughterhouses should be questioned before bullfighting in order to block all possible progress in the matter. This, nonetheless, has had a double edge to it. On one hand, as intended, in comparison with the horrors of the newborn chickens whose beaks are being cut off, the fight of the bulls seems beautiful. On the other hand, however, this argument has slowly turned attention to the hidden evils of today's meat industry.

Apes

The question of the rights of humans and animals lies at the heart of biopolitics. Not all animal rights activists are fighting for animal rights, but most of them are convinced that some kind of legal framework is necessary for effective protection. Without the possibility of penalizing cruelty, it is hard to stop some people from torturing animals simply for fun. Antonio Lafora's *El trato de los animales en España* (The Treatment of Animals in Spain, 2004) contains a catalogue of cruelties committed every year by young males in search of entertainment. Without a legal framework, it is close to impossible to expect that animal farms invest in better conditions for chickens, cows, and pigs that are crowded into exceptionally small spaces, surrounded by their own excrement, and unable to move. European Union norms have been applied vigorously for cows in 2007, for chickens in 2012, and for pigs in 2013. They regulate the density of animals in cages, the minimum and maximum temperature of the animals' housing, and the hygiene conditions that farm factories need to fulfill. As a penalty, those who have not met the new requirements are not able to export their products. In Spain, they are sold for less money and marked appropriately; for example, eggs are stained with a red dot (Maté, 2012, n. pag.).[13] The widest and most interesting debate on the legal frameworks regulating animal (and human) existence was, however, motivated by the proposition to grant limited human rights to great apes.

The members of the *Proyecto Gran Simio* (Great Ape Project), internationally led by philosophers Paola Cavalieri and Peter Singer and in Spain by Paula Casal and Jesús Mosterín, by asking for limited human rights for the nonhuman great apes (i.e., chimpanzees, gorillas, orangutans, and bonobos) demand no more than protecting them from death, torture, and imprisonment. Casal (2012) argues that due to great similarities with human beings, these apes should be viewed as a newly discovered human race and treated accordingly. She argues that their human-like capacities predispose them to human-like suffering when they are imprisoned and tortured. Like humans, they have an extensive long-term memory and metacognitive capacities.[14] The 2014 manifesto entitled "El quinto simio somos nostros" (We Are the Fifth Ape) written by Paula Casal and Jorge Riechmann and signed by various other intellectuals internationally, argues that great apes possess all fifteen attributes defining a human, according to Joseph Fletcher (1966): intelligence, self-consciousness, self-control, sense of the time, sense of the future, sense of the past, capacity to establish relationships with others, capacity to take care of others and worry about them, communication skills, capacity to commit suicide, curiosity, perception and capacity to change, idiosyncrasy, and neocortex activity.

In 2006, Francisco Garrido, on behalf of the activists of the *Proyecto Gran Simio*, presented the Spanish Parliament with a resolution to grant great apes the right not to be killed, tortured, or arbitrarily imprisoned. The resolution was voted in favor, but because it has never been ratified, it has not become law. The most recent manifesto by the *Proyecto Gran Simio* from June 2014 again requests granting to the great apes a legal personhood, which in absence of the law protecting their rights, would help to argue for liberation of particular individuals at the court. The manifesto suggests that great apes

---

13. Some other legal changes and the anticruelty measures are discussed in chapter 6.
14. "Pensar acerca de pensar no está limitado a los humanos" Casal and Riechmann 2014.

deserve to be considered nonhuman persons more than corporations and thoughtfully connects the movement of empathy toward nonhuman animals to the ecological crisis experienced by the contemporary world. Casal and Riechmann write the following:

> Por otra parte, sin dar un salto en la difusión social de valores como la biofilia y la sustentabilidad, las perspectivas de futuro de nuestra propia especie son muy sombrías en un mundo sometido a la severa crisis ecológico-social que hemos causado nosotros mismos. Ampliar la comunidad moral más allá de la barrera de nuestra especie, no sólo sobre la base del reconocimiento de capacidades de los grandes simios, sino también atendiendo a la obligación moral de respetar la vida de los animales sintientes, que son sujetos de su propia vida, y de no dañar a los seres que pueden ser dañados, supondría un avance decisivo en ese deseable cambio valorativo. (n. pag.)

> On the other hand, if we do not popularize social values such as biophilia and sustainability, the perspective of the future of our own species is very grim in a world immersed in the severe crisis of ecological and social nature, caused by ourselves. Opening the moral community beyond our species, not only on the basis of the recognition of the capacities of the great apes, but also due to the moral obligation to respect all sentient animals who are subjects of their own lives, not to hurt any beings that can be hurt, would constitute substantial progress in this desired change of values.

Hardly anything exemplifies better the search for what Campbell calls "affirmative biopolitics" that these words of the two leading Spanish philosophers, but the legal aspects of the question have appeared more debatable than the ethical message of the manifiesto. The resolutions and manifiestos of the *Proyecto Gran Simio* prompted a debate on the human/animal divide in connection to the nature of rights that were in various cases considered as the unique prerogative of humans.

In his article "Derecho de los iguales" (The Law of the Equals, 2007), Carlos Pérez Vaquero, professor of constitutional law, sums up the debate on the limited human rights for great apes, pointing to the most frequently mentioned arguments. The most frequent reason to oppose granting rights has been, according to him, the great deal of work that remains on the human front. Humans whose rights are not respected will not benefit in any way, however, from great apes' lack of rights. According to Pérez Vaquero, it is also unfair that concerns about animal well-being be postponed until all human problems are solved.

Pérez Vaquero presents the paradox that medical experiments are done with great apes because of their human similarity, while this very similarity is not used as an argument against using them in such experiments. The author ends his article suggesting that in the not-so-distant future, people will be surprised that there ever were any doubts about the great apes' rights in the same way that it is hard to believe for us today that some objected that women or other races deserved them.

Cortina (2009) argues that human beings have rights because they are aware of them. If they are not treated fairly, their self-esteem is lowered and they may become depressed. Cortina finds no reason why animals deserve rights because they are not responsible and cannot learn how to read. Casal and Riechmann (2014) explain that this imperfection is in fact a condition of a large part of humanity: "Los humanos con ciertas enfermedades, o en coma, carecen total o parcialmente de los atributos de las personas que en cambio exhiben, aunque en distinto grado, todos los homínidos o grandes simios" (n. pag.) (Humans with certain illnesses or in comas totally lack attributes of personhood with which all hominids or great apes are equipped to a different degree). The cognitive abilities of the great apes are above the level of very young children. Today we do not question children's rights to be free of torture and to fully develop their potential because they are irresponsible, do not understand politics, and cannot fulfill their duties. In Casal's (2012) argument, those who cannot defend themselves due to lack of capacities deserve a special protection from the society. In this aspect, animals are like children.[15]

In reference to Arendt, Hunt (2011) argues that animal rights are as natural as human. Reading Arendt's *The Origins of Totalitarianism* (1951), Hunt observes that she is uncomfortable with the phrase of the Universal Declaration of Human Rights that states, "All human beings are born free and equal in dignity and rights" because it places emphasis on the fact of birth and the natural existence rather than on "the artifice of speaking and acting as a member of an organized community, for example a citizen of a nation-state, something quite different from being a member of a biological species" (223). This biologist assumption of birth in the Universal Declaration of Human Rights allows Hunt to argue, in contrast to Cortina (2009), that the platform of rights removes humans from the artificial circumstances

---

15. Sue Donaldson and Will Kymlicka (2011) bring to equation the handicapped, the mentally ill and other kinds of people whose rights may vary depending on their capacities but are never totally taken away. These two authors propose that a law consider separately different kinds of animals (pets, denizens and wild animals) and grant them different sorts of legal status.

of society and abandons them into animal bodies. In Hunt's reading, if the birth is the only condition mentioned for the right to have rights, it is not the human but the animal who is the bearer of rights.

In the article entitled "The Perplexities of the Right of Man" (2013), Arendt reflects on human rights in a deeply pessimistic fashion, arguing that their innate character is in fact a fantasy or, in a more positive reading, a never-realized ideal. She points to the fact that Burke's criticism of the Universal Declaration of Human Rights for the emptiness of its "Universal" character has been proven correct by history. According to Burke, it is the citizen and not the human who is the bearer of rights because the rights are guaranteed by the government of the nation to which the individual belongs. Arendt laments that people without a nation are unprotected as "savages" and destroyed as "beasts," which was the case of the Jews during the Holocaust. The Rights of Man, which humans acquire by birth and which are announced as "inalienable," are in fact "unenforceable" (85) outside of a nation. The philosopher pessimistically states: "We are not born equal; we become equal as members of a group on the strength of our decision to guarantee ourselves mutually equal rights" (94). She notes that these groups often insist on their ethnic homogeneity because differences "arouse hatred, mistrust and discrimination" (94). By attracting readers' attention to the fact that equality as a platform of rights is a social construct and functions only in particular national contexts, Arendt establishes the rights of humans as a question of politics rather than nature. At the same time, however, she notes that the political and ethical ideal, which should ideally be realized in the future, is that these rights exist as if they were granted by nature and by birth, independently from citizenship. If that was so, however, or, when it becomes so, "whenever a civilization succeeds in eliminating or reducing to a minimum the dark background of a difference" (95), animals would bear rights as well.

Esposito similarly searches for a framework that would eliminate the difference that effectively erases the right to life. He writes that *sensu stricte* biopolitics is about *zoé* (pure animal life) rather than *bios* (political life), but he also argues that this distinction is false, because on the one hand there is no pure life, all life being transformed by human politics, but on the other there is the idea "of [the] impossibility of a true overcoming of the natural state [that politics] is anything but the negation of nature, the political is nothing else but the continuation of nature at another level and therefore destined to incorporate and reproduce nature's original characteristics" (2008, 17). In Esposito's view, this idea, although powerful, is not correct and his writings suggest that deconstructing the difference between *zoé* and *bios* may help construct a framework more friendly towards life.

Savater (2010) considers the animal rights movement as a great danger for human civilization, which, as he rightly states, "is based on mistreatment of animals" (n. pag). Granting animals the right to life and to be free from torture would radically change the human life style, taking away a large part of the privileges of those well-off humans. But as the reading of Arendt and Esposito shows, these privileges are held not only at the expense of animals but also at the expense of animalized humans and maintained together with "the dark background of difference" that is the basis of wars and ethnic cleansings. It is plausible to think that both humans and nonhumans would benefit in a civilization granting rights by birth. According to Riechmann (2003), the transformation of the system of food production that would be necessary if animals bear the right to life would help in solving not only world hunger problems but would also slow down climate change.[16] If only ten to fifteen percent of the grain today consumed in husbandry was destined for human populations suffering from hunger, the consumption all over the world would increase and rise above the levels of malnutrition (25). Limiting or completely giving up husbandry would increase the amount of terrain available for crops that could feed more people, thus eliminating one of the main causes of poverty.[17] According to Ron Bowman's (2008) film titled *Six Degrees Can Change the World* consumption of cheeseburgers in the United States contributes more to global warming than all the American SUVs. Thus the suffering of animals is intimately connected with the suffering of humans, and they need to be targeted together. As Riechmann (2005 and elsewhere) argues, limiting or even completely stopping our consumption of meat is not only an ethical question but an environmental one, a *sine qua non* condition for stretching the duration of life on Earth unless, as some venture, the power of science can help to overcome the crises.[18]

Genes

If war, capital punishment, and other kinds of socially regulated killings are examples of necropolitics (Mbembe) and a market-regulated process of letting the poor die constitute an operating way of necro-economics (Montag,

---

16. See also Steinfeld et al. (2006).

17. Riechmann bases his statement on a number of English language studies. See also Thompson, *Spirit of the Soil* (1995); Blatz, *Ethics and Agriculture* (1990); Norton, "Agricultural Development" (1985).

18. The first hamburgers from in *vitro* meat grown in a lab have been already produced and consumed ("World's First Lab-Grown Burger Is Eaten in London," BBC, 2014).

2013), then bioeconomy, based on genetic modification, appears as a compulsory overgrowth of life that is transformed into something else (not-quite-life or more-than-life) to better function as capital. Even if most of the genetic experiments are carried out on plants and animals, humans do share the risks as their consumers and cohabitants and also because genetic modification becomes possible in humans. However, in all these forms of mastery over life by state, market, or science, there seems to be a common denominator.

In the very first pages of *Tiempo de silencio* (*Time of Silence*, 1962), Luis Martín-Santos compares laboratory genetic research to bullfighting, "como si fuera una lidia" (2) (as if it were a bullfight). Both the scientist and the bullfighter experience the thrill of being able to master life and death, both may fail to do so. Pedro, a young researcher from *Tiempo de silencio*, analyzed in chapter 3, wants to prove that viruses and not genes cause cancer, because then cancer could be prevented through immunizations. When he sees Madrid's slums, however, he decides that poverty is the main factor of human illness and that perhaps he should be researching human suffering in this environmental context rather than leading laboratory experiments. Martín-Santos's novel constructs an antideterministic, antigene discourse; it argues that it is not only genes that determine health but also people's attitudes shaped by politics and language. Martín-Santos's point of view has not become outdated in the twenty-first century when science has become capable of modifying genes of living organisms. A group of intellectuals such as Riechmann (2011b), Salvador López Arnal (2006), Carlos Amorín (2000), and Martínez Castillo (2008) argue, like Pedro from *Tiempo de silencio*, that instead of leading laboratory experiments on genes to overcome the crisis of life on earth, we should focus on protecting the life that has not been yet destroyed.

In the neoliberal framework, where the environmental problem is often reduced to the insufficiency of resources, biotechnology and the new discoveries related to DNA are thought to be able to infinitely stretch the capacity of the biosphere to feed a growing humanity. According to Sheenan and Tegart (1998) this means a new stage of capitalism where what is exploited is not human labor, but rather the generative and regenerative capacity of live organisms. Genetically modified organisms (GMOs) are integrated into the cycle of production and commercialization of the market, revamping it through various crafty strategies.

For example, genetically modified salmon grow larger and can multiply all around the year, thus remedying the progressive destruction of fisheries. Genetically modified corn or rice, after just one cycle of growth, requires additional purchases of activators to retain its genetically modified (GM) feature, forcing a close relationship between agriculture and chemical corporations and blurring the distinction between agricultural and industrial production. Genetically modified organisms constitute then not only an environmental risk and a risk to consumers' health but also a biopolitical risk of the privatization of life, which might be administered according to corporate financial interests rather than human needs.

Genetic modifications of crops, especially corn and soy, have been an object of debate for many years, and there has been no agreement in respect to their influence on human health. In spite of the pressure of the large multinational corporations and of the U.S. administration, most of the EU countries did not admit GMOs. Seven of them forbade cultivating genetically modified Monsanto corn. But in this aspect, Spain has been different again, as it possesses the greatest quantity of GMO crops in the EU (90 percent, according to Corinne Lepage, 2013). In Spain, GMOs have been cultivated since the late 1990s, when the PSOE government admitted Monsanto GM corn and allowed it to spread over 76,000 hectares, amounting to 21 percent of all corn in Spain ("España: El cultivo de maíz transgénico vs. ecológico," n.d., n. pag.).

Even though science cannot foresee the long-term consequences of GMOs on human health, the scientific character of GMOs has been for a number of Spanish intellectuals a sufficient argument to defend them, because science has been symbolically connected to progress previously denied to Spain because of past regimes. The desire to make up for lost time often leads to an uncritical enthusiasm for anything with a label of "science" or "progress" attached to it. Those who warn against excessive enthusiasm for everything that science condones are frequently accused of irrationality or of being retrograde. However, as some articles argue, unlike science funded by corporations, independent science has occasionally managed to show that the effects of GMOs on human and animal health could be very serious (Séralini, 2009; Lepage, 2013).[19]

While the debate on the genetic modification of animals and plants continues among philosophers, politicians, and scientists, a new issue has already arisen: genetic modification of humans. While the production of humanoids through cloning is still mainly a subject for science fiction, various types

---

19. According to Serralini's research GMO corn that is fed to chickens, cows, and pigs and that is also a common ingredient in sodas, sweets, and breads has produced the following results in experiments on rats: "Aumento de grasa en sangre (del 20% al 40%), de azúcar (10%), desajustes urinarios, problemas de riñones y de hígado, precisamente los órganos de desintoxicación." (n. pag.) (Increase of fat in blood from 20% to 40%, increase in sugar of 10%, urinary problems, problems with kidneys, and with liver, which are the organs of detoxification.)

of research in cutting-edge laboratories, carried out for medical purposes, could be also employed to "enhance" certain human capacities and to modify human bodies in many ways—the line between therapy and enhancement is blurry.

Clouds

A new study sponsored by NASA Goddard Space Flight Center and to soon be published in the peer-reviewed *Ecological Economics* shows a growing likeliness that the current civilization might collapse as soon as in the next fifteen years due to growing levels of economic stratification in the human population, which is directly linked to the overconsumption of resources. By investigating the dynamics of collapse in past civilizations, the study identified the most important factors, which explain the decline and converge on two salient features of social systems on their way to catastrophe: the stretching of resources beyond the associated ecosystem's capacity and the stratification of the society into overconsuming elites and increasingly poor masses. The study puts into doubt theories that technological innovations may in fact solve these problems. Due to their ecological backpacks—that is, all the materials needed for their production—these innovations significantly raise the per capita consumption of resources, leading to further depletion of the planet and, in turn, further prompting the growth of poverty. The study ends, however, by advising that collapse can be still avoidable if the depletion is reduced to sustainable levels and resources are distributed in a more equitable fashion, but structural changes need to be made immediately.

This means an urgent need for an alternative biopolitics, a new way of administering life that would repair ecosystems at the expense of economic growth and profit. Programs and visions of such new politics of life appear in the works of economists focused on environmental justice and degrowth, initiated among many by the writings of Catalan economist Joan Martínez Alier (2003) and by the alternative economies that emerged in various Spanish localities as a reaction to this deep crisis. These are environmentalisms for the times of Anthropocene, conscious of human responsibility for the planet and interested in doing what is possible for survival in times of climate change, contamination, and loss of health.

Those who do not share the optimism of the science and technology enthusiasts have reason to worry. On 12 December 2013, *El País* published a forecast of the World Energy Outlook that warned that by 2035 CO_2 emissions would increase by another twenty percent, elevating the world's median temperature by almost four degrees, which could mean much greater increases in hotter areas (as well as lower temperatures in some colder areas), possibly making them inhabitable. The United Nations Conference on Climate Change in Durban in 2011 and the 2013 Warsaw Summit on Climate Change have brought little or no progress in this matter, and the most important environmental organizations, such as Greenpeace, the World Wide Fund for Nature (WWF), Action Aid, and Amigos de la Tierra, left the summit in protest. Juana Viúdez, reporting for *El País*, shared with her readers the mood of helplessness and disappointment. In an article entitled "Llegaremos a tiempo?" Lola Arpa Villalonga (2013) quotes Stephen Emmots's supposition that if we found out that Earth would be hit by a huge asteroid, we would concentrate all resources and energies to prevent it. A change of climate will bring a comparable disaster by the end of the twenty-first century, but we cannot prevent it because, to echo Villalonga, in this case the problem is not external, but it is *us*, that we are reluctant to change. Climate change is a slow-motion disaster that still could be prevented to a certain extent but may not be due to our constant procrastination and inability to compromise our pleasures. It is thus a disaster that results from human culture, yet because it comes from the outside (i.e., the weather), it requires new strategies to confront it. To a great extent, the change may need the collaboration of the humanities and social sciences, as new discourses on the human and the human relation with the environment need to be elaborated. Riechmann (2001) stresses that humans should not imagine themselves as conquerors of nature—this is precisely the attitude that is leading to the destruction of Earth—but rather should realize their vulnerability and desperate dependence on the air, water, sun, sunlight, range of temperatures, stable ground, and soil that produces grain, all guaranteed by a particular chemical balance in the atmosphere that is now threatened. In other words, the nonhuman should appear not as an antagonist to be destroyed but rather as an extension of our physical body, our home, as in the performance in front of Bilbao's Guggenheim Museum.

The principle of precaution that Riechmann (2011b), López Arnal (2006), and Amorín (2000) defend, arguing against a wide adoption of genetically modified crops, fish, and domestic animals, is one of the main principles of ecological thinking. Because of this, it is in a certain sense a conservative attitude in contrast to the apparently progressive enthusiasm for novelty, and this leads to criticism of the environmental movement. For example, Martín Caparrós (2010) declares in an interview with Ima Sanchís that "el ecologismo es una forma presentable, cool, elegante, del conservadurismo" (n. pag.) (environmentalism is a form of conservative attitude, presented as cool and elegant), and he argues that it transforms the religious discourse on an apocalypse and mobilizes people to prevent distant dangers while there are many immediate ones that they thus neglect. Indeed, the question of time, the indefinite future of the announced catastrophe—since the clouds are not yet permanently over one's head—has been one of the greatest challenges for mobilizing environmental action. In certain parts of the world, however, the future has already arrived, and clouds have exploded, producing death and destruction. In his dramatic speech during the United Nations Climate Summit, Filipino delegate Yeb Sano, still emotionally affected by the 2013 hurricane that devastated his country, turned to those who think that climate change is a question for the future:

> I dare you to get off your ivory tower and away from the comfort of your armchair. I dare you to go to the islands of the Pacific, the islands of the Caribbean and the islands of the Indian Ocean and see the impact of rising sea levels . . . where climate change has likewise become a matter of life and death as food and water become scarce. . . . If that is not enough, you may want to pay a visit to the Philippines right now. (Sano, 2013)

As this talk shows, environmentalism and the alternative biopolitics that it mandates means not only protecting the future, but also dealing with the aftermath of catastrophes that have already happened and preparing for those that are on their way. As Morton and Rosa Montero state independently, in the global ecological vision everything appears interconnected through the material particles and the information. All and each action and event has an impact on the general state of things. Ecological vision is overtly rational in its precautionary approach, but it is also quasi-marvelous in its perception of all life as if it formed part of one complex single living organism, as in J. E. Lovelock's *Gaia: A New Look at Life on Earth* (1987). Ecological vision requires the capacity to skillfully imagine the movement between the bird's view and a view from below, a dialogic connection between the big picture and a localized predicament where individual and local are partaking in the processes taking shape in a planetary scale and where there is a correspondence between each cell and the body of the world. The seemingly irrational fear that a wound of the smallest critter can contribute to the loss of equilibrium on the planet may in fact be justified.

The greatest wound and largest environmental disaster that awoke ecological awareness in Spain was the oil spill of 20 million U.S. gallons from the oil tanker *Prestige* just off the coast of Galicia in 2002. It caused "la marea negra" (a black tide) as the whole sea was covered with a thick layer of black oil that killed hundreds of thousands of fish, birds, and other creatures. The polluted coast and destruction of life produced a sense of loss, mourning, and a consciousness of a threat, and it was compared in Fernández Mallo's *Nocilla experience* (2008) to the end of the world. The disaster caused despair, but it also mobilized a great number of volunteers who came to clean the oil from all over the Iberian Península and abroad. Juan López de Uralde (2010) called them "la marea blanca" (53; a white tide), since most volunteers wore white cloths, sharply contrasting with the black oil. Activists gathered around the platform "Nunca Mais," which called for protests attended by thousands of people and demanded that Galicia be recognized as a catastrophe zone and for a disaster prevention system to stop further catastrophes from occurring. But it also mobilized an awareness that black tides will keep occurring as long as energy production depends on the oil and thus provided an incentive to search for alternative sources of energy.

José Luis Rodríguez Zapatero's government (2004–11) significantly subsidized Spanish production of solar photovoltaic panels, leading to a boom in its implementation. Spain moved to be the fourth largest manufacturer of world solar power, and it became the world leader in concentrated solar power (CSP). In 2012 Spain was also the world's fourth largest provider of wind power, and the wind farms in the mountains have slowly become a part of the Iberian countryside. Wind is the third most important source of energy in the country, and as of 2012, it has covered sixteen percent of the demand. As a result of the economic crisis and subsequent change of government in 2011, however, the subsidies for solar panel production were removed, retroactively leading to great losses as well as many lawsuits where the legality of such changes was questioned. Several hundreds of photovoltaic plant operators faced bankruptcy. The economic and political crisis has not proved to be an opportunity for development of those technologies that were not promising an immediate profit but rather only an environmental benefit. It has stopped funding research and promoting the implementation of solar energy. The crisis became an excuse to intensify the economic measures and processes that are responsible for the collapse of the international economy and the destruction of the environment. In that sense it had begun earlier than 2008.

In her *Environmental Culture: The Ecological Crisis of Reason* (2002), Val Plumwood employs the metaphor of the Titanic to talk about "the technological hubris" (1) of contemporary capitalism that receives the warning of an iceberg, but instead of slowing down, it decides to go "Full Speed Ahead" (1) in order to avoid a loss for the business. Riechmann believes that it is not possible to stop the destruction of Earth given existing structures of government and property. While alarming discourses of environmentalists have not brought sufficient change, sociologists have already identified the phenomenon of "ecofatigue"—coined by professor of psychology José Antonio Corraliza and popularized by media (Montalbán, 2011)—which refers to the weariness many feel with all these announcements of future catastrophes. In this sense, the economic crisis may have proven to be an opportunity for change as it prompted the emergence of a number of alternative movements that slowly renewed political consciousness.

On 15 May 2011, a protest movement known as the "Indignados" (Outraged) occupied the central Plaza del Sol of Madrid, where they stayed debating and organizing themselves for several weeks. Various participants remain active in regularly meeting assemblies even today. The Indignados's manifestos ask for the development of green energies, closing of nuclear plants, sustainable means of transportation, and development of community life and public spaces. In various publications coming from the movement, the general failure of the system has been diagnosed. The Manifesto of Indignados from 15 May 2011 reads as follows:

> El obsoleto y antinatural modelo económico vigente bloquea la maquinaria social en una espiral que se consume a sí misma, enriqueciendo a unos pocos y sumiendo en la pobreza y la escasez al resto. Hasta el colapso. La voluntad y el fin del sistema es la acumulación del dinero, primándola por encima de la eficacia y del bienestar de la sociedad. Despilfarrando recursos, destruyendo el planeta, generando el desempleo y consumidores infelices. (10)

> The current economic model, obsolete and unnatural, blocks the social mechanism in a spiral that consumes itself. While few get rich, the rest sink into poverty. It will collapse. The will and the end of this system is an accumulation of money, which is more important than the efficacy and well-being of society. It wastes resources, destroys the planet, causes unemployment and unhappy consumers.

The Indignados have been one of the first ample social movements on the Iberian Peninsula, displaying an awareness that the decline in quality of life experienced by many is not only due to the neoliberal economy's failure to distribute wealth in a fair way but also to its destruction of the environment, especially in the poorer communities. There is a widely spread consciousness that injustice is responsible not only for the social crisis but also for the environmental one, consequently making justice an environmental value.

Also, Spain has been home to the degrowth economic theories that argue for the need to stop the infinite economic growth postulated by neoliberal capitalism.[20] In March 2010, Barcelona hosted an International Degrowth Conference with five hundred participants from all over the world. The basic assumption of the movement is that infinite growth is impossible due to the limited amount of resources on Earth. According to Peter Brown (2012), "Degrowth is an attempt to decolonize the mind" and make people see that our consumption ends up consuming our lives. Joan Martínez Alier (2003) argues that Western world societies have a false image of the economy as based on Growth Domestic Products (GDPs), which does not take into consideration the damages caused to the environment—including human health—by economic growth (externalities). According to Martínez Alíer, the real economy is that which deals with energy sources and the environment. The Indignados have been inspired by these debates on alternative economy, which are also entering academic curricula in Spanish Universities.[21]

The Indignados movement has been criticized for its lack of a political program. Such a program, however, could not have existed in the movement's initial moments because it rejected the political strategies and conceptual frameworks that were available, instead opting to work toward revising the most elementary forms of politics and the unquestioned truths of a capitalist economy. The Indignados called alternatively themselves "Democracia Real Ya" (Real Democracy Now) and have attempted to develop a local, organic, participative democracy as an alternative to the political system that disappointed them. Among alternative economic projects, they have proposed to limit the workweek to only thirty hours not only to provide for more jobs but also to give people time for political participation. As the discontent driving millions of Indignados connects to alternative discourses of other new social movements, a network of new visions of ethics and politics of life synergizes, and a change has begun to happen.

In Paula Casal's *Martina y el mar* (2007), a teenage Martina gets lost on a beach and is taken care of by Captain Gunnar, who shows her his boat and tells her about dolphins and whales. Martina falls asleep on board the ship, and in her dreams, she hears the voices of whales that got stuck in ice and are calling for help. In Martina's dream, they are in the North Pole where it is even

---

20. Riechmann, deeply engaged in these debates, prefers the term "autoncontención" (self-containment), which implies the need for self-limiting consumption on the individual as well as the systemic level, including the nationalization of banks and various other parts of the economy.

21. In his blog, tratarde.org, during the last few months of 2013, Riechmann announced three courses on degrowth and on ecological economy that were offered for free.

colder than usual since the sky is covered by thick contaminated clouds that had been formed in the excessively heated areas of the Earth. While as a result of climate change, parts of Earth are becoming very hot, other places become even colder. Animals stuck in ice die a slow painful death, crying with the full strength of their lungs. But it is only very few that can hear them. As in his poetic oversensitivity, Lorca heard the piercing bellowing of mechanically milked cows in the factory farms hidden from New York dwellers, Martina in her dreams hears the whales calling for help from the distant North Pole. If Araujo is right, and all the important changes are brought by movements that were marginal at first, perhaps Lorca's and Martina's sensitivity will spread, and important changes will be implemented by new generations.

HISPANISM AND NECROPOETICS OF BULLFIGHTING

> Some of the hardest people to reach are those who work to a well-developed theoretical framework, with whose assumptions your findings do not concur.
> —George Monbiot

Among the various examples of writings by great artists and intellectuals of the Second Republic whose ideas have become a sort of New Testament of Hispanism is Lorca. This inspiration, however, didn't come from Lorca's complaints about New York slaughterhouses but rather his enthusiasm for the particular Spanish capacity to move between life and death and to give life to death, as is played out in the bullfighting ring in his *La teoría y el juego del duende* (Theory and Play of the Duende, 1932) and other works. Lorca's excitement about life in death provides for a series of poems and theater dramas of thrilling beauty. It is also to an extent responsible for Hispanists' approval for bullfighting. Lorca wrote poems about bullfighters and bullfights, and declared bullfighting to be the human treasure of Spain, which should be exploited by artists and writers. The artists and scholars, admiring Lorca, have followed his call. There are plenty of scholarly articles and books analyzing the meaning of death and of bullfighting for Lorca and other poets of his generation—many of which are far from critical toward the necropoetics of those artists, which aestheticizes the necropolitics of the state.[22]

---

22. See, for example, *The Tragic Myth; Lorca and cante hondo* by Edward F. Stanton, The University Press of Kentucky, 1978; or Carl W. Cobb, *The Bullfighter Sánchez Mejías as Elegized by Lorca, Alberti and Diego*, Spanish Literature Publications, 1993, as well as its review by Salvatore J. Poeta in *Revista Hispánica Moderna*.

A Lorca-like vision of bullfighting's relation to Spanish culture appears in *Música callada del toreo* (1989) and other works by José Bergamín, a great friend of Lorca, who survived him by many years, was similarly fond of bullfighting, and was a key intermediary between the world of literature and scholarship during the early years of the postwar Hispanism. During the time of intense debate on bullfighting, prompted by Spain's entrance into the EU, Bergamín, just as Lorca, connected bullfighting to the Gypsy way of being. Just as Lorca (and also, more recently, Prada and Grandes), he described bullfighting art as understandable only to those equipped with a special sensibility, and he connected this sensibility to the spiritual superiority of a quasi-religious trance that bullfighting elicited for him, "este estado de posesión divina—o diabólica" (30) (this state of divine or diabolic possession). For Bergamín, the Andalusian bullfighting style defines Spanishness together with the contradictory spirits of Kierkegaardian Christianity, Don Quijote, *cante hondo*, flamenco, and the baroque.

Bergamín, just as Lorca, belonged to the famous Generation 27. He was also a disciple of Miguel de Unamuno and a political activist of the Spanish Second Republic. During the Spanish Civil War, he presided over Alianza de Intelectuales Antifascistas (Alliance of Antifascist Intellectuals) and, in besieged Madrid, next to Rafael Alberti, Miguel Hernández, Luis Cernuda, Manuel Altoaguirre, and Vicente Aleixandre, published in *El mono azul*. During his exile in Mexico, he founded the journal *España peregrina* as well as a publishing house, named Seneca, which printed literary and critical works forbidden in Franco's Spain. These contained the paradigms of today's Hispanism in which we are still to some extent located. Bergamín himself was adopted as an emissary of Spanish culture by the first generations of postwar American Hispanists. In 1941, for example, David Lord wrote in *Books Abroad* an homage to Bergamín, praising Bergamín's physical appearance, his Catholicism, as well as his work on "Spanish music, drama and the authentic Spanish sport of bullfighting" (409). He presented the Spanish exile as a living example of that spiritual superiority of Spain over the rest of the Western world that Lorca and Bergamín had voiced and that still resounds in many articles in the field. In the words of Lord:

> There is probably no adequate historical study of the problem, owing to the fact that the Western world largely denied its existence, but the present crisis of society is bringing the matter again to the forefront, and Spain's role in the spiritual development of Western men will soon become widely recognized.
> ... It is as a bearer of ... [the] spiritual health of Spain that Bergamín comes now to the new world. (407)

Once again, for Lord as for Lorca and Bergamín, bullfighting is the truest expression of Spanish soul.

Sebastiaan Faber (2005) analyzes the ideologies of three journals that had an important role in the foundation of the postwar U.S. Hispanism: *Revista Iberoamericana, Romance,* and *España peregrina,* the latter founded by Bergamín. He notices that in spite of their differences, they all share the belief in the spiritual mission that originates in the "madre patria" and that represents a set of values "without which the world would not be able to survive" (82). Not so radically different is Carlos Fuentes's vision in *El espejo enterrado* (The Buried Mirror, 1997), which the Univeristy of Wisconsin–Madison, together with various other American universities, still uses as a textbook to teach its Hispanic cultures course. Fuentes, like Bergamín, claims that the essence of Hispanic civilization is intimately connected to the symbolism of the bull and bullfighting. The spirituality that bullfighting represents for Fuentes reveals the *true* relation between human and nonhuman nature, as opposed to the hypocritical concealment of this relation in Anglo-Saxon cultures.

The opposition of the "spiritual health of Spain" to the crisis of Western civilization that Lord brings into his article on Bergamín appears also in Américo Castro's inaugural lecture at Princeton in December 1940, entitled "The Meaning of Spanish Civilization," where this famous writer states that "the way that Spanish life has realized itself in history is different from what we observe in other great peoples of the West" in that instead of material and technological development, it opted for confronting "the ultimate problems of life and death" (Margaretten and Rubia Barcia, 1976, 25). Later he adds: "To live or to die are for [the Spaniard] equivalent points of departure . . . [and] today it seems certain that only those countries able to face death will be able to survive" (39). In 1940, Castro lectures on "the crisis of the European civilization," (Ibid, 26) which he supposes might be brought about by its excessively materialistic and technocratic tendencies, and he suggests that the cure can be found in the Spanish way of life. Once again the vision of Spanish culture's superiority, due to its intimate relation with death, is presented as its most relevant feature. Gonzalo Pasamar (2010), basing his assumption on John Beverly's research, claims that Castro's influence in U.S. academia and especially in U.S. Hispanism was enormous due to his affinity with "the primary assumption of the American liberalism during the cold war" (218). According to Pasamar, "Castro's influence must be regarded as an ideology of North American academic Hispanism" (218). This influence was so strong because Castro provided American Hispanists with an updated version of Spanish history that was, however, framed and phrased by concepts that were crucial for his vision of Spanish culture as not materially oriented but with a spirituality focused on death.

If, according to Mbembe, Hegel defines the "life of the Spirit" forming a human subject as different from an animal in terms of "upholding the work of Death" (2013, 164) in a struggle against it, it may be argued that Spanish bullfighting is the most classical performance of necropolitics' spirituality. Additionally, Hispanists defining Spanish national spirituality as enthusiastic acceptance and commitment to necropolitics are themselves committed to these critical conceptualizations. Intellectuals willing to serve national culture develop discourses that reveal interiorization of the state necropolitics that may be deeper than their commitment to a particular form of the state. Lorca, Bergamín, Castro, as well as Luis Cernuda, Rafael Alberti, and so many others were abhorred by the violence of the state but attracted by the sublime and spiritual allure of death caused by this violence. They did not even see this as an inconsistency or take note of the fact that this spirituality of necropolitcs helps to build, maintain, and justify oppressive regimes that kill and let die. The cultivation of this spirituality among Hispanists may be the reason for relatively little interest in alternative biopolitics in our field. Alternative perspectives that propose to protect life, like those found in animal studies or environmental studies, seem not essentially Spanish.

A celebratory approach to heroic death encounters such as is found in a bullring still appears in literary criticism published in professional journals in the twenty-first century. For example, María G. Hernández, in her article "Matador: El deseo transgresivo en Rafael Alberti and Pedro Almodóvar" (Matador: Transgressive Desire in Rafael Alberti and Pedro Almodóvar, 2002) published in *Lenguaje y Textos,* not only aestheticizes the violence of the bullfight but also considers the passions expressed in bullfighting as an answer to ethical problems:

> Ellos usan la irracionalidad como arma arrojadiza contra una moral en crisis. Rompen las barreras que separan lo racional de lo irracional para conducirnos a una intrahistoria donde se alberga el deseo, los sueños y la pasión. Alberti con su poema "Matador" y Almodóvar con su película del mismo nombre dan cuenta de una fuerza pasional sin encasillamientos. Los roles fijos desaparecen y sólo existe el fluir de la vida y el arte. Los papeles del toro y el torero se intercambian: el toro mata al torero, el toro es mujer, el toro es hombre, el matador es animal y, al final, toda diferencia y barrera entre lo creíble e increíble se disuelve en la estocada que mata. De esta manera, este deseo irracional, constructivo o destructivo, que persigue la plenitud, encuentra su profunda realidad en la muerte. (65)

> They use irrationality as a weapon against moral crisis. They break barriers that separate the rational and the irrational in order to conduct us into an

"intrahistory" inhabited by desires, dreams and passions. Alberti with his poem "Matador" and Almodóvar in his film with the same title tell us about limitless passions. The fixed roles disappear and there is only a flow of life and art. The roles of the bull and bullfighter switch: the bull kills the bullfighter, the bull is a woman, the bull is a man, the matador is an animal and, in the end, all differences and barriers between the credible and incredible are dissolved in the thrust of the sword that kills. In this way the irrational desire, constructive or destructive, that peruses the plenitude, finds its profound reality in death.

This long quote reflects the vision of the particularity of Spanish culture where death is represented as an erotic object, a realm of plenitude, and the ultimate reality that makes earthly existence irrelevant. Hernández's analysis arises directly from the cultural theory by Lorca as well as Alberti's text. We can trace the origins of Hernández's discourse on "moral crisis" resolved through irrational passions leading to death in Castro's writings on the "spiritual health of Spain," as opposed to the "crisis of the society" in the Western world ("The Meaning of Spanish Civilization," 1942/1977). The interpretation of Almodóvar's *Matador,* however, seems to be insensitive to the fact that his representation of bullfighting in the film lies halfway between parody and eulogy.

In Hernández's analysis supplied by the passage above, all differences, including those between human and nonhuman, feminine and masculine, the victim and the victimizer, even the constructive and the deconstructive, dissolve in the symbolic arena of death, rendering all possible inequalities and injustices irrelevant. This irrelevance of injustice and inequalities, as Hernández herself notes, was inherent in Unamuno's idea of *intrahistoria,* which envisions a life that persists on the back burner, mostly in poverty and uneducated, unaffected by the events that push history forward, as an essence of the national culture.[23] As a result, Unamuno suggests that no action, no intervention, and no reform are necessary to improve the life of those people. Thus even if he claims to dislike the performance of necropolitics in bullfighting, in his philosophy, Unamuno is also focused on death although he favors dying peacefully, which is made possible through the dying of necro-economics.

According to Terry Eagleton's widely read *Literary Theory* (1983), literature and literary criticism in the modern nation take on the connected function of religion and entertainment. Eagleton presents the foundation of

---

23. See, for example, the characterization of the life of *el pueblo* (people) as connected to land, almost a part of it, never changing, and far away from the worldly politics in *Don Manuel Bueno, mártir* (1933) or in "Las hurdes" (1914).

English literary criticism as a substitute for religion in its mission of turning people, especially the middle and lower classes, into good citizens. That mission was to be accomplished through propagating literature as ideology, creating symbolic ties to both the state, with its right to violence, and the nation, with its need to be defended. In this early stage, the symbol, as it is defined in Romantic aesthetic theory, "becomes the panacea for all problems. Within it, a whole set of conflicts which were felt to be insoluble in ordinary life—between subject and object, the universal and the particular, the sensuous and the conceptual, material and spiritual, order and spontaneity—could be magically resolved" (19). Nation-building has been a foundational mission of literary criticism not only in England but, in a different context, also in nineteenth and early twentieth-century Spain as well as after the Spanish Civil War, both in Spain and in exile. In its nation-building mission, literary criticism largely recurs to symbols. Symbols, as Eagleton explains, are crucial for nationalistic identifications which override real cultural and class differences. Symbols subtly divert readers' attention from the immediate realms of life and direct it toward the transcendental domains, where all become one. This is what happens in bullfighting, and it is exactly the process described and enacted in Hernández's passage quoted earlier. Bullfighting, as the all-encompassing symbolic "difference" of Spanishness, resolves tensions between real differences of life (also diverse cultural forms of Iberian life as well as the conflicts of history) in "the deep reality of death," which constitutes the foundation of the "deep tradition" (65). Given its symbolic reach, it is not surprising that Hispanism displays a fascination with the necropoetics of bullfighting where all differences resolve in death and a unified nation emerges as a result.

As a mark of difference, a number of deeply reverential analyses of bullfighting have been produced by anthropologists who adopted the lenses of cultural relativism. Anthropologists have generally privileged tradition and treated the anti-bullfighting movement as an insignificant or, as it has grown strong, a troubling aspect of bullfighting culture, mentioning it only in passing. Anthropology as a field emerged in the midst of a romantic rebellion (Shweder, 1991) that found meaning in the traditions criticized by enlightenment philosophers. Traditions as repositories of cultural differences have been respected for the sake of diversity as well as, from the ethical perspective, realms of otherness. The relativistic attitude within the academy, and especially in the field of anthropology, understands *objective analysis* as a balanced view and respects difference, that is, tradition rather than movements criticizing them, and in doing so it promotes production of discourses that are often neither objective nor progressive, but rather prejudiced against progressive social movements opposing oppressive local traditions. Paradoxically, these discourses that claim to be objective because they adopt the other's perspective and to be ethical because they protect otherness may sometimes be subservient to the political ideologies, which use traditions to promote nationalism and social injustice.

Bullfighting and National Identities in Spain by Carrie Douglas (1999) is an example of analysis accomplished from an anthropological perspective. Douglas admits the existence of heated debate on bullfighting, but she privileges a perspective according to which bullfighting is still, as Franco's regime wanted, a unifying element of Spain, where geography and languages change, but the bull is everywhere the same: "It makes the contemporary Spanish state possible by integrating its various parts" (8). This prejudice in favor of bullfighting is even more pronounced in the article by Alberto Bouroncle (2000) where anti-bullfighting is only mentioned *passim*. Bouroncle's vision of bullfighting as a factor of all provincial allegiance to the Spanish state is not only ideologically subservient to the remnants of Francoism but also, in the year of the publication of the article, is simply not true as a fact. Mitchell's *Blood Sport* (1991), although already quite dated, is perhaps the best researched study of the cultural context of bullfighting. It presents an in-depth analysis of the "cultural complex of beliefs and behaviors that sustains [bullfighting]" (2). His research leads him to assert that "bullfighting has been nothing less than a microcosm of the Spanish social order," replicating "almost every feature of the Spanish political system" (132). Mitchell avoids ethical judgments or, having made them, dilutes them through relativistic considerations. He states, for example, that "it is entirely possible that bullfighting is immoral or unethical in some way or in every way. Extreme caution must be exercised, however, when applying personal or abstract moral standards to specific aspects of cultural performances" (2). He explains that if bullfighting is evaluated in terms of the taurine subculture itself, it is impossible to condemn it. Mitchell maintains this ambiguous attitude toward his topic through his whole book, moving back and forth between anti-bullfighting arguments and the alleged "objectivity" of his perspective. His immersion in the "planet of bulls" (2), as he calls the taurine subculture, visibly prevents him from appreciating the change introduced by the growing anti-bullfighting movement, which in 1991 should have been already visible. Like Douglas, he expresses unsubstantiated certainty that for every Spaniard critical of bullfighting there are hundreds who enjoy it. In spite of all this prudence, he is strongly criticized by Sarah Pink, who in *Women and Bullfighting: Gender, Sex and the Consumption of the Tradition* (1997) accuses him of essentializing and orientalizing Spanish character through its inscription to bullfighting.

The efforts undertaken by various Spanish provinces, and especially Catalonia, to disassociate themselves symbolically from the bull are analyzed in the insightful article by the renown anthropologist Stanley Brandes, "Taurophiles and Taurophobes" (2009). Although Brandes devotes attention mainly to anti-bullfighting as a means of asserting Catalan and other separatisms, he describes the search for a symbolic mascot different from the bull in the Spanish autonomous provinces, recognizing the broader significance of the growing moral objections of the people and the effects of activism in the animal rights movement as important components of changes in Spaniards' self-perception. Brandes's article is to the best of my knowledge the only publication to date on anti-bullfighting that presents the dynamics of change of attitudes towards bullfighting in Spain written by an American anthropologist. He claims that for a majority of the contemporary Spaniards, "a blood sport like bullfighting seems particularly anachronistic" because "to a growing segment of the Spanish population, anything that sets the country apart from the rest of the continent is undeserving of preservation" (789).

Not only Franco's cultural propaganda but also Hispanism critical of Franco's politics and relativistic anthropology have constructed the vision of "wild" Spain that most of today's Spaniards want to get rid of. Hand in hand with certain discourses of Spanish politics both on the right and the left, artists such as Lorca, Alberti, and possibly Almodóvar, as well as most literary critics like Hernández who glorify their visions, still propagate Spanish culture as one of bullfighting, violent passions, irrationality, desires, and death as a mark of Spanish difference understood as a sort of spiritual superiority. At the same time, however, Spain has undergone a great number of changes. If it is debatable to what extent its reality had ever corresponded to the Prescott paradigm or Castro's theories, nowadays there are even more reasons to question these past discourses of the national culture in Spain. It seems inappropriate that in American classrooms students are still being informed that bullfighting is a true expression of Spanish soul, while in Spain over seventy percent express no interest in bullfighting (Gallup, 2002; "Comparativa ICSA-Gallup," 2009; Lafora, 2004; Abend and Pingree, 2007; and others).[24] A number of grassroots organizations against animal cruelty are founded by regular citizens, and anti-bullfighting protests happen all over Spain. The discourses of the field have had a limited sensitivity to these changes in the culture.

Giorgina Dopico Black (2010) paid attention to the anti-bullfighting debate but reduced its significance to that of Catalan nationalism. Dopico Black's article in the *Journal of Spanish Cultural Studies* is nonetheless signifi-

---

24. See note 12 for more on polls on bullfighting in Spain.

cant as a call for attention to the question of the animal in Spanish cultural studies. Two other scholars note the significance of the anti-bullfighting movement in Spanish culture: John Beusterien and William Viestenz. Beusterien's book *Canines in Cervantes and Velázquez: An Animal Studies Reading of Early Modern Spain* elegantly highlights the significance of the animal question in the sixteenth century and shows that even then there existed certain opposition to bullfighting. Viestenz's article, entitled "Sins of the Flesh: Bullfighting as a Model of Power" (2013), surveys recent debates on bullfighting in Spain and does so in reference to Agamben's and La Capra's theoretical frameworks. The article discusses Martín-Santos' *Tiempo de silencio* (1962) and Juan Goytisolo's *Señas de identidad* (1966) as reflections on traumatic experiences of violence connected to bullfighting in Spanish postwar history. It concludes with analysis of Salvador Espriu's collection of poetry *La pell de brau* (1960), which calls for revising the symbolic association with the bull in Iberian artistic production. In his forthcoming article "The Bull Also Rises: The Political Redemption of the Beast in *La pell de brau* by Salvador Espriu" (*Hispanic Issues*), Viestenz connects "redeeming the bull" from the category of "the beast" to that of "a political animal" to the consideration of a shift from the discourse of Hispanism to "Iberian Studies." Iberian cultural studies with its focus on diversity of the cultural conceptualizations of life and languages on the Iberian Peninsula questions the paradigmatic discourse of Hispanism related to spirituality, death, and bullfighting. Scholars related to this reforming movement in the field show more openness toward the significance of the nonhuman life in general. Also beyond Iberian Studies, in recent years various young scholars such as Luis Martín-Estudillo, Paul Begin, Sara Brenneis, Luis Prádanos, Daniel Ares López, and others have undertaken criticism that includes the perspectives of animal studies and environmental humanities.[25]

Jesús Torrecilla's *España exótica* (2004) deserves a special mention as a pioneering effort not only to recognize bullfighting as a cultural problem but also to openly take a stand in analyzing it. Torrecilla's sympathies are visibly with the anti-bullfighting movement as he boldly states that "las corridas de toros son un espectáculo anacrónico y primitivo. . . . El nacionalismo valora lo que considera propio y lo hace sin restricciones ni distingos" (151) (bullfighting is an anachronistic and primitive spectacle. . . . Nationalism values its own without any restrictions and distinctions). Torrecilla evokes

---

25. *Hispanic Issues* has recently published a volume of essays written from environmental perspectives and focused on human/animal relations. A Northeast Modern Language Association (2014) panel on environmental approaches to Spanish culture received so many graduate students' abstracts that it was changed into a roundtable seminar.

an article by Galdós published in *La Nación* in 1868 where the famous novelist considers bullfighting within the dichotomy between nationalism and modernization. After considering keeping bullfighting as the last resort of Spanish authenticity threatened by foreign fashion, Galdós rejects this possibility, stating that "más vale parecer extranjeros en España que bárbaros en Europa" (qtd. in Torrecilla, 2004, 144) (It is better to seem a stranger in Spain than a barbarian in Europe). Torrecilla comments that this dichotomy is debatably false, as the modernity has various faces, and he gives more meaning to the fact that the defense of bullfighting has turned into a "national crusade" because at its heart lies wounded national pride and resentment toward the rest of Europe (153). The popularity of bullfighting was a result of an "inferiority complex" that must be given due attention. In *Le phantom de la liberté* (1974), Buñuel develops a similar intuition of Spaniards' traumatic relation with France as comparable to the predicament of animals. It is to this surprising comparison that I turn now.

CHAPTER 1

BULLFIGHTING BIOPOLITICS AND DEBATES ON NATIONAL CULTURE

THE FIRST SCENE of Luis Buñuel's film *Le phântome de la liberté* (The Phantom of Liberty, 1974) evokes Goya's painting "Los fusilamientos del 3 de mayo" (Executions of the 3rd of May, 1808), which is one of the common references for political debates on Spanish national identity and marks the beginning of one of the most important periods in Spanish modern history. The last scene of the film represents a rebellion in the Paris zoo at the end of the twentieth century, when a Parisian youth attempt to free animals from their cages, but this idea is transformed to suggest a rebellion of the animals themselves. As the camera shows animals in their cages, we hear the echo from the first scene of a famous slogan—¡Vivan las caenas! (Long live the chains!)—shouted by Spaniards welcoming Ferdinand VII. Even though this happened in 1814 after the end of the War of Independence against the French, Buñuel anachronistically puts the words in the mouths of the Spanish patriots executed in the beginning of that war in 1808.[1] This comparison of the Spanish rebellion against the French invaders to that of animals against their human

---

1. The paintings "El 2 de mayo de 1808" and "El 3 de mayo de 1808" (or "Fusilamientos") were financed by Ferdinand VII and were destined to commemorate rebellion against the French and celebrate the return of the Bourbon King in 1814, which is precisely when the slogan "¡Vivan las caenas!" was launched. This may explain the fact that Buñuel hears it shouted by the patriots in Goya's "Fusilamientos" (Shootings).

oppressors (who also happen to be French) is more than a surrealist joke; it points to certain parallels in the logic of animalization and subjugation as well as the subsequent loss and search for freedom.

Animalization was played out in those segments of Spanish culture that had rejected French models of modern citizenship and had instead adopted a seemingly more primordial outlook on life, characterized by a particular intimacy with violence and death. While the rejection of the French model of civilization was dictated in the particular historical moment of 1808 by the desire to be free from the French, that is, a desire for liberty, the culture that developed as a result of that rejection may have led, as Buñuel sees it, to a much greater loss of freedom in the social and political realm. This rejection of civilization was a gesture of rebellion (and thus for freedom) against foreign military forces as well as foreign norms that seemed to be trading spontaneity for correctness as, for instance in Mariano José de Larra's "Castellano Viejo". While the intuitive sense that civilization was taking away freedoms was likely correct, the manner in which these freedoms were taken back may have occurred in the error of considering violence as its most important aspect. This romantic illusion, however, that violence could somehow open the way to freedom was hardly unique to Spanish culture. In his essay on "Necropolitics," Achille Mbembe (2013) attributes to Hegel the construction of "spirituality" that arises in proximity to death:

> Within the Hegelian paradigm, human death is essentially voluntary. It is the result of risks consciously assumed by the subject. According to Hegel, in these risks the "animal" that constitutes the human subject's natural being is defeated.
>
> In other words, the human being truly *becomes a subject*—that is, separated from the animal—in the struggle and the work through which he or she confronts death. (164–65)

This description coincides with Federico García Lorca's vision of the Spanish intimacy with death acted out in *los toros,* discussed in detail later in this chapter. On the other hand, the subjectivity that arises from dealing with death has its effects on the body. Bullfighters are known to discipline their bodies, repressing pain and neglecting suffering. In this way, bullfighting culture could be seen as what Michel Foucault (2013) calls a "calculated management of life" (44) that includes both the mind and the body. Mbembe explains that the spirituality shaped by and for dealing with death serves the purposes of the state. Governing a nation of bullfighters, the state does not need to divide the population into those who are destined to live and those who have to die because a population will divide itself, with some volunteering to die if needed due to a spiritual need to do so. The relation between the spirituality shaped by death and the state biopolitics that takes advantage of it explains why the illusion of sovereignty and subjectivity that the bullfighting model provided ended up dehumanizing its subjects, turning them back to a bare animal-like life.

Persistent complaints in modern Spanish literature about the lack of political freedom, poverty, and the cult of bullfighter-like masculinity invited comparisons of Spanish people to animals in narratives by Emilia Pardo Bazán, Leopoldo Alas (Clarín), Luis Martín-Santos, and many others.[2] Paradoxically, the persistent construction of the inferiority of the lower classes in the political realm was simultaneously masked by an illusion to the contrary in the domain of the bullfighting spectacle. The bullfighter had an aura of defiant freedom and superior humanity. He was seen as a man who was allowed to transform into a beast, a werewolf moving effortlessly back and forth between the city and the primordial forest. He appeared to be free from the mores of civilization and from the hierarchies of his society. There was an important angle, however: when in front of the bull and watched by thousands, he was simply killable, like a Roman gladiator, a slave of the public, indeed like the bull that he confronted. The institutionalization of bullfighting under Ferdinand VII, and later contracts forcing matadors to undertake a certain number of fights per year, strengthened this slave-like aspect of the bullfighter's career. This loss of freedom was particularly dramatized in the case of Manolete, the most famous Spanish bullfighter of the postwar period and truly a national hero, who famously had to wait for the bullfighting season to end in order to retire, which he never managed to do as he died from being gored by a bull. What Lorca and other probullfighting intellectuals represented as a transgressive search for freedom through a pursuit of an animal-like humanity meant for many bullfighters a loss of freedom. While in the popular understanding, bullfighting displayed the superiority of human over animal, in fact the bullfighter was reduced to an animal-like predicament. Just as the bulls that he fought, he was well fed and well treated only to later be turned into a bloody mass for the public's morbid pleasure. It is from the perspective of this morbid pleasure that Lorca describes bullfighting as a spectacle that displays intimate relationship of the Spanish with death and that inspires a particular darkly erotic perception of life. This morbid aspect of Spanishness, connecting the violent and the erotic, has appeared enormously attractive to the rest of the world.

---

2. These are discussed in chapters 3 and 4.

This chapter analyzes Spanish debates on the aspects of the national culture that connect violence to the erotic, as Lorca defines it in his lecture, "La teoría y el juego del duende" (Theory and Play of the Duende, 1932). While Lorca envisions bullfighting as an exercise of freedom and search for the sacred, in the eyes of its critics this kind of search leads to animalization and a loss of freedom. Buñuel's *Le phântome de la liberté* reflects on this predicament in the context of Spanish relations with the French. In *Las Hurdes: Tierra sin pan* (Land without Bread, 1933), he parodies Lorca's theory on the Spanish people's intimacy with death, showing that it is due to extreme poverty rather than heroic bullfighting adventures. In the following section, Pedro Almodóvar's *Matador* (1986) is analyzed as a darkly humorous reflection on the erotic implications of the bullfighting's intimacy with death that focuses on the both bullfighter's and filmmaker's dependence on the public's taste. To end the chapter, the discussion of Pablo Berger's *Blancanieves* (2012) connects the aestheticization of violence and death in bullfighting to a social context where the poor and simple are easily taken advantage of by the greedy and manipulative.

LORCA AND BUÑUEL ON ANIMALIZATION, KILLABILITY, AND THE SACRED

In 1807 Napoleon pushed 100,000 French troops into Spanish territory under the pretext of attacking Portugal. Once the French army occupied Spain in 1808, Napoleon placed his brother Joseph on the Spanish throne by deposing the Spanish legal monarch as inept, causing the Spanish people to rebel. The ensuing war that started in May of that year and was to last four years, took a half of a million victims, and passed into history under the name of "The Spanish War of Independence." On Goya's famous painting "Fusilamientos del 3 de mayo," which pays homage to the first victims of the war, a French execution squad fires at the Spanish rebels, one of whom, dressed in a white shirt and with his arms spread in a shape of a cross, according to Buñuel, cries: "¡Vivan las caenas!" (*Le phântome de la liberté*). These famous words of welcome for the absolutist monarch in 1814 in Buñuel's film connote a symbolic rejection of French "civilization," which came to Spain as a violent imposition in 1808. According to Timothy Mitchell (1991) and Jesús Torrecilla (2004) this rejection significantly marked Spanish identity during the following century, and it had already been perceptible in the previous one. In *España exótica* (Exotic Spain), Torrecilla explains that it was as early as the eighteenth century, when the Spanish throne was occupied by the French

Bourbon dynasty, that Spaniards began to turn the complex of inferiority into a mark of their national identity.

According to Torrecilla, in reaction to French ideas of progress, some Spaniards deliberately posed as being less civilized, identifying themselves with exotic, peculiar, wild, and lower-class ways of being. One of the symbolic performances extremely significant in this context was bullfighting. However, the connection could be extended to a number of other performances, such as small town festivities involving bull-baiting, donkey teasing, or cock neck wresting, the last portrayed in the opening sequences of Buñuel's *Las Hurdes* (1933). It could also be extended to the construction of a macho-masculinity that is capable of killing with flippant indifference, which all of these practices required. Consequently, it could also be applied to cruelty toward humans. Mitchell (1991) and Torrecilla (2004) coincide in observing that after the War of Independence these reactionary ways of being became progressively and more self-consciously Spanish in a number of ways. They constituted performances against the Enlightenment's dislike for spectacles of torture and cruelty, against the more feminine masculinity constructed by the French, and against elites influenced by the French culture. This challenge to the French sympathies of the elite in popular entertainment was connected to the fact that bullfighting spectacles as well as other similar shows were in the eighteenth century performed mostly by the lower classes. Under Ferdinand VII, however, bullfighting became transformed from a spontaneous festivity into a spectacle with a great number of rituals designed to enact a connection between the people and their ruler and to form people's attitudes toward the state and toward violence. It linked people and the state in a display of state power and especially the sovereign's right to violence. This particular connection of the Spanish state to the bullfighting spectacle was cultivated by the monarchy, subsequent dictatorships, and even by the post-1978 semifederalist (but still centralist) Spanish democracy, when Enrique Tierno Galván characterized it as proof of the exceptional character of Spanish men (1988, 74–75). Bullfighting has become a symbolic spectacle of the Spanish state's biopolitical discourses with animals and the animalized as its central figures.

Buñuel's representation of the national insurrection in 1808 in *Le phântome de la liberté* can be interpreted as a reflection on history that universalizes the Spanish people's predicament during the French invasion and transcends binaries between the self and the other through surprising insights, which consist of human/animal comparisons. In other words, he compares the human loss of freedom to the similar loss suffered by animals. Animals appear in various scenes of the film: an ostrich and a cock are wandering in the bedroom of M. Foucault; a nurse driving to see her father encounters policemen hunting for foxes. Later a stuffed fox looks at us from a wall as his glass eyes shine with an eerie glow. M. Legendre disrupts the arrangement of his mantle piece with a large spider in a glass frame, stating that he is tired of symmetry, and the surrealist poet curses all the bastards who mistreat animals just before he shoots at the crowd. One could say that the ghosts of animals hunted and caged by humans haunt them in this film, every day, in all places, as if they were the ghosts of freedom lost. The significance of animals for the question of liberty is revealed in the last, absurd but significantly climactic sequence of the film, when we find out that an animal rebellion has shaken the Parisian zoo. As the two police chiefs, surrounded by armed guards, cross the threshold of the zoo's gate, the camera's circular traveling accelerates through the zoo's space, all images blur, and the camera stops on the face of an anxious-looking ostrich that, in the words of Linda Williams (1981), "cranes his neck and darts his eyes wildly about in a manner similar to Goya's bug-eyed, stretched-neck Spanish prisoner at the moment of execution" (176). Then we hear again the same cry of Spaniards welcoming backwardness rather than civilization, but now voiced by animals: "¡Vivan las caenas!" and the shooting. The literal blurring of the space between humans and animals through the camera lens brings out the symmetry between the execution of Spanish patriots and the massacre of animals. This symmetry between humans and animals, of which M. Legendre, a French bourgeois who will soon be diagnosed with cancer, is essential for the question of freedom in Buñuel's vision.

The comparison of the Spanish people's rebellion against the French invasion, which came with the rhetoric of progress, to the zoo animals' rebellion against different but equally violent aspects of progress, is surprising but justified. For both Spaniards and animals, progress came as an imposition from elsewhere with slaughter as the result. For Spaniards, progress brought by the French army resulted in the bloodiest war to date. For animals, technologies brought about by the civilization of reason resulted in slaughtering them in more and more efficient ways. While Spain was turned into a land of bullfighters and flamenco dancers by the tourist industry during the nineteenth and twentieth centuries, for animals, the Enlightenment resulted in their zooification in urban spaces (Friese, 2013). The early 1970s, when the *Le phântome de la liberté* was made, brought the boom of bullfighting and flamenco tourism in Spain. It was, perhaps, that present danger of turning Spain into a "theme park" (Gabilondo, 2008, 19) that inspired the surrealist director to represent his nation's modern history as a process of animalization and zooification. In drawing this pessimistic picture of the human loss of freedom, Buñuel coincided with some of the brightest philosophers of his times.

In *The Eclipse of Reason,* (1947) Max Horkheimer writes that reason is corrupted because it was born as an urge to dominate nature. The social apparatus, which is required to administer control over nature, backfires for "the power of the system over human beings increases with every step they take away from the power of nature. Enlarged, collective domination over nature is matched by a comparably heightened domination by some people over others" (24). In other words, Horkheimer sees the parallels between the domination and the consequent loss of freedom in human and nonhuman lives as well as in those realms that are simultaneously human and animal, that is, peoples who are animalized and rendered uncivilized. Buñuel's film seems to be reflecting on these processes in the context of Spanish-French cultural opposition. It represents civilization as a process of depriving freedom for both the civilized and civilizing. This loss of freedom leads to desperate attempts to regain it, ranging from sexual transgressions to random shootings at the crowd by a surrealist poet, all of which are doomed to fail but at the same time are understandable and ubiquitous. The action of the poet who shot at the crowd becomes a great public attraction and the accused leaves the courtroom free and signing autographs. Even M. Legrande's smoking can be seen as a transgression against the science of modern medicine, unavoidably leading him to cancer. Buñuel never talks about bullfighting, but it could easily be added to the chain of violations of the civilized habits that he portrays. The title, *Le phântome de la liberté,* indicates that as freedom is lost, it becomes transformed into a ghostly ideal, an omnipresent and powerful illusion, indeed a phantom provoking rebellions and transgressions of all sorts, but one that has little prospects of being realized.

Although Buñuel portrays French civilization as curtailing freedom, he suggests that the Spanish rebellion against it led to an even greater loss of freedom for Spaniards. The political culture promoted by the restored Spanish monarchy turned out to be deeply oppressive and manipulative by complex strategies, one of which was its sponsorship of the bullfighting culture. According to critics of the nineteenth-century culture, such as José Mariano de Larra, and at the end of the century, Eugenio Noel, the growing popularity of bullfighting under Ferdinand VII as well as other animal-centered entertainments where animals were defeated by humans had a profound effect. The obsessive comparison of humanity against animality in these festivities revealed a defensive attitude, the consciousness of inferiority and a subsequent need to perform superiority. According to Larra and Noel, instead of a superior humanity, these performances point to a progressive dehumanization through a real loss of political freedom, as well as a high death toll resulting from all sorts of violent conflicts. In sum, in the realm of the bull-

FIGURE 1.1
Cover of Eugenio Noel's *La novela de un pueblo en capea*

fighting culture, the inferiority of the lower classes constructed by the politics of the monarchy was masked as superiority. The cover of one of Noel's short anti-bullfighting novels (1926) sums it up bluntly: bullfighting culture means animalization. Because of this, as Noel points out, governments find bullfighting culture particularly helpful for their control of the population, that is, for manipulating citizens to renounce their right to life and sacrifice for the sake of the *status quo.*

The concept of "sacrifice" that often appears in the context of bullfighting is puzzling if the purpose of this sacrifice is just the public's morbid pleasure. It is also puzzling if *sacrifice* is to be understood according to Giorgio Agamben (1998) as being connected to citizenship and opposed to *killability.* The bullfighter, killable and sacred, seems to be in fact an example of *homo sacer,* an individual taken outside of the law, which would preclude the possibility of sacrifice for him. It may be helpful to notice, however, that if we consider Agamben's vision, the concept of sacrifice seem to be similarly misused in war, where soldiers in a battle are also simply killable. In both cases the sacrifice takes place beforehand, when a young man or woman decides to give up their right to life and, predisposed by spirituality shaped by necropolitics of the state, volunteers for a bullfighting career or for a war. (In the case of a popular draft, there is never a sacrifice but always killability and therefore no citizenship as Agamben understands it). The parallelism between those processes of renunciation of rights in bullfighting and in war is not accidental, since bullfighting had historically been a sort of an exercise for war. This extension of sacrifice in the context of bullfighting and in war is obviously manipulative and mystifying in the way it gives the meaning of sacrifice to killing. A similar slippage is recorded in *Le phântome de la liberté,* where the 1808 Spanish rebellion against the French, marked by sacrifice, is exposed as animalization and slaughter in the last sequence of the film.

Ferdinand VII saw in the popular anti-French zeal that led to the enthusiastic welcome of "the chains" the possibility for structuring his own dictatorial power. In Antonio Buero Vallejo's *El sueño de la razón* (The Dream of Reason, 1970), Goya's friend Doctor Arrieta complains that Ferdinand VII closed universities and opened schools of bullfighting. The king sponsored bullfighting rings, bull-breeding, and actual bullfights while saving money on education. The bullfighting spectacle contained a number of rituals whose purpose was to connect the authoritarian monarchy with its people and the people with the bullfighter. Bullfighters performed patriotic salutations to those representing the nation's power, and the people imitated bullfighters. Slowly bullfighting was transformed from a popular form of mass entertainment to a mark of Spanish national culture. The spectacle received the significant name of "La Fiesta Nacional" (The National Feast) and was celebrated during state holidays and on other culturally significant dates. Local fights for symbolic power through a display of skill and courage in performances of killing animals (so called *capeas* in the villages) became institutionalized, built into the system of state power. Seeing that moment in history as a purposeful manipulation of bullfighting festivities for the interests of the reactionary monarchy's biopolitics allows today's intellectuals, allied with the animal rights movement, to argue against a view that bullfighting is an "authentic" Spanish tradition.

This construction of national culture was not unique to Spain. In his analysis of Chinese history, Prasenjit Duara (1995, 1999) describes the state's construction of national traditions as means of assuring stability of power for a governing class as *othering* of the nation (xx). The *othering* of a nation seems to be a process parallel to what Torrecilla calls "nacionalismo defensivo" (23) (defensive nationalism).[3] It is fueled by a state's desire to shape a citizenry capable of sacrifice and insensitive to death. The mythification of violence legitimizes repression inside and aggression outside of the state's borders.[4] In this way, defensive self-othering of nations is unavoidably reductive of citizenship privileges. In welcoming the absolutist rule of Ferdinand VII, the Spanish people were subjected to indiscriminate state violence that indeed reduced many to the status of a bullfighter. Although not exactly equal, the human and the animal enacted the script of what the absolutist king, and others who followed, chose to represent as the unchangeable dynamics of domination and destruction of nonhuman by human, of the weaker by the stronger. Ignacio Sánchez Mejías's famous allegory sums it up succinctly: "El mundo entero es una enorme plaza de toros donde el que no torea embiste" ("La Tauromaquia" n. pag.) (The world is a bullfighting arena and people can be bullfighters or bulls). In the bullfighting worldview, all people are condemned to fight each other to death. Sánchez Mejías's metaphor displays the

---

3. Torrecilla (2004) defines and explains the concept in chapter 1.

4. Tim Cross in "Othering the National Self: 'Japan the Beautiful' as Lethal Aesthetic" argues,

> During the eighties the impenetrable cultural distinctiveness of Japan was employed as a strategic response to American demands to open Japanese markets.... An aestheticised cult of death was a significant presence in the Showa formation of "the authentic" Japan. Cherry blossoms were employed by political elites to advocate death in the name of the Japanese nation. The ideological application of literary and visual texts also implicates cultural practices like tea in the creation of patriotic fervor. (Cross, 2012, n. pag.)

Abeer Najjar (2010) tells a similar story in Palestinian context in "Othering the Self: Palestinians Narrating the War on Gaza in the Social Media."

awareness that in a bullfighting culture, the public shares bullfighter's predicament. The "life exposed to death" in the bullfighting ring can therefore be contemplated as a model of national identity established by the state biopolitics. "The technology of power centered on life" (2013, 48) that Foucault considers as a mark of modernity is in the Spanish case to a great extent spectacular and performative.

The slippery movement from a voluntary sacrifice to killability can be also observed in the art accompanying the Spanish War of Independence. The sacrifice that is envisioned as voluntary in Jovellanos's poem "Canto guerrero para los asturianos" (The War Song for the Asturians, 1812), which summons Spaniards to give their lives for the country, is represented as killability in Francisco Goya's series of *Desastres de la Guerra* and especially in his "Naturaleza muerta," produced between 1810 and 1812, which constitute one of his most pessimistic reflections on war. Goya paints a ram's carcass with its head pulled away from its skin, with one visible eye retaining a shade of lost life—surprisingly human. In this animal's butchered flesh with a human-like eye, Goya finds the true expression of war. In *Desastres,* killability is represented as dehumanization. In the words of J. M. Matilla "Víctimas pierden su humanidad, cuando se convierten en pedazos de carne desmembrada, donde Goya alcanza la cumbre expresiva de la brutalidad" (n. pag.) (Victims lose their humanity when they turn into pieces of dismembered flesh, where Goya reaches the expressive heights of brutality). In Goya's *Tauromaquia,* a similar obsession with the animalization of humans is notable, so striking also in his *Caprichos* and in *Desastres.* The first eleven prints of *Tauromaquia* are illustrations for Nicolás Fernández de Moratín's *Carta Histórica sobre el origen y progresos de las Fiestas de Toros en España* (Historical Letter on Origins and Development of Bullfighting Festivities in Spain, 1801), engraved in the beginning years of the nineteenth century. The work was interrupted by the War of Independence, whose horrors influenced the mood of the later etchings. According to Alfonso E. Pérez Sánchez and Julian Gallego (1997), after the end of the war, Goya's style in *Tauromaquia* became closely related to that with how he represented the war in *Desastres.* The bullfights and the war fused in various scenes. For example, as Lafuente Ferrari (1966) points out, in order to depict the origins of bullfighting in Arab times, the artist uses Arab and French soldiers as models, called in Spain *Mamelucos,* who formed part of Napoleon's army in Spain. The bodies of a bull's victims lying on the ground in the etching 12 remind the viewer of the cadavers in *Desastres.* J. Blas Benito asks in his "Prólogo: la tauromaquia de Goya" (Introduction to Goya's *Tauromaquia,* 2001): "Cuál es la diferencia estética entre las cruentas matanzas de los *Desastres* y los mortales impactos del caballo, el picador y el toro de las estampas 26, 28, y 32 d la *Tauromaquia*?" (n. pag) (What is the aesthetic difference between the cruel killings from *Desastres* and the deadly encounters between the horse, the matador and the bull in the etchings 26, 28, and 32 of *Taurmoaquia*?). In other words, the war changed Goya's way of looking at bullfights; instead of the spectacular and artistic, he could only focus on the horrific. The critic supposes that most of the etchings done between 1814 and 1816 were hard to sell precisely because they reflected the trauma of war and were nothing like the traditional, colorful, anecdotal, and distanced bullfighting scenes that people liked to buy and hang over the dining table.

Lorca's idea that Spaniards have a very particular, intimate relationship with death was inspired by bullfighting. In his lecture, "Juego y teoría del duende," *el duende* is thought of as "el espíritu oculto de la dolorida España" (the hidden spirit of the suffering Spain), which comes from blood and, similar to Unamuno, from the "deep tradition" (viejísima cultura) and blood's connection to the earth as "el espíritu de la sierra" (the spirit of the mountains).[5] Lorca's theory of the Spanish soul identifies for the most part with the emotions of Andalusian Gypsies and the Castilian cultural tradition. Lorca claims blood superiority for the Gypsy community of Cádiz and explains its ritual connection to the mythical past of the sacrifice of a bull to Gerión.[6] The concept of *el duende* relates to the violence and darkness justified by this mythical sacrifice, which "es un luchar y no un pensar" (is a fight not a thought), "quema sangre" (it burns one's blood) and guarantees "la vida de la muerte," (life in death), which defines Spanish culture according to Lorca. The bullfighter is the best example of someone who is more alive after death. In this nine-page lecture, the word "bull" in the context of bullfighting appears fifteen times. It is in the context of bullfighting that Lorca calls the Spanish people "contempladores de la muerte" (spectators of death). It would be fair to say that Lorca's theory of the Spanish soul is to a great extent based on an analysis of the bullfighting spectacle. Because *el duende* only comes when death is close, "En los toros adquiere sus acentos más impresionantes" (In bullfighting it reaches its most impressive tonalities). The last page and a half of the lecture is devoted to the aesthetics of bullfighting and it is there that Lorca pronounces his widely quoted sentence: "España es el único país

---

5. There are no page numbers in the online version of the lecture.
6. He says that "Manuel Torres [es] el hombre de mayor cultura en la sangre" (Manuel Torres carries more culture in his blood) and that "Eloísa, la caliente aristócrata, ramera de Sevilla, descendiente directa de Soledad Vargas, . . . en el treinta no se quiso casar con un Rothschild porque no la igualaba en sangre" (Eloisa, the hot aristocrat, prostitute from Seville, direct descendent from Soledad Vargas, . . . in the 1930s did not want to marry a Rothschild because he was of an inferior blood).

donde la muerte es el espectáculo nacional" (Spain is the only country where death is the national spectacle).

In the same lecture, Lorca establishes a hurried European and Spanish literary canon in which he claims Spanish Castiliano-Andalusian primacy. *El duende* is superior to "el ángel" (the angel) and "la musa" (the muse) because it guarantees Spanish art's superiority over Italian and German art. Within Spain, it is Castilian productions that are best as they are inspired by *el duende* as opposed to Galician ones, under the influence of an angel, and Catalonia is the kingdom of the muse. At the same time, the poet proclaims the superiority of Jorge Manrique over Gonzalo de Berceo and then goes on with the names of Arcipreste de la Hita and Juan de la Cruz as superior to Góngora and Garcilaso, preferring always those who exhibit the presence of *el duende* that is a fascination with death. This appropriation of Andalusian folklore, and especially its Gypsy elements, to represent Spain became a common feature of the postwar period, although at the same time it was accompanied by discrimination against real Gypsies. The idea that the sacrifice of the bull and the culture of aestheticization of death that surrounds it is a reason for the superiority of the Spanish culture over non-Spanish ones had also been exploited in Falangist discourses during and after the Spanish Civil War period. Even during the period of democracy, it appeared in various pro-bullfighting publications, such as those by Enrique Gil Calvo (1989), Enrique Tierno Galván (published after his death in 1988), Andrés Amorós, (1987) and José Bergamín (1989).

Lorca's lecture can be read as a manifestation of cultural nationalism that vindicates its own exceptionalism and superiority, manifesting at the same time that this superiority can be only understood by those endowed with a special sensibility able to establish communication with *el duende*. Focusing on intimacy with death as an important element of national spirituality, Lorca and various other intellectuals, contributed *nolens volens* to the state necropolitics that they themselves criticized and personally suffered from in other contexts. Similar to Goya in his still life, Buñuel ridicules this grandiose vision of a Spanish intimacy with death in his early documentary, *Las Hurdes: Tierra sin Pan* (Land without Bread, 1933), where he also links the cruelty perpetrated on animals to poverty. As Jeffrey Rouff (1998) notes, "*Land without Bread* reverses the anthropomorphism of the nature documentary and instead treats its human subjects like animals" (52). Rouff sees this film as a sort of "black comedy" (49), ridiculing first ethnographic documentaries such as Robert Flaherty's *Nanook of the North* (1922), noting that in representing the people's fight against nature, Buñuel takes away the nobility of such an enterprise, which was essential for Flaherty's work.

Buñuel's parody also ridicules the representations of Spanish culture where, in Lorca's words, death is a national spectacle, showing how uninspiring a desperate fight for survival of the flesh can be. Thus it constructs a parody of Spanish intimacy with death where bullfighting is significantly substituted by cock neck wresting, and heroic sacrifice is reduced to the suffering sicknesses and dying of bare flesh. In his documentary, Buñuel relies heavily on Maurice Legendre's ethnographic work (2006) on *Las Hurdes*. Legendre writes that the earth beneath the Hurdanos lies ever ready "to open up like a tomb" (78) and that sickness is "the ordinary state" of the local inhabitants (104).

The film opens with a scene of a little town in Las Hurdes, Alberca, celebrating its annual festival. A rope is stretched between housetops and a live rooster is tied by his legs and hung upside down from the rope. A number of young men prepare to prove their masculinity in a competition. They will ride on horseback under the rope and jump to wrest the head of the rooster off its body. The camera takes the point of view of the hanging cock, showing in a high angle the people crowded around in eager anticipation. For a second, it focuses on a young man's face looking up at the camera, which films him approximately from the point of view of the cock hanging upside down. The man is caught with a baffled, cruel smile as if taken from one of Goya's *caprichos*. It is evident from the critical camera angles, its dehumanizing close-ups, (e.g., old women combing their hair or the silver charms hanging on a baby and the swaying movement that alludes to people's drunkenness) that, in spite of the apparent scientific distance, Buñuel does not hide his dislike of the spectacle.

In the commentary that Buñuel had written together with Pierre Unik, read with an "insolent indifference" (Rouff, 50) during the film's premiere, the abhorrence of the traditional festivity of the people of Alberca and of the sickening poverty of the inhabitants of the villages in Las Hurdes is mediated by dark humor. Rouff notes a number of gaps between the images and the commentary. For example, the girl's throat does not seem inflamed, the goat that falls down from the cliff is visibly shot by the film crew, the children that go to school may look poor but their hair is combed. These gaps are, in Rouff's interpretation, purposefully offensive toward the bourgeois public who would fall into the trap of the documentary convention and take every word at its face value without questioning it like a "choir of idiots" (51). Following Rouff's commentary, the exaggeration of the voice-over in the film may be viewed as an effort to blur the opposition between the animalized objects of the documentary and the authority of the subjects, studying and viewing what's on the screen, implying the similarity in their dehumanization. In *Las Hurdes*, the parody of the scientific discourse, built into the convention of the ethnographic documentary, subverts the presupposed superiority of the ethnographer as well as the superiority of the public viewing the film. As a result, instead of building hierarchies and stating the superiority of civilization over the animalized poor, the film in fact does the opposite; it equalizes across the board. If *Las Hurdes* is viewed as a black comedy, as Rouff suggests it should be, we need to remember that such black humor has always been a way of talking about disasters, a way of deforming them in order to ease their reception, but it never aims at negating them. As Juan Egea (2013) argues in his book on dark laughter, it signals that in fact there is nothing to laugh about. The half-baffled and half-cruel smile of the men watching the cock hanging upside down on the rope in Alberca is the facial expression that Buñuel may be projecting to his bourgeois public, similarly indifferent and curious to hear about the death hanging over Hurdanos.

MATADOR, MANOLETE, AND THE PUBLIC

Foucault writes that as a mark of modernity, the regulation of life by the rule of blood is replaced by an obsession with sex. While, according to Foucault, concern with blood led to the obsession with death, transgression, and sovereignty, in contrast "sexuality was on the side of the norm, knowledge, life, meaning the discipline, and regulations" (2013, 51). But while, in the nineteenth century, sex became the crucial target of power, it has not happened without overlapping with the dynamic of blood. For example, "in the second half of the nineteenth century, the thematic of blood was sometimes called on to lend its entire historical weight toward revitalizing the type of political power that was exercised through the devices of sexuality" (2013, 52). The bullfighter's sexuality could be viewed as an example of this kind of overlap, as it gave more importance to transgression and shedding blood than to educating children or building a life-sustaining environment.

According to Foucault, an interesting ambiguity affects the relation between sexuality and power. While sexuality "is a result and an instrument of power's designs," sex is the power's "other" (2013, 54). Even if the myth of sex is somehow a biopolitical production for the sake of managing the population, the moment of experiencing it allows us to resist and subvert the power that we are subjected to. This is what Almodóvar attempts to imagine in *Matador* (1986), a film that grows out of bullfighting metaphors, equalizing killing and dying with the sexual act and orgasm, and comparing women to bulls. Ultimately, however, bullfighting eroticism is disappointing (or not). While, according to Foucault, in modern sexuality, "sex is worth dying for"

(2013, 57), in the bullfighting version, it is only death that is worth dying for, and the only desirable sex is one that ends in death.

While in *Las Hurdes*, Buñuel deconstructs the myth of a grandiose intimacy with death, bringing it down to the dreaded reality of poverty, exaggerated by both the camera and commentary, in *Matador*, Almodóvar portrays the bullfighter's sex as a killing orgy. He also connects it to the story of Manolete, Spain's most famous postwar bullfighter. According to Emilio Maille's documentary *Manolete, la leyenda*, "La propaganda franquista lo ha convertido en un símbolo" (Franco-era propaganda made him into a symbol), so much that he would be identified as one of the most sacred (and therefore most killable) icons for the state's ideology: "era en parte un santo" (he was part saint). Rumors that Manolete "se entrenaba estacando con prisioneros" (trained using prisoners tethered on stakes) challenged this sainthood. Or, perhaps, not, since, according to Maille, these rumors were consistent with the bullfighter's ethos of a killer and only confirmed that Manolete was the regime's favorite hero.

The obsession with death is an integral part of both Manolete's and *Matador*'s stories. Maille's documentary starts with a close-up from Abel Gance's film (never completed) in which Manolete himself was going to act before the cameras.[7] This shot is followed with photos of deceased bullfighters as they were photographed in the moment of death while the voice-over announces Manolete's killability, "Cualquier día en algún lugar un toro le matará de una cornada, como aquella que recibió Granero en 1922" (At any moment and anywhere a bull could kill him with one thrust, like that which Granero received in 1922). Granero's gored face appears on screen and, a moment later, Manolete is shown in the bullring gazing into space as if he had a premonition of his own death, as if he were on his way to the martyrdom that was going to convert him into another morbid, sacred icon.[8] In Almodóvar's *Matador*, a very similar shot of Nacho Martínez's face appears when Diego enters a movie theater and looks at the screen where in *Duel in the Sun*, the two lovers kill each other and then die in each other's arms. The position of the face on the screen, the light, and the expression in the man's eyes fixed somewhere beyond the immediate reality, his hair combed back and sticking out a bit, and the face itself, angular, sad, halfway between life and death, are the same in both movies.

The only woman in the mythologized life of Manolete was Lupe Sino. Lupe competed with Manolete's mother, Doña Angustias, and with Manolete's

---

7. I owe this discovery to a conference conversation with José del Pino who led me to discover "Manolete, una película desconocida de Abel Gance" (1985).

8. Possibly also a shot of Gance.

public to win the bullfighter's heart, and according to many testimonies, she lost the fight. The real love of the bullfighter was his public. While at Lupe's side, Manolete had planned "según confiesa a algunos íntimos, volverse a encontrar con el hombre sencillo que hay en él" (according to intimate friends, to rediscover the simple man inside himself), Maille shows his relation with the public was tense and marked by contradictory emotions, passionate and antagonistic. Articles covering Manolete's death, again and again, almost obsessively, insist that in the last hours of his life he made no remark to or about Lupe. In the article "Manolete el mártir expoliado" (Manolete, the Exploited Martyr), K'Hito (2004) writes, "Pocos minutes antes de morir, la principal preocupación de Manolete fue la reacción del público. 'Maté al toro de la estocada? Y no me han dejado ni una oreja siquiera? Fue lo primero que preguntó al recuperarse parcialmente del choque. Al saber por su apoderado, Camará, que le dieron las dos orejas y el rabo, se sonrió, él, que nunca sonreía, lo hizo al borde de la muerte" (n.pag.) (A few minutes before his passing, Manolete's main worry was the public's reaction. "Did the bull die from my blow? And they didn't even give me the ear?" was the first thing he asked upon partially recovering from shock. When he learned from his manager Camara that he had been awarded both ears and the tail, the man who never smiled, smiled, on the verge of death). *El Mundo de Andalucia* remembers that before his death, Manolete also said, "Álvaro, tráme mis medallas" (n. pag.) (Alvaro, bring me my medals); he was informed that Lupe was waiting in the hallway, but he did not answer. Manolete had a premonition that he would only receive the public's sought-after "unconditional love at death" ("Manolete, Doña Angustias y Lupe Sino," 1998, n. pag.). In an interview reprinted in *La Nación*, he confessed in a theatrical tone: "Tengo mucho miedo, pero el público, que paga, pide: 'Más cerca!' Yo me aguanto el miedo, que tiene mucho más filo que la espada que debo hundir en la diana del morrillo del animal" ("Un mito del toreo, 1997, n. pag.") ("I'm so afraid, but the public, who pays, demands, 'Nearer!' I swallow my fear, which is much sharper than the sword I must sink in the 'bull's eye' of the animal's snout"). According to Maille's documentary, at the end of Manolete's career, "una parte del publico se ha convertido en un toro peligroso" (a part of his public had become a dangerous bull). They whistled at him because he did not approach the bull closely enough, because the tickets for his fights were too expensive, and because his performances had become repetitive. His public threatened to betray the love they professed for him and applauded other bullfighters with more vigor. Lupe thought that he was killed by his public. After his death, she said: "They didn't let him retire until he was left dead" ("Manolete, Doña Angustias y Lupe Sino," 1998, n. pag.).

Maille describes Manolete's passion as not directed toward women but toward the public, which is compared to a bull and which brings death. This is also the reading of a bullfighter's love life in Almodóvar's *Matador*. Diego is a retired bullfighter, who transfers his passion for killing from bulls to women, whom he strangles during lovemaking and buries in his garden. He does that until he meets María, a bullfighting fan, who has a similar morbid passion. She becomes Diego's lover in order to kill him and to shoot herself in the sexual climax. María, however, is not a real woman, but rather, she symbolically represents Diego's public. The battle for Manolete's heart fought between the public who wanted him dead and Lupe who wanted him safe is dramatized in *Matador* by the character of Eva, who wants to save Diego, and María, who wants to kill him and die with him. Eva, full-faced and full-bodied, even physically resembles Lupe, and María, with flamenco features and dramatic angles, reminds us of the woman from the 1947 poster announcing the Linares bullfight at the fair where Manolete died. The manner in which María worships Diego corresponds to the way the Spanish public revered Manolete. She only surrenders to her idol when she is certain he "will continue being a *matador*"; her love, like the public's love, is conditioned by his "performances" for her. When Diego kisses the television screen where María's features can be distinguished among the crowd, the bullfighter is passionately kissing the face of his public.

Almodóvar personifies the metonymic chain of signifiers—public-bull-death—through the monstrous character of María who has aspects of each. Whereas Lupe blamed the public in Manolete's drama for his death, in *Matador* Eva blames María for what she perceives as Diego's falling prey to the obsession with death haunting him since his accident, when he was gored by a bull. Like Lupe, Eva does not matter in the end. From the moment Diego sees María cheering him on from the bottom of the bullring, crowded with people, María personifies the public and becomes Diego's only desire, even if he realizes that she wants "to see him die," or maybe precisely because of this.[9] When Diego, the bullfighter, and María, his public, meet, they determine to kill each other during a sex without wasting any time beyond what is necessary to prepare the staging for their love and death scene. The exaggerated urgency of mutual killing perceived by both characters is a darkly humorous outcome of the symbolic dynamics of bullfighting.

---

9. This realization appears also in Luis Martín-Santos's short story entitled "Tauromaquia" (1970). It is a scene of an encounter and a conversation between a bullfighter and his lover in whose climax he realizes that she comes to bullfights due to a dark desire to see him die.

As Buñuel before him, Almodóvar with tongue in cheek compares humans to animals. A televised commentary regarding an eclipse establishes the aesthetic context of the deadly desire of the bullfighter and his public, María. As the lovers prepare their suicide, the solar eclipse is approaching. A television speaker announces:

> La luna ocultará el disco solar. El cielo se oscurecerá. Aparecerán las estrellas haciéndose visible el disco solar. Las aves volarán a sus nidos y las alimañas se precipitarán sobre sus madrigueras creyendo que llegó súbitamente la noche sin sospechar que a los pocos minutos el sol volverá a brillar con el consiguiente trastorno para estos pobres seres carentes de raciocinio y esclavos de sus instintos.

> The moon will hide the solar disc. The sky will darken. Stars will appear, making it possible to see the solar disc. Birds will fly to their nests and vermin will crawl to their dens thinking that it's night without suspecting that a few minutes later the sun will again shine, causing consternation in those poor creatures that lack reason and are slaves to their instincts.

Even the television commentator shows her fear of the sun's eclipse, although she claims that humans are different from nonhuman creatures devoid of reason. In a dark parody of bullfighting heroism, Almodóvar elevates his two monstrous protagonists, because, in contrast to all little animals and little people that hide in their burrows, Diego and María will use the moment of eclipse to embrace their urge to kill and die while copulating.

Almódovar's attitude toward the myth of Spanish intimacy with death seems ambiguous. While he recognizes the erotic attraction of bullfighting for the national sensibility, he rejects the political discourses that use bullfighting, spirituality, and sexuality for their own purposes. In other words, he aestheticizes bullfighting sexuality, but condemns biopolitics that produces this sexuality. If *Matador* is the least successful of Almodóvar's films, as he admitted in an interview with Frederick Strauss (1998), it is perhaps due to this inconsistency. In attempting to recreate bullfighting sexuality, however, Almodóvar, not unlike Buñuel, fuses it with its own parody. Right before the murderous sexual orgy organized by Diego and María, Diego buys flowers from a Gypsy who attempts to tell his fortune and stops, terrified as he sees death written in his hand. In response to her facial expression, Diego winks his eye as the camera closes up on his face. This seems to be a sign for the viewers of the film to take what is coming as tongue in cheek. As in *Las*

Hurdes, the dark humor does not change the meaning, but rather it places the responsibility with the public for desiring violent shows.

The film reflects on the relationship between art and ideology that serves state's biopolitics. *Matador* suggests that the bullfights and the violent cinema arise out of public demand and that the film director as well as the bullfighter find themselves obliged to satisfy this desire if they want to be applauded. The initial scenes of *Matador*, where we see Diego masturbating while viewing killings in Mario Bava's *Blood and Black Lace*, establish a parallel between the violence and eroticism of a bullfight and the violence and eroticism of cinema, and consequently, also between the bullfighter and the director and their respective audiences, which, in María's words, expect to get some good shows out of them. The film director, like the bullfighter, takes pains to satisfy the audience's taste, and it is precisely this taste that condemns him to privilege violence. Almodóvar is one of the artists who attempt to "save" bullfighting for democracy while building his symbolic capital for the bullfight-loving public. While Almodóvar's parodies often partake in Buñuel's dark humor, he also shares José Ortega y Gasset's and Lorca's desires to be located in the center of what he imagines as Spanish taste. But the Spanish public may not be as enamored by bullfighting as Almodóvar seems to believe.[10]

BLANCANIEVES AND THE QUESTION OF TASTE

Pablo Berger's silent film *Blancanieves* (Snow White, 2012) goes much further in its cultural critique. Berger subverts the bullfighting spectacle by drastically deforming the aesthetics of the silver age of Spanish culture where silent films were produced and bullfights were most popular. While the sublime aesthetics in the first part of the film goes awry and transforms into a horror and then into a freak show, the bullfighter, an admired hero, becomes vulnerable, helpless and taken advantage of "bare flesh" (Agamben, 1998), in the final scenes substituted by a wax-like figure displayed to deceive people for money. Apart from this degradation of bullfighting aesthetics, the film tells a story of a tragedy caused by bullfighting, and its happiest moment is when the public indults the bull instead of killing him.

This film, which has received nine Goyas and some international awards, begins when a famous bullfighter is gored by a bull, and his pregnant wife,

---

10. The Gallup polls show that the number of bullfighting fans fell from fifty percent in 1980 to only ten percent in 2000, while twenty percent of Spaniards declared only a partial interest. (Instituto Gallup, 2002). See introduction, note 12, and chapter 5 for analysis of these data.

scared by the accident, goes into childbirth and dies, leaving a daughter. The father survives though paralyzed, but is manipulated by an evil nurse into abandoning his daughter. Encarna (Maribel Verdú), who seduces him for his money, makes sure that his daughter's life is like that of the saddest Grimm brothers' stories. Carmencita, the daughter of the bullfighter, lives in animal-like conditions, and her only joy comes from a friendship with an animal, her cock, Pepé, who follows her everywhere making her smile. Her loving grandmother also dies, and Carmencita has to dye her white dress into black as a sign of mourning. Soon afterwards, her beautiful hair is cut, and her pet cock killed by her stepmother, Encarna, who also kills Carmencita's father and attempts to murder the girl. Carmencita miraculously escapes death, saved by a dwarf who is a comic bullfighter. Traumatized, she forgets who she is and learns bullfighting by accompanying seven dwarfs in their comic shows. In the second part of the movie, the grandiose scenery of Sevilla, its bullfighting plaza, and Villalta's mansion cede to disheveled little towns' plazas, a dwarf circus, and deformed bodies. While still indulging in the melodramatic emotions elicited by the bullfighting *pasodobles*, the second part of the movie warns that the stories that accompany the music are dark and painful. The childhood illusion of the beauty of the bullfighting culture appears terribly deformed as this illusion destroys its subjects. Yet, even if bullfighting is the cause of all the terrible things that happen to Carmen and her loved ones, even if it is compared to a freak show, the real "freaks" in this film are those who take advantage of the bullfighting dynamics to get rich and achieve power.[11] Soon Carmencita, called now by the dwarves Blancanieves, becomes famous. She is manipulated into signing a contract for life and is invited to perform in Sevilla, where Encarna poisons her with an apple.

Somewhat similar to Buñuel's work in *Las Hurdes* and Almodóvar's in *Matador*, Berger's film distorts what Lorca praised as the Spanish intimacy with death by creating a sequence of spectacles, in decrescendo, that go from a classic bullfight to a morbid photographic session featuring a dead bullfighter surrounded by his smiling family and friends, through a comic bullfight to a freak show. Berger compares these spectacles to real life: the attempted murder and rape of Blancanieves, the killing of her father by Encarna, and Encarna's subsequent deadly attack on her docile lover whom she animalized in sexual play. The sexual habits, domestic violence, lack of family protection, and human/animal relations are parts of one cruel prime-

---

11. In this as well as various other aspects, *Blancanieves* establishes dialogue with *Freaks* (1932) directed by Tod Browning. I owe this observation to Juan Egea. Some of the scenes are also inspired by the remake of *Freaks*.

val dynamic where the stronger and the smarter have no pity on those over whom they have power, like in bullfighting.

The happiest moment of the film, when the whole bullfighting arena fills with white scarves demanding the bull's pardon, minimally suggests that suspending the killing is the most enjoyable element of the tradition. Berger also makes the point, however, that it is not enough to ban bullfighting, because those who intentionally benefit from the blood of others—the vain stepmother and the bullfighters' agent—will find ways to continue doing so. When the agent takes out the contract to be signed by the girl who cannot even read it, the tip of his pen flies out like a head of nuclear rocket, signaling the desire for domination and destruction not only "until the end of her life" as the contract claims, but even beyond it, which is as it eventually comes to be. Even in death, she is not free from the greediness of the manager who exploits her beautiful cadaver, selling her kisses to the necrophiliac public.

While indulging in the beauty of the music, dress, buildings, countryside, and the faces of iconic ladies looking on at the bullfight, the film develops a deep concern with the human/animal relations acted out in bullfighting. There are memorable scenes of hunting and the cutting of a cock's head, which connect to the nausea evoked by dinner scenes when chickens and other small birds are served. In one of its sequences, the film has five consecutive scenes that portray humans in their relation to different animals and suggest the parallelisms that they have with interspecies violence. In the first scene, Encarna's ill-fated lover takes the place of a dog at her feet, posing for a painting on all fours. The next opens with Blancanieves pointing her sword at a bull in her father's room, to be followed by an image of the nose of a bunny that is hunted by Encarna's dogs. In the scene that follows, Encarna leads her dogs by a leash to the bedroom, where Blancanieves is secretly playing with her father, only to discover her pet cock, Pepe, hidden by the curtain when he comes out to protect Carmencita. Encarna then orders the cock to be killed and served to the girl for dinner. Her following victims will be human. The bunny's nose moving in fear, as in Carlos Saura's *La caza* (1965), has already alerted the viewer to this possibility. *La caza* establishes a series of parallels between humans killings animals and other humans, while comparing the hunt to war and criticizing the fascist ideas of the hunters.[12] Berger pays homage to this classic film, adopting its discursive coherence in the construction of inferiority and the consecutive destruction of subaltern humans and animals as in fascist biopolitics. Animalization as a construction of inferiority in sexual relations and as a form of oppression is a constant

---

12. The film is analyzed in chapter 4.

motif of the film. Innocent and unaware of the injustice inflicted upon her, Carmencita suffers like an animal, and she is treated like one by Encarna, who tortures her for a long time before ultimately killing her.

In contrast, other human/animal relations are deeply meaningful in the film. Blancanieves's forgotten identity literally leaps out of the eyes of the bull that later avenges her death. After the attempted assassination by her stepmother's lover, Blancanieves loses her memory when bullfighting dwarves come to take care of her. She only begins to remember at the sight of a cooked bird on her plate, served during one of the dwarfs' dinners.[13] The memory of Blancanieves is triggered twice by the depth of her bond with an animal. First it is her beloved cock Pepe, whose death makes her incapable of eating bird meat. (The scene in which Encarna eats Pepe's thigh in front of the girl, terrorizing her, is one of the most torturous moments of the film). As a cultural commentary, revising myths and stereotypes, Berger leads the viewer to see that the natural relationship between people and animals is that of sympathy rather than cruelty and killing, and the film not only questions bullfighting but also hunting and eating meat. The critical eye of the camera challenges cultural *doxa*, for example, when the jovial smile of the old cook acquires a tone of cruelty at the moment she cuts off the cock's head.[14]

The second time Blancanieves's memory returns, this time scene by scene, is when she looks into the eyes of the bull that stands in front of her in the bullfighting ring of Seville. A series of accelerated flashbacks of the most important moments of her life may be interpreted as a proposition about her identity. She experiences this hallucination-like trance of returning memory during the few seconds she has as she faces the bull that may mean her death. That bull is not a thoughtless fierce beast, however, but rather a creature that understands somehow his situation and establishes a relationship with the human being that he confronts. He stops in front of Carmencita and does not attack her as she looks into his eyes. Through the images that run through Blancanieves's head, projected onto the eyes of the bull, she

---

13. This scene establishes a dialogue with *Freaks* (1932), but even more directly with its remake *Freak Show*, (2007) in which a group of criminals hides in a circus, composed of a number of monstrous humans, similar to the original film, to act as circus security. The girlfriend of one of the criminals seduces and attempts to marry the director of the circus. In the scene of the engagement, she sits at a table with all the circus "family" and is invited to drink from the same bowl as they do. At that moment, she feels unable to perform as nausea overcomes her. The meaning of the nausea in *Blancanieves* is different, however, as in Berger's film, the girl is not really reacting to the monstrosity of her companions, which she does not perceive as such, but rather to the monstrosity of her culture's cruel treatment of animals. As she looks at her plate, the cooked bird appears for a moment alive and reminds her of her childhood pet.

14. "Doxa" according to Pierre Bourdieu (1977) are unquestioned cultural beliefs.

realizes who she is. Among various scenes of her life, there appears her cock, Pepe, who returns to her as her dead father does to stand by her in the most difficult trance. The cock and the father, as if coming out from inside of the bull, form Blancanieves's identity as those that she loved and who became parts of her.

While on one level, the meaning of the film arises from the progressive aesthetic deformation, on the other level, it results from the contrast between the spectacular and the everyday reality. The scenes of spectacular violence, such as the goring of Villalta, are followed by shots of pain and sorrow as they lead to loss of health and love. Berger is as sensitive to the suffering flesh as Almodóvar, especially in his later movies, but in *Blancanieves* love does not thrive in proximity to death, but rather to the contrary, it is destroyed by it. After Villalta's tragic goring by the sixth bull of the *corrida*, his wife dies during a premature birth brought about by the shock, and he becomes incapable of loving and protecting his daughter. Perhaps, the most important figure of the film is the parallel between the story of Blancanieves, who suffers as a child devoid of father's protection and the bullfighting context. By fusing the plot of the Grimm brothers' story on Blancanieves with the bullfighting love script, Berger arrives at a vision where the heroic death of a matador is substituted by a narrative of illness, disability, and abandonment. The paralyzed Villalta spends his time in a wheelchair, unable to move, not feeling his arms and feet, sad and lonely while his child is suffering from lack of protection. According to the story, it is the violence, the erotic attraction to death that characterizes the bullfighting culture that is responsible for the loss of a father's love and his incapacity to give care. In commenting on this loss of the father, the film can be seen in dialogue with Belén Gópegui's novel *El padre de Blancanieves*, (2007) which debates the responsibility of the middle class for having abandoned the fight for social justice in Spain. According to Gópegui, it is the middle class, absorbed with building its own security, pleasures, and luxuries that should be held responsible for the damages done to the weakest members of the society. This egoism, greediness, and obsession with one's own looks are features of Encarna, the most evil character of the film, who can be seen as one of those who gain power and riches at the expense of the lower classes and benefitting from their destructive engagement in the bullfighting dynamic. Seen in this way, the story of Blancanieves can be interpreted as a generalized commentary on Spanish society, whose weakest members are abandoned by its "fathers," who are busy killing bulls (and elephants).

In an essay on time in film, and writing on Buñuel's *El ángel exterminador* (Exterminating Angel), Gille Deleuze (1986) explains that a repetition with a difference is a way out of trauma (131). Through a series of repetitions with a difference, Berger's film reflects on the possibility of an alternative ideology of life, one that would do away with the eternal repetition of violence, though it is negated by the evil speculations of the greedy, vain, and power thirsty. *Blancanieves* is constructed through a series of repetitions, some of which are obviously traumatic, where the possibility of liberation from trauma is debated. When the sixth bull takes Villalta on its horns and then throws him up and down, the camera follows this torturous spectacle for a purposefully extended length of time. Halfway to the end, there is another similar sequence when a scene of a comic bullfight is acted out by the dwarfs who saved and adopted Blancanieves. The dwarf acting as a bullfighter, Jesusín, is butted and pushed around the ring by a small bull, and the public laughs and cherishes this. "Nadie va a hacer nada?" (Nobody is going to do anything?) asks Blancanieves, and someone responds: "Esto les encanta a la gente" (The crowd loves this). This is when the girl takes the cape to fight the bull for the first time. It is then, against the public's desire for the bullfighter's suffering and death, that she fights, still once again, she does it through violence. One more repetition with a difference is about to occur when Blancanieves faces the bull in la Colosal in Seville, when the public decides to pardon the bull. While the girl's performance as bullfighter plausibly constitutes a traumatic repetition in reacting to her father's death (the cultural value of this repetition is explicit when in a Bollywood fashion the ghost of Villalta appears in the sky to talk to his daughter), it is only the public who can point her a way out of the vicious circle of violence. When they do so, the filmmaker portrays the resulting happiness, which, however, ends prematurely when the girl is poisoned by her stepmother with an apple given to her from the public.

It is not enough to stop bullfighting to avoid suffering and injustice. For one thing, those in the principal lodge who represent the government who forgive their killable subjects only affirm their right not to do so, even if they do so under popular pressure. Villalta did not die the day that the bull took him on his horns, but rather several years later, pushed down the stairs by Encarna. Bullfighting, in light of this film, is a social practice that makes it easier for the evil to be done because it is responsible for facilitating a cultural dynamic that normalizes the killability and intimacy with death in real life.

There are two sequences portraying obsessive intimacy with death. The first one takes place after Villalta is pushed down the stairs. The dead Villalta is dressed in his bullfighting costume, set on the sofa, and photographers take a series of pictures with him surrounded by past companions, wife, domestic servants, and finally next to his daughter. Everyone celebrates this grim moment, smiling widely; only Carmencita cries. This same fetishization of death appears in an even more disturbing fashion in the final scene of the film, when Blancanieves, poisoned by the apple, lies in a glass coffin, exposed in a small circus-like show that is led by her bullfighting agent, who not even after her death releases her from the contract to act for him. For ten cents, anyone can try to wake the beauty up with a kiss. Rafita (Sergio Dorado), the dwarf who saved her and loved her, turns his eyes away, not being able to bear when one of the young men takes a handkerchief out to clean the mouth of Blancanieves, dirty from so many kisses she has received. It is chilling when her bed suddenly moves upward, and Blancanieves sits up and opens her eyes. While the terrified man who seemingly woke her up with a kiss escapes in a panic, Rafita moves the spring of the bed down, rearranges Blancanieves's face with love, repaints her lips, and lies in the coffin right next to her. This last scene with a dead woman dressed like a saint and a dwarf picador lying in a coffin next to her subtly reveals the morbidity of the national intimacy with death.

As in all the bullfighters' love stories previously mentioned, Manolete's and Diego Montes's from *Matador*, and in tune with Lorca's vision of the bullfighter's "life in death," also in *Blancanieves*, the love script is focused on the moments of deprivation, loss, and mourning. As a result, except for the brief time when Blancanieves manages to play with Pepe while secretly visiting her paralyzed father and there are some hopeful exchanges of looks between her and Rafita, there is no happy love in bullfighting culture as it is portrayed here. There are only fetishes of the dead that once and again appear on the screen: the dressed-up body of the deceased bullfighter, his portrait, his hat, the medallion with the portrait of his wife, and, finally, the fetishized body of Blancanieves, who resembles a church's wax saint. In Berger's film, the beauty of death is deeply corroded, darkened, and perverted. The collapsing body of the dead bullfighter on the sofa during the photographic session and the heavily made-up face of a dead Blancanieves in a glass coffin are grotesque and monstrous, reminding viewers of Fernando Arrabal's and Alejandro Jodorowsky's cinematic aesthetics.[15] In spite of the sweetly orientalizing camera focus corresponding to the seductive music accompanying bullfights and dances, this film is much less an homage to the national tradition and much more its reevaluation. Berger points out that bullfighting maintains a spirituality and aesthetics that lead to a sadomasochistic dynamic in an abusive family, unjust social hierarchies, and lack of education. In other words, bullfighting is deeply connected to the necropolitcs and necro-economics of an unjust state.

---

15. In particular Arrabal's *Viva la muerte* (1971) and Jodorowsky's *Santa Sangre* (1989).

The last words that appear on the screen are "milagro o maldición?" (a miracle or a curse?). They reappear in bold, enlarged letters to stress their significance to the conclusion. The question concerns the reality of "life in death," that, according to Lorca, defines Spanish spirituality. According to his famous lecture "Juego y teoría del duende," Spanish bullfighters are more alive after they are dead. The ending of *Blancanieves* is ironic in the sense that it is contrary to what it announces. The final word: "Milagro!" (Miracle) announces that the bullfighter is in fact alive and woken up by a kiss from the most cynical man in the public. After the simulation of the miracle is revealed when Rafita fixes the mechanism on the bed, one more proof of "life in death" appears on the screen. This time it is visible only to the film viewers: as the dwarf has joined Blancanieves in her bed, a tear that comes out of her dead eye and runs down her cheek toward the bottom of the screen. Just as in the previous instance when the dead bullfighter seemed to have woken up and metaphoric life in death was staged, in this last image of the film, the realization of the metaphor of "life in death" renders it powerless and deconstructs it. The tear seems to be one last element of the spectacle that consists of clichés that are part of the national archive and that this film subtly readapts for a new vision of cultural dynamics. A dead bullfighter may be crying, but she does not come back to life.

CONCLUSION

Lorca defines Spanish national culture through a particular sense of intimacy with death, which he connects to bullfighting and which constitutes the basis for pride or even feeling of superiority over other cultures. For Buñuel, this discourse of Spanishness masks poverty, suffering, vulnerability, and animalization. Almodóvar's *Matador* builds on Lorca's ideas while at the same time distancing itself from them; it recreates the necrophiliac sensibility with its erotic implications and parodies it. Berger also recreates bullfighting aesthetics, but he immediately subverts it by showing that on the other side of glamour hides a reality of pain, injustice, animalization, abandonment, and abuse. In the last scenes of *Blancanieves,* he also deconstructs the bullfighting aesthetics, showing that its attraction resides in violence and death, and amounts to a morbid kind of eroticism.

Buñuel's and Berger's criticism of bullfighting and their decision to animalize their Spanish characters do not imply their anti-Spanishness, however. When in the last scene of *Le phântome de la liberté,* French police shoot at the animals in the zoo, Buñuel's camera stops at the ostrich's face as if taking in its perspective on the events. When, as Williams supposes, he implicitly compares the ostrich to the Spanish patriot from Goya's famous painting, he animalizes the patriot, or perhaps humanizes the ostrich, in order to comment on Spanish culture but also in order to stress the community of human and nonhuman animals in their vulnerability in a moment of violence. Buñuel empathizes with the animalized Spaniards, although he regrets the process of animalization. While ridiculing the beastly practice of the festivity in Alberca in *Las Hurdes,* where men wrest the heads off live cocks hanging on a rope, the director repudiates the self-animalizing tendency in Spanish culture that leads to cruelty toward animals. At the same time, however, he empathizes with the animalized poor in their victimhood, in sickness, and at the edge of death.

In Berger's film, the image of a heavily made-up Blancanieves in the glass coffin with artificial hair and eyelashes leaves us in the place where the construction of animality and the sacred coincide. In the previous scene, Blancanieves makes wild faces in front of the bull, equally killable and representing "bare life" confronting death. Her animality is assumed as a national tradition passed to her by her father. She was killed not by the bull but, as in *Manolete* and *Matador,* by the public and was immediately turned into a sacred icon. In this sense, she is a moving figure. While the world surrounding her is to blame for her suffering, it is hard not to empathize with the heroine.

If, as Foucault writes, biopolitics is administered by modeling conduct and manipulating attitudes, artists have an important role to play in it. In *Blancanieves,* the relation between the public and the bullfighter who needs their applause is intimately connected to the tension between the contradictory desires of the artist to satisfy and to transform the national taste. This tension is also commented on in *Matador,* where Almodóvar compares the thrills and morbid pleasure of watching bullfighting to those of violent cinema and, consequently, a bullfighter to a film director. However, while *Matador* ends up blaming the public for the violence created by artists and performers who feel forced to satisfy the national taste, *Blancanieves,* while pointing to this dynamic, deplores it and searches for ways to subvert it. Blancanieves first became a bullfighter in reaction to the public's sadism when, during the comic *corrida,* people laughed at Jesusín's goring. At the end of her career, however, she ends up satisfying the same morbid need of the public against which she initially rebelled. The directors of the films about bullfighting, even those deeply aware of its problems and willing to mention them, as Berger does, run a similar risk. But the last scenes of Berger's film, where bullfighting is compared to a freak show, can be read as a refusal to indulge in the nostalgic *pasodobles* that the film begins with, a decision not only to reveal the other side of the glamour of the bullfighting spectacle but to stay on the other side.

The works discussed in this chapter show that the human/animal dynamic has been at the foundation of Spanish modern biopolitics, and it is at the core of the "spirituality" that has accompanied this biopolitics and facilitated its administration. Animality is implied in the definition of Spanish national character and in the definition of Spanish difference and superiority over other cultures. On one hand, humanity in Spanish culture is defined in bullfighting as a superiority over the animal, on the other, the Spanish alleged superiority over other cultures is due to the Spaniard's being able and willing to be like an animal, to be *bare flesh* in front of death. Complexity is the right term to characterize an animal's ubiquity and functions in Spanish narratives, but the contradictory dynamics described in this chapter point to the essential significance of the animal in Spanish culture. The next two chapters analyze works and life trajectories of intellectuals who, like Berger, attempt to transform Spanish culture's relationship with animality, challenging its public's taste and ethics and unambiguously opposing a spirituality that is subservient to the politics of death.

CHAPTER 2

ANTI-BULLFIGHTING ACTIVISTS AS MARGINAL INTELLECTUALS

CHAPTER 1 REVEALS that in order to achieve success, some artists, like Pedro Almodóvar in *Matador* (1986), connected with the death-obsessed spirituality that Unamuno and Federico García Lorca had earlier announced as a mark of Spanish culture. It also discusses how other artists, such as Luis Buñuel and Pablo Berger, attempt to deconstruct this aesthetic, revealing its other side where human sickness, animalization, pain, and death do not appear spiritually uplifting or attractive. This chapter shows how the debate on bullfighting, biopolitics, and spirituality changes if the anti-bullfighting point of view is adopted instead. While it begins by briefly discussing the paradigmatic case of José Ortega y Gasset as an intellectual who strategically planned his success and popularity as a function of his presumed Spanishness that in turn led him to praise bullfighting, it develops a longer commentary on those who adopted an opposite strategy. Mariano José de Larra and Eugenio Noel are marginal intellectuals who are willing to compromise their success and even their livelihood so as not to compromise their views and ideas. By maintaining painfully critical positions on the national tastes and customs that they see as favoring necropolitics and necro-economics, Larra, and even more visibly Noel, establish a model of an intellectual activist who tells truth to power and assume the consequences. Even if they were not celebrated sufficiently during their lifetimes, their discourses became foundations for the rhetoric of contemporary social movements.

There can hardly be two more distinct intellectuals than Noel and his contemporary Ortega. Noel was a son of a servant and mostly self-educated. Ortega was a son of the director of an important Spanish journal, owned by his mother's family, and he was educated at the best universities in Spain and Germany. Noel had never had a job while Ortega was appointed a professor upon his return from his studies in Marburg and soon promoted to full professor. While Noel was considered by many an outlandish jester, Ortega was respected, due to his social origins, education, and position. All these circumstances must have contributed to the reception of their respective ideas.

Ortega (1962 [1944]) said that it is impossible to understand Spanish history since 1650 without knowledge of bullfighting and its role in Spanish culture. Ortega's argument, gathered from various sources, is that problematic class relations in Spain have been shaped by bullfighting dynamics.[1] When the elite did not convincingly provide role models for the nation, the vacuum was filled by the lower class, which in his opinion had led to anarchy, made Spain impossible to govern, and contributed to the country's decay. The lower classes took over the repository of the national "soul" in part due to the protagonism that they assumed in bullfighting. When the higher classes began to abandon bullfighting under the French influence coming from the court after the dynastic change in the eighteenth century, the lower classes subsequently took the lead in it, inventing so-called "modern" bullfighting on foot (for they had no horses to waste). When sometime the nineteenth century the aristocracy came back to the roots of the "national" (so the argument goes) they had to do so by imitating the people's way, which caused an imbalance in social hierarchies, making simple people believe that they deserved the same or more than those above them. Ortega writes in *España invertebrada* (Invertebrate Spain, 1921) "En suma: donde no hay una minoría que actúa sobre una masa colectiva, y una masa que sabe aceptar el influjo de una minoría, no hay sociedad, o se está muy cerca de que no la haya" (89) (In sum, where there is no minority that could direct a collective mass and where a mass is able to accept the influence of the minority, there is no society, or there is a risk of losing it). Bullfighting as a privilege of the lower classes, so the argument goes, increased their self-esteem and contributed to an emergence of

---

1. I base this on Ortega's *La caza y los toros* (1944), *España invertebrada* (1921), *La rebelión de las masas* (1937) as well as on works commenting on Ortega such as Enrique Gil Calvo's *La función de los toros* (1989) and Timothy Mitchell's all-embracing work *Blood Sport* (1991).

majismo, the cult of an outlaw and of the bullfighter. But also, as Mitchell (1991) puts it in paraphrasing Ortega, not without irony, "an invertebrate and directionless Spain collapsed upon itself, to writhe in economic backwardness and sterile civil wars throughout the nineteenth century" (83). Given this analysis, in which Ortega contributes his view of the inversion of the class and culture dynamics in Spain that bullfighting reflects and that he deplores, it is surprising that he becomes one of the patrons of what Mitchell calls jokingly "the planet of bulls" or "bullfighting sub-culture" (2). This inconsistency or, perhaps, even hypocrisy of the famous philosopher has been noticed by various commentators[2] and mercilessly ridiculed by Luis Martín-Santos in *Tiempo de silencio* (1962), which is discussed in detail in the next chapter.

Ortega's enthusiasm toward bullfighting can be connected to his strategic accommodation to the national culture. An *ABC* article from 16 January 2011, suggests that Ortega did not criticize bullfighting because he was not critical of it, but rather because he was afraid of adding yet another reason for being accused of anti-Spanishness due to his long stay and intellectual formation in Germany. The article reports on the discovery of the unpublished notes for the book on bullfighting that Ortega had promised to write, but never had: "Llama la atención un manuscrito titulado 'Sobre las corridas de toros o Secretos de España,' en el que juega irónicamente Ortega con los que le tildan de filósofo 'extranjerizado y extranjerizante': paradójicamente—dice—es este pensador tan poco castizo el que viene a llamar la atención de los españoles ante el hecho de que las corridas modernas cumplan su segundo centenario." (Notably in a manuscript entitled "On Bullfighting or the Secrets of Spain" [n. pag.], Ortega establishes an ironic exchange with those who accused him of being "under foreign influences," saying that paradoxically, this philosopher accused of lacking Spanishness is the one who reminds the Spaniards of the significance of the second centenary of modern bullfighting). Ortega's centralism, his view of Catalan and Basque separatism as a "tumors" (1921, 79) can be, perhaps, similarly attributed to the philosopher's choice of his intellectual location in the center of Spanish taste. In spite of the evolution of Ortega's political ideas from socialism in 1909–14 and hesitant republicanism in the 1920s and early 1930s to the partial assimilation to Franco's dictatorship during the postwar era, Ortega's support of the centralist politics have remained unchanged.[3]

---

2. For example, Gil Calvo's *La función de los toros* (1989) suggests that Ortega's theory of bullfighting is not different than that emerging from its greatest critics among regenerationists, and that "it is almost as if Ortega reveled in decadence, entertained himself in degradation, and congratulated himself for impotence" (144).

3. I comment on this in detail in my previous book, *Del infierno al cuerpo,* (2007, 125–32).

In *Representations of the Intellectual* (1993), while deliberating if an intellectual is obliged to maintain solidarity with his nation and his people, Edward Said opposes Julien Benda's *The Treason of the Intellectuals* (1927) against Matthew Arnold's *Culture and Anarchy* (1869) and distances himself from both. Arnold's thesis is that the state is the nation's best self, and a national culture is the expression of the very best that had been said or thought, and it is the duty of an intellectual to be "helping a national community to feel more of a sense of a common identity" (31). Benda criticizes precisely this kind of commitment, which he perceives as a betrayal of the universal spiritual values that, in his view, an intellectual should serve regardless of his national and class origin. Ortega's views fall in the crossroads of both these approaches, revealing that in spite of their apparent opposition, there is a point where they meet. Ortega believes in the ethical mission of the elites, who are viewed as defenders of universal civilizing values, while at the same time he supports nation, central state, and bullfighting. The universal character of the values Benda defends is problematized in light of Ortega's views, which reveal that these values are posited as universal in national languages dominated by national values and are thus national universals. Said criticizes both Arnold's and Benda's ideas and takes inspiration from Walter Benjamin and Frantz Fanon. Benjamin was one of those who best understood the dangers of nationalism founded on "blood and soil" as in Alfred Rosenberg's *Blut unde Ehre* (1935), and Fanon knew how to transcend the antagonism between the nation and "its enemies" (Said, 242). Said praises Fanon for his way of "universalizing the crisis" of the Algerian nation, saying: "It is inadequate only to affirm that a people was dispossessed, oppressed or slaughtered, denied its rights and its political existence, without at the same time doing what Fanon did during the Algerian war, affiliating those horrors with the similar afflictions of other people" (1967, 44).

While Ortega is right in noting that bullfighting has been gaining symbolic significance in Spanish culture during the last three centuries, it should also be noted that the significance of anti-bullfighting has grown alongside it. As Mitchell (1991) writes, "The most resolute adversaries of bullfighting have always been Spaniards. There has never been a period of Spanish history, in fact from the Middle Ages to the present day, that has not witnessed a clash between the apologists of bullfighting in one form or another and those who would abolish it for good" (82). He later proposes to see bullfighting "against the familiar gestalt of *las dos Españas*, 'the two Spains'" (83). A similar suggestion appears in Carrie Douglas's *Bullfighting and National Identities in Spain* (1999). Although the two Spains that confront each other in Machado's famous poem are undoubtedly different from bullfighting and anti-bullfighting Spain; the anti-bullfighting Spain calls for its examination as an important aspect of national identity.[4] The persistence of anti-bullfighting in Spanish debates on culture shows that it constitutes a significant undercurrent, an always-alive opposition. It contains the critical reformist spirit that carries a framework for an alternative biopolitics to which contemporary Spain owes many of its transformations. This argument is then related and opposite to Ortega's at the same time, namely that it is impossible to understand Spanish contemporary culture without the knowledge of anti-bullfighting and other manifestations of sensibility to animal pain because they represent reactions to the centuries-long debates on national culture centered around bullfighting, bullfighting masculinities, and bullfighting love scripts, in sum, centered around of what could be called a bullfighting biopolitics.

NOEL IN LARRA'S FOOTSTEPS

The most interesting insights into the role of bullfighting and its role in the national culture come together with the most important anti-bullfighting ideas. They are often elaborated outside of academia by activist-intellectuals such as Eugenio Noel (1885–1936). Activism is not an obstacle for understanding as is sometimes believed, but rather helps to achieve a deeper insight precisely because it leads to production of a "situated knowledge" (Haraway, 1991, 183). While certainty of one's research objectivity tends to be a mistake or a camouflage, the so-called objective research often amounts to a strategically structured discourse which, instead of analyzing reality, limits itself to paraphrasing discourses previously established and recognized as legitimate knowledge of a given subject (Bourdieu, 1991). Conscious political engagement announces its prejudice and passion for change openly. In order to show the need for change, the politically engaged intellectual challenges the established knowledge of a problem by contrasting this knowledge with an alternative, often marginalized perspective of reality. The passion for change becomes a motivation for getting things right as opposed to the desire for symbolic capital within the field that strategically allies with mainstream discourses. Among American academic publications, for example, Mitchell's book (1991) shows the deepest understanding of the role of bullfighting in Spanish culture. Even if dispassionate and cynical itself, as the author himself admits, the book owes a great deal of its understanding to Noel's work, who, as Mitchell affirms, knew everything about bullfighting and spent all his life fighting against it. The insights come from Noel's activist engagement.

---

4. Poema LIII appears in Machado's "Proverbios y cantares de *Campos de Castilla*" (n.d., n. pag.) and is often used in discussions of the Spanish Civil War.

Noel's visions of Spanish culture and even his style of writing have at times a lot in common with Larra's. Noel's best production, like Larra's, can be classified as journalism bordering on literature. Both Larra's and Noel's writings are nourished by travel, observation, and interviews; both become irritated by the same characteristics of Spanish life—laziness, arrogance, harmful masculinities, and unwillingness to learn from one's mistakes—which they judge as backward. Both develop a love-hate relationship with their *patria*. A tone of tender bitterness can be identified in their articles when they compare the Spaniards' ways of life to that of other European nations.

Noel, like Larra, was a "marginal intellectual" according to Said's definition of the term in his *Representations of the Intellectual* (1993) as one that "remains outside the mainstream, unaccommodated, unco-opted, resistant" (52). He compares the marginal intellectual to the "ranting Thersites" (53) whose "curmudgeonly disagreeableness can become not only a style of thought but also . . . an habitation" (53). In Said's view, it is only this kind of antisystemic positioning that allows for a permanent check on reality and functions as a sort of security against falling prey to prejudice, following fame, gain, and other kinds of worldly rewards. Said quotes Adorno to the effect that "it is part of morality not to be at home in one's home" (57). This kind of ethos of self-exile from "self-congratulating honor societies that routinely exclude embarrassing troublemakers who do not tow the party line" (59) allows one to see what can only be seen from the margin. But the reason most people do not situate themselves on the margins, but rather aim at the center of their respective cultural milieus, is that marginality, even consciously chosen, tends to be a very painful position and one that rarely brings any kind of success.

Paradoxically, Larra's marginality was manageable for him as long as he felt public support for it, and when that feeling was no longer there, he found it unbearable. Larra's was a strategically compromised marginality, first by financial success and popularity that Noel never had, and then by his decision to become a politician, presenting his candidacy for the Spanish Parliament on behalf of the *Moderados* party. As Donald E. Schurlknight (2009) succinctly summarizes, "Having succumbed to ambition, having staked his reputation on the prospect of political power, and having lost, Larra evidently believed his professional life to be in shambles" (156).[5] As Kirkpatrick notes, one motive for his frustrated attempt to become a politician, was disappointment with his power as a journalist writing from the

---

5. Larra was elected, but the rebellion of la Granja, which errupted one day after the elections had been announced, annulled their results.

margins of power. When the liberal public that cherished his articles fell into opposing factions, Larra felt too marginalized to continue in his position. His was thus in a very complex relationship with marginality, which is reflected in his works. Even though his articles became part of the canon of Hispanic literature and are regularly read in the classroom, those most appreciated are the ones in which his critical voice breaks down to subjective, self-critical, and self-mourning tone and remarks, especially "El Día de los Difuntos de 1836" and "La Nochebuena de 1836." In spite of his resistance and convincing criticism of the cult of death, which he observes already in the practices of his time, in these last articles Larra succumbs to it. In his last drama review ("*Los Amantes* de Hartzenbusch"), which appeared in *El Español* on 22 January 1837, just a few days before his suicide, he talks about death as a means of proving the value of love. In his "Exequias del conde de Campo-Alangue" from 16 January of the same year, he praises death as a way to prevent "death in life" (Beilin, 60). These last articles by Larra, in which despair is so clearly perceptible, do not merely play with the gothic tastes of his readers but should be read as an abandonment of his reason-based worldview, which was critical of destructive habits, and a preparation for a full adoption of the death-ridden culture that surrounded him by shooting himself in the head on 13 February 1837.[6] "La Nochebuena de 1836" in particular can be read as a self-deprecating recognition of the inappropriateness of his previous biting critiques, a sort of apology to Spanish culture, which he recognized he was in no position to criticize. After Larra gives up his position as a marginal intellectual, and after the Granja rebellion shuts him off from politics and an amorous disappointment takes away the possibility of love, Larra finds no further reason to live.[7] It is possible to consider this act of rendition of his passionately critical marginal position as what ultimately gained Larra his place in the Hispanic canon. In other words, had Larra persisted in his critiques, driven by the spirit of the Enlightenment, rather than succumbing to the Spanish flirtation with death and then to death itself, he may have earned himself the status of a minor author and not of a great one. Susan Kirkpatrick (1977) seems to mean something similar when she says that his work would not be so important if it did not reflect his personal experience (206). His adoption by Generation 27 may have resulted from his positioning himself as a protagonist of those last writings but also as someone who is inside and one with his culture, as he confesses in "El Día

---

6. See Rebecca Haidt's "Gothic Larra" (2004) for analysis of gothic as a literary strategy in Larra.

7. This occurs through an accumulation of various circumstances in Larra's life, as I explain at greater length in my previous book, *Del infierno al cuerpo* (2007, 56–61).

de Difuntos de 1836" that the death permeating Madrid's life is one that eats his heart.

Noel's trajectory was different, although his critical engagement with Spanish culture was similar. At the heart of Noel's search for the improvement of Spain was his anti-bullfighting obsession. For this *sui generis* intellectual, activist, and bohemian writer, bullfights were Spain's greatest problem because they nourished and fostered the worst features of what he, along with various other thinkers of his times, called "la raza" (literally, "race" but with the meaning that today we give to culture). In Noel's view, Spanish popular art, especially flamenco and *cante hondo*, was similar to bullfighting and deeply connected to it, conductive to unhealthy emotions and political tragedies. Just as Larra, Noel was worried by the excitement with death in Spanish popular culture that, around the same time, Lorca chose to promote. Noel coined the term *flamenquismo*, which meant "the culture of bullfighting and flamenco," and he spent a great deal of his 50-year-life campaigning and writing against it. His essays, articles, and novels are all about Spanish culture and Spanish character, which is analyzed often brilliantly in a Larra-like style through descriptions of customs, festivities, attitudes, art, and language. His writings produced a modest source of income and, even though he managed to achieve considerable fame in his time, he was invariably lacking money and at times desperately so. This constant poverty was an integral circumstance of his work, which is impossible to understand without reading his *Diario íntimo* (1968), where economy often replaces inspiration. These are two thick volumes of accounting for incomes, understood as financial gains, immaterial success, and simple feelings of satisfaction as well as expenditures—efforts and sacrifices. They tell the story of his life, torn between his anti-bullfighting activism, his love of his country, his desire for fame as a writer and intellectual, his family (most of the time deprived of his company and without money to pay the bills), and his weakness for bodily pleasures: good food, drink, and elegant clothes, which he often had to sell in order to eat. His contemporaries, including Ortega and Unamuno recognized his greatness publicly (with Ortega claiming once that he was the best of his generation), but they did little or nothing to help him to succeed. His biographers systematically point out that he failed in his aspirations, as if this were the main reason they talk about him. One of them, J. García Mecadal, adopts a preaching tone, stating that Noel lacked practical sense and did not organize his life in a wise enough way to achieve success (1967, 16). His constant travel and lack of regular income made it very hard for him to write literary works, limiting him to brief essays and articles. Mecadal suggests that Noel should have gotten a job, one of these positions that would give him a steady income while not requiring much dedication. Perhaps, however, for Noel this unwillingness to accommodate himself and his decision to persist in his quasi-missionary anti-bullfighting campaign was the essence of his critical attitude toward the culture in which he was immersed, and this made his work possible but unsuccessful at the same time.

In contrast to Larra, Noel's faithfulness to his critical marginality, guaranteed by his obsessive anti-bullfighting stance, let him live almost twice as long and build the foundations of the discourse of today's anti-bullfighting activism, but it may have also turned him into a *persona non grata* in the literary canon of Hispanism and an unlikely object of research. The Modern Language Association (MLA) database shows fewer than ten scholarly articles focused on him and, despite the great intellectual quality of his writing, various very erudite Hispanists have never heard his name. Several of those articles focused on Noel are negative criticisms. Sánchez Reboredo (1985), for example, reproaches Noel for his eccentric style of dress and writing, specifying that instead of elaborating on others' theories and building his own on the top as all critics do, Noel just mentions names of authorities chaotically as a "nouveau riche." Reboredo (1985) accuses him of "sensationalism" and "superficiality" in contrast to "atildamiento de Ortega" (155) (Ortega's smartness). Ángeles Prado (1969) claims that Noel's works reflect on the author's confusion (3). The most interesting article on Noel, and the only one devoid of negative criticism, does not even appear in the MLA database. "El ideario costista en Eugenio Noel" (Eugenio Noel, Following on Ideas of Costa) by Jesús Vicente Herrero (2003) shows how Noel puts into practice and expands on the ideals of the Institución Libre de Enseñanza and in particular those of Joaquín Costa (1846–1911). Following Costa between Paris and Madrid and Noel's trips across Spain, Herrero's analytical voice constantly changes its point of view, not only geographically but also as he alternates between Costa's and Noel's critical visions. In contrast to Hernández's article on the death of matadors, which blurs all differences discussed in the introduction, Herrero's approach describes a complex reality, and it shows great sensibility toward all the nuances of Costa's and Noel's approaches, differentiating them from each other and, additionally, from the reactions that their writings elicited from their contemporaries. In Herrero's view, even though the critical framework that both Noel and Costa elaborate is structurally turned against institutionalized national symbols (such as *caciquismo* in the case of Costa and bullfighting in Noel), it is full of praise for Spanish people and some of their traditions, such as communitarian cooking or the use of certain plants. They evaluate traditions looking at their practical outcomes.

Both Larra and Noel hesitantly include themselves in the culture they criticize through an occasionally emotionally ridden use of "we" or first-person narrations that reach the deepest tones of romantic pessimism in Larra and are suddenly interrupted when the enthusiasm runs out in Noel. Their approach toward their own culture verges on reverse ethnocentrism.[8] They look up to other European countries as they analyze Spanish culture, but at the same time they privilege it emotionally. Even if they believe that other nations' ways of life and social organizations are in some respects better than theirs, they do not abandon their culture but rather display a quasi-apostolic zeal of reformism that proves that their criticism is rooted in their love for their origins. They fight to include in Spanish culture the subversive, marginal forms of thought that they themselves practice, which they promote as a key to a desired change. This strategy will also be chosen by Juan Goytisolo in the second half of the twentieth century.

The rhetoric of self-marginalization that Larra and Noel use is fairly similar. In a revealing paragraph on bullfighting, Larra implicitly includes himself in "una clase de entes [que] no va a estas funciones: esa bandada de sentimentales" (Larra, 1828, 151) (a class of beings that do not go to these functions; a band of sentimentals), implying that he is one of those called "afrancesados" (French-like) who voluntarily distinguish themselves from their compatriots, having changed their tastes, values, and behaviors and adopted French ways that they believe to be superior. Yet in the description of the group of dissidents to which he belongs, Larra sarcastically uses the tone that "normal" Spaniards, would use while talking about him. Through the acceptance of the perspective of those who do not respect him, Larra attempts to include his marginal views, here characterized by anti-bullfighting, in the national culture. Noel uses the same strategy:

> España es un país delicioso para quien sabe entenderle. Ahora bien: hay mamarrachos disfrazados de intelectuales . . . los cuales andan por ahí gritando como endemoniados que España es una pandereta pintada de toreros, chulas, corridas, flamencos y gente de bronce. (1913, 122)

> Spain is a sweet country for those who know how to understand it. Now, there are some fools presuming to be intellectuals . . . who go around announcing that Spain is a tambourine painted with bullfighters, cuties, bullfights, flamencos and people made of bronze.

---

8. Judgment of another culture(s) as superior to one's own.

The tone of sarcastic self-inclusion into the group he mocks is significant here. Self-deprecation becomes the price of acceptance. Self-categorizing as a marginal and not well-respected intellectual serves as a bridge toward the ideological adversary. In that rhetorical maneuver, both writers construct their national culture as a mixture of competing but mutually tolerant ideologies and approaches to life. But simultaneously, they also accept the negative perception that comes as a result of their marginality, with which each of them dealt differently.

Larra and Noel want their criticism of Spanish culture to be a part of Spain, to be in fact taken as a form of love for their country. This sense of anti-Spanishness is similarly displayed by the anti-bullfighting movement today. Baltasar Porcel (2004) argues in *Vanguardia* that "la posición antitaurina conlleva un antiespañolismo. Desde luego . . ." (n. pag.) (the anti-bullfighting position carries with it anti-Spanishness. Of course . . .). But not in the sense of vindication of Catalan nationalism, but rather "porque este españolismo basado en el sostenella y no enmendalla, en el nacionalcatolicismo, en el 'por cojones,' en el culto a la muerte, el punto de honor, en el España es diferente, indigna a mucha gente" (n. pag.) (because this Spanishness based on "keep it, don't fix it," on national Catholicism, on "because I say so," on the cult of the dead, on honor, on "Spain is different," infuriates many people). In this sense, Larra's and Noel's critiques have served as models for the critiques of the contemporary anti-bullfighting movement not only in Cataluña but every place of Spain.[9]

Larra's and Noel's marginal intellectual position, like those of Martín-Santos and Goytisolo in more recent years, permit culture to be seen from a distance as a set of connected discursive realms. For these writers, who share a deep dislike of bullfighting, the fight between bullfighter and the bull that excites the public simultaneously represents numerous aspects of cultural relations, all characterized by cruelty, aggression, domination, and destruction. Works by anti-bullfighting writers consistently display sensitivity toward human and animal suffering and the destruction of nature, as well as criticism of state violence, whether direct or inscribed in structures of oppression. They are critical toward gender violence and war, and they attempt to deconstruct emotions, aesthetics, and discourses that excite them. Last but not least, the anti-bullfighting writers consistently learn of reality as a result of experience; they are traveling researchers or journalists, and their analyses are place-sensitive, rather than generalizing. In Noel's and Larra's writings, all these connections appear very distinctly. The following pages map the anti-

---

9. See introduction and chapter 7 for more on these critiques.

bullfighting sensitivity as an inspiring alternative to the cultural excitement of death.

MAPPING THE CONNECTIONS: ANTI-BULLFIGHTING AS A CRITIQUE OF STATE VIOLENCE

The most interesting aspect of Larra's criticism of bullfighting is his observation that the state's power promotes a bloodthirsty society whose dynamics is acted out in the ring. As the writer builds the history of Spanish bullfighting through its relationship to national heroes, he shows how those most courageous and violent men accumulated power and became models for the Spanish people, merely by their use of aggression. His reflections coincide with Charles Tilly's work on how the national discourses provide a manipulative legitimization of might.[10] For Larra, the Cid's *gesta* of killing the bull alone, as he was legendarily the first one to do so, shows how brutal force assisted by cleverness is mystified as a virtue that grants a right to power, and this mystification is registered and ritualized in bullfighting. In Larra's view, bullfighting not only symbolizes brutal political power but is connected to it in a historical way. Larra points out that there are various correspondences between the changing ways of bullfighting and the transformation of the state. The intensification of bullfighting corresponds invariably to an escalation of political violence and is often followed by wars. For example, as Larra tells us, bullfighting really flourished in Spain during the reign of Charles V, when most of American lands were conquered and thousands of Indians killed. It was Charles V who suffocated the rebellions of the commons in 1521 and who himself killed bulls at the ring. Bullfighting decreased under Felipe II and then revived under Felipe III, who expelled the Moriscos, and especially under Felipe IV, who himself "además de alancear y matar los toros, quitó la vida a más de 400 jabalíes con estoque, lanzón y horquilla" (Larra, 1828, 250) (besides throwing pikes and killing bulls, took the lives of more than 400 boars with a sword, spear, and yoke).

In his survey of the history of bullfighting Larra points out that the Spanish noblemen learned it from the Arab aristocracy in the times when they cohabited the peninsula. In this way, he further deconstructs the argument

---

10. According to Tilly (1985), the line separating criminal violence from legitimate violence is blurry. Historically they come from the same origins, and even today, might still makes right. Tilly writes that in premodern Europe, "a king's best source of armed supporters was sometimes the world of outlaws" and that mythical conversion of Robin Hood to a royal archer "records a common practice" (173).

that bullfighting is somehow connected to Spanishness at its origins. Larra debunks and mocks this vision as he mentions that the most popular entertainment among Christians during their festivities in medieval times was in fact a spectacle in which blind men hit pigs with sticks. This comparison of bullfighting to pig-fighting ridicules all sorts of animal circuses while also exposing the falseness of the national myths. According to Larra, the popularity of bullfighting among the lower classes started as a way of imitating the nobles, while the nobles practiced and cherished bullfighting if their rulers did so as well. To this extent, Larra cites Voltaire's verse "*Quand Auguste avait bu, la Pologne était ivre*" (1828, 250) (When August drinks, Poland is drunk). Popular tastes and customs, he argues, are hardly "popular" but rather result from the influence of the elite. Larra states that while in distant times bullfighting was justifiable as an exercise for war, a daily reality in the lives of noblemen, in his own historical moment, which he would like to see as less violent, there is no reason to continue it.

In *Pan y toros* (1913), Noel similarly justifies the Spaniards' passion for bullfighting with Spanish history:

> Muchos siglos de guerras exteriores y civiles nos han dejado en el estado lastimoso de nuestro flamenquismo. La emoción nos está vedada si no viene directamente de sangre vertida. Cuando nos hacemos paladinos de alguna idea, lo primero que compramos para imponerla es un revólver y los que van a oírnos lo llevan también. (88)

> Many centuries of wars immersed us in the deplorable state of our *flamenquismo*. We are not able to experience emotions if blood is not shed. When we become convinced of an idea, the first thing we do is to buy a gun to impose this idea on others, but those whom we target are similarly armed.

In these brief phrases, Noel ventures a historical analysis of the cultural atmosphere of his time as a warning. He is wary of ideological excitement in a society lacking the skills to resolve conflicts peacefully, where only bloodshed guarantees emotional fulfillment. While he derives the people's fascination with bullfighting from the Spanish history of wars, his words seem to announce the most horrid war, which Noel fortunately did not see, as he died of pneumonia in 1936.

Noel's personal life trajectory points to a connection between bullfighting, wars, and the construction of masculinity. At the age of twenty-four, he was advised by Ortega, who praised his writings but deplored their lack of connection to the nation, to volunteer for the war in Morocco to "hacerse hombre allí" (1968, 212) (make himself a man there). While, as Noel confesses in his diary, he followed this advice for financial reasons and because of curiosity, in Morocco he acquired a deeply negative vision of both war and warrior masculinities, as well as of the social hierarchies connected to its exercise. In his "Notas de un voluntario," (*Diario íntimo*, 1968, 338–70) written in 1908 in Morocco, he describes the abyss that separated the war life of a marquis, for whom the war was "a magnificent sport," (341) who seldom took risks, lived in luxury, and received royal decorations for his service to the country, and the regular soldiers who paid for those decorations with their flesh. The series of articles on these themes that Noel published after returning to Spain from the war in *España Nueva* condemned him to nearly two months in prison in 1909. There, he kept writing, progressively acquiring a more intense antiwar focus. He finally suggests that Moroccans and all colonial powers involved in the conflict should negotiate rather than fight. Right after his release from prison, Noel began his "campaña antiflamenca," (anti-bullfighting and anti-flamenco campaign). His ideas appear in 1914 in his best known book, primarily because of the title, *Pan y toros*. The order of his experiences suggests that it was war that turned him into an anti-bullfighting activist.

Although it was a direct experience of the reality of violence that conditioned him for the rest of his life, Noel's worldview was also a result of his readings, especially those of Joaquín Costa. Costa had also been critical of bullfighting in the context of the disaster of 1898, when Spain lost its last colonies in the war with United States. Noel's article "Los toros de Carabanchel y el año del desastre" (*Diario íntimo*, 1968, 155) (The Bullfights in Carabanchel and the Year of the Disaster), whose aura reminds the reader somewhat of Larra's "El Día de los Difuntos de 1836," had been written a few years before the author's participation in the Moroccan war. In this ingenious essay, Noel connects bullfighting and the naval disaster of 1898 through subtle images, sounds, and thought-provoking anecdotes. The war and the bullfight first become connected by music:

> Se la oye de lejos y salta el corazón. Viene tocando un pasodoble que enardece la sangre. Esos somos nosotros, los españoles. Chispean todos los ojos, se marcializan los movimientos, andan los cuerpos solos. Es ¡*La Giralda*, de Juarranz. Bravo! ¡Un viva España! Formidable atruena el espacio y La Guerra es recordada. La Guerra y la música están endiabladamente unidas. ¡Pobres soldaditos de Santiago de Cuba! ¿Que será de los barcos? . . . Las proas de las naves guerreras llevan por mascarón una cabeza de toro y los estandartes y enseñas de almirantes son fajas de toreros celebres. Eso es . . . la victoria es nuestra, no puede menos de ser así. Somos los hombres más

valientes del mundo. ¿Qué país del Universo tiene esta música tan alegre, tan viva, tan torera? (161)

We can hear it from far and our hearts jump. It is a *pasodoble* that inflames our blood. This is us, the Spaniards. Our eyes shine, our movements become solemn, our bodies move by themselves. This is *La Giralda* by Jurranz. Bravo! Long live Spain! There is an echo that evokes the War. The war and the music are so intimately connected. Poor soldiers from Santiago de Cuba! What will happen with their ships? . . . The prows of the war vessels carry heads of bulls and banners; the admirals decorate themselves with belts of famous bullfighters. This is . . . the victory is ours, it has to be. We are the most courageous men of the world. Which country of the Universe has a music so joyful, so lively so bullfighting?

The bullfighting spectacle projects itself over the spectacle of war, and the bullfighting mood brings certainty that the war must end in victory. The illusions of heroism and greatness created by the music and by the bullfighter's nearly inevitable destruction of the bull take away the sense of reality. At the same time, while the movement of the people toward the bullfighting plaza paralleled that of the navy toward its defeat at the sea, Noel points out that no two events could be more antithetical. The echo of the faraway disastrous maritime battle resounds in the cheerful music of the bullfighting spectacle. The military defeat throws an ominous shade over the illusory triumph of the bullfighter Mico Chico. This repressed consciousness of failure becomes a ghostly projection contaminating the bullfighting reality. Noel comments on the defective construction of the bullfighting plaza, which, as the architects stated, had not fallen down during all the years of its exploitation only thanks to a miracle. The loss of the empire in Noel's text is somehow caused by the abuse of old horses that are about to expire while pulling the tramway full of people up the hill to the bullfighting plaza of Carabanchel.

The horses pulling up the tramway merge with the "heroes of 1898," "viejos, estrechos, bajos, sucios, feos y héroes. Quién podía explicarse cómo no se hundían bajo el peso tremendo que soportaban?" (158–59) (old, narrow, short, dirty, ugly and heroic. Who could explain that they did not fall under the tremendous weight that they had to bear). The description of the horses (which stand for the soldiers) extends for half a page. Noel notices signs of abuse in the bodies of the animals: tensed tendons in their legs, their protruding ribs, and an expression of anger suppressed by a mortal tiredness. As the tramway slowly climbs up the hill, the tension in the horses' bodies grows and a catastrophe imposes itself: "Y quien no recuerda aquellos latigazos del mayoral . . . cuando había que abordar un cruce, un cruce de aquellos en los que el tranvía se salía de los rieles o raíles con un ruido macabro de hierros viejos y cadenas y cristales hechos añicos y ruedas rechinantes?" (159) (And who does not remember the whips of the coachman . . . when a turn was approaching, one of those turns where the tramway was falling out of the rails with a macabre noise of the old steel and chains and crystals broken into small pieces and squeaking wheels). The anticipation of a street accident brings the premonition of the national disaster as the author uses an overburdened horse's body to talk about the soldiers. On the other hand, this description of overly exploited animals that pull a crowded tram metaphorically renders the cruelty and inefficacy of the national biopolitics where unbearable burdens are deposed on those who are lowest in the hierarchy of national life.

The description of the office of the director of the bullfighting plaza contains another memorable allegory of national necropolitics. Romero, the director, is like a god disposing of the lives of a never-ending line of boys, tall, slim, and pale, who arrive from all parts of the country and do not ask for money but only for the privilege of fighting with a bull at Carabanchel. Romero's office is thus "una funeraria en eso de no cerrarse jamás ni de día ni de noche" (164) (a funeral home open day and night). For Noel, however, Romero, rather than a god is in fact a politician, an important functionary of the state who administers human lives and deaths for the national spectacle that maintains the status quo of the powerful. The state appears as a life-sucking machine, and bullfighting and war are its preferred modes of exploitation of the energy of the flesh. Culture promoted by a state whose symbolic spectacle is bullfighting contains manipulative schemata that make all these young boys desire their own destruction. Sacrifice, which, according to René Girard (1972) is an unquestionable foundation of a state, is viewed by Noel as a deceit that guarantees the dominion of the powerful over all the others through the *mimetic desire* that both Larra and Noel describe in their articles on bullfighting.[11] Noel sarcastically characterizes "las síntesis de la raza" (the synthesis of the race) that has been attributed to bullfighting in the following terms: "Entraba sangre de la estirpe y salía el alcohol de las almas españolas, ese alcohol que ha desencadenado hoy por la Penísnula el delirio taurino" (165) (The blood of the stock entered and the alcohol for Spanish souls came out, that alcohol that today spreads in the peninsula the bullfight-

---

11. In his *Mensonge romantique et vérité Romanesque* (1961), Girard analyzes the structure of desire in great European novels and explains that it is inherently "mimetic," that is, it arises as an imitation of the Other's desire.

ing delirium). The use of the word "estirpe," which can refer to both human and animal lineage, suggests that in the bullfighting biopolitics people and animals are similarly used for power maintenance. Noel complains that the excitement produced by the closeness of death intoxicates Spaniards, making them incapable of seeing this dynamic, indeed incapable of dealing critically with the problems of social life. Noel believes that overall social progress requires the abolition of the spectacle of bullfighting and a consequent revision of the patterns of sacrifice, which should not be measured in liters of blood poured in the ring and in the war but rather in a resourceful activity for the common good. The article "Los toros de Carabanchel y el año del desastre" ends by counterposing the colossal triumph of the bullfighter Mico Chico with a total defeat of the Spanish army in Manila, Santiago de Cuba, and Cavite.

Noel also challenges the state's right to dispose of people's lives in the penal system. Like Larra, he criticizes the death penalty, and his cunning critique in the article "Ante la cuna de una infanta" (*Diario íntimo*, 1968, 269) (By the Cradle of the Princess) manages to save the lives of the six political prisoners sentenced to death who were to be executed the same day that the king's daughter was born. Noel calls upon the king to pardon the lives of the sentenced prisoners, arguing that otherwise their death would weigh over the little princess's life but also threatening the king that one day Republicans may attain power and similarly decide to pardon the life of the king. The king commutes their sentences, but Noel himself is punished by imprisonment for his audacity. This article, which saves the lives of prisoners but is paid for with a piece of the author's life, once again exemplifies Noel's condition as a marginal intellectual, putting in evidence an intimate correspondence between writing, political reality, and the writer's life. For both Larra and Noel, writing is an instrument of social change, and like bullfighting, it requires courage and self-sacrifice, but unlike bullfighting, it seeks to subvert power and to save the victims of state violence.

In "Los toros de Carabanchel y el año del desastre," written very early in his life, Noel's tone is similar to the Generation of 98's bitter regret over the loss of the colonies, and the article thus unavoidably displays an imperialist nostalgia. His later writings, however, became progressively more and more antiwar and anti-empire. If "Notas de un voluntario"(*Diario íntimo*) consists mostly of a first-person narration of the war—dinner with the commanding elite, conversations with soldiers, battles, suffering, thirst, wounds, withdrawal, always hinting at major mistakes by the commanders, and the general purposelessness of the war—his most mature reflections come from the second stage of the Moroccan war, where he participates as a journalist in 1913. Observing soldiers departing from a pier in Cadiz to Morocco, he complains that thousands of young boys are thrown into a war to risk their lives without adequate arms and preparation. The ship that will carry them to war is too small and so overloaded that it promises to sink before reaching the other shore. Watching the departure, Noel regrets that "por causa de la negra honrilla, nosotros arrojamos al abismo de la Guerra los millones de las escuelas por crear y puertos por hacer" (340) (because of stupid honor, we are letting the war consume resources that could otherwise be used for millions of schools and harbors that we could build). He complains that since the war is an essential exercise of statehood, the protests against the war are punished with prison. The associations between war and bullfighting impose themselves again through the music played at the ceremony of the farewell to the soldiers deployed to Morocco: "Esta música es tan viva y torera, que en oyéndola cualquiera se flamenquiza" (344) (This music is so lively and so reminiscent of bullfighting that while listening to it anyone would fall into the bullfighting mood). What Noel understands as "flamenquizarse," and that for lack of a corresponding expression is translated as "falling into the bullfighting mood," means falling prey to those melodramatic emotions that demand a spectacular self-sacrifice for the sake of the nation.

Like Ortega in the first few articles on the war of 1913, Noel deplores Spain's lack of colonial success and does not challenge Spain's right to dominate Morocco, but soon his attitude changes. His descriptions of Moroccan soldiers, Moroccan men and women at the market or sitting by their houses watching him pass evoke deep respect or even that sort of fascination that would later appear in Goytisolo's novels. He calls the Moroccans "heroes" and finds them impossible to conquer. He also recognizes the harm that Europe causes them by imposing its civilization on them, and he recognizes the Moroccans' right "de ser como les da la gana" (366) (to live as they wish). He notes that Spain fights the Moroccans, forgetting that Spain itself had been invaded for imperial reasons and that Moroccan people defend their freedom in that very same way that Spaniards had defended and would again defend theirs. In these last series of reflections, Noel moves quite far away from the tone of imperial nostalgia. As in bullfighting thus also in war: ultimately he must identify with the victim and his pen turns against the discourse of the aggressor. He concludes that Spain and Morocco are so close that they should understand each other as neighbors rather than fight.

Masculinity

While Larra despises bullfighting for aesthetic reasons, citing Jovellanos' description of a feast, which contains a deep distaste for the noisiness, dirt, vulgarity and aggression among the viewers, his main arguments are ethical. Bullfighting and also hunting, which he criticizes in a later article, are remnants and enactments of war. They are ruled by a war mentality, the sad need to be passionate about killing. Both of Larra's articles, the one on bullfighting from 1828 and the one on hunting from 1835, imply that the pursuit of these entertainments perpetuates violence between men and animals as well as among men, and even more notably between men and women.

For Larra there is an intimate connection between the cruelties in bullfighting and the self-othering construction of Spanish masculinity as impervious to the fear of death and insensitive to the suffering of others. He also notes that love for bullfighting goes together with love of wine, and it inspires domestic violence. The bullfighting fans, "estos parcos españoles se contentan con ser dichosos el domingo y el lunes, y reservan para los demás días, en que ya no hay harina en casa, el trabajar la obra y las bien cuidadas costillas de su mujer, como si quisiera indemnizarse en su pellejo del dinero mal gastado" (Larra, 1828, 252) (these frugal Spaniards are happy to enjoy Sundays and Mondays and the other days, when there is already no dough in the house, they work the ribs of their women as if this was a good way to compensate for the wasted money). Tauromaquia also transforms the sensibility of women, who faint at the sight of blood from a finger when they pinch themselves while sewing, but who do not mind watching horses tread on their own intestines during bullfights. Larra criticizes women who leave bullfighting rings unsatisfied if they have not seen sufficient amounts of interior organs and blood splashed on the sand. In fact as Martín-Santos and Almodóvar would show later, women's sensibility shapes the gaze of the public. It is they who require that matadors infinitely perform their heroic masculinity at the expense of bulls and other weaker creatures that appear in their way, ironically including women themselves.

In his brilliant analysis of the hunt in "La caza", Larra (1835) connects the killing of animals and other forms of violence with further insights: "La afición a la caza es como el amor, que donde está ha de dominar. Es como ciertas enfermedades que se apoderan hasta de los huesos del enfermo: el cazador es todo caza. Una puerta cerrada de golpe es un tiro para él" (441) (The passion for hunting is like love, which always needs to dominate. It is as some sicknesses which take over the body completely: the hunter is all about hunting. A closed door is an invitation to shoot). Larra observes that the passion for conquering and killing animals is transported to other domains of life, especially that of love, where women are treated as prey. But he also notes that the zeal of shooting translates to violence between neighbors as well as generalized cruelty toward animals, even those domesticated in human service: "Un compañero que bulle entre la jara es un ciervo: y el burro del ganadero que corre espantado de los tiros entre las encinas, recibe más de una vez una posta que se le dispara" (441) (A pal who moves in the roses turns into a deer: a farmer's donkey that runs terrified by the shooting through the oak forest, receives more than one pellet shot at him). Bullfights and hunting are thus portrayed by Larra as workshops of violence that develop violent instincts, desensitize people to blood, and promote that intimacy with death that Lorca proudly presents as a grandiose feature of Spanish culture in his lecture on el duende and that Larra finds troubling.

Larra sees himself as alienated from the Spanish way of life by the formative years that he spent in France that made him appreciate different tastes and formed him into a more feminine man. He looks at his country's reality from the distance of a quasi-foreigner, where familiarity mixes with estrangement. Not without a grain of admiration but otherwise with a strong critical distance, Larra creates a caricature of a Spanish man as he portrays a hunter: "El cazador generalmente es infatigable: a la larga le sucede siempre alguna avería, o pierde un ojo o un dedo . . . pero todo ello es nada a sus ojos. Hay que matar, y vamos viviendo. En eso se parece al militar . . . (441) (A hunter is generally tireless: eventually something always happens to him, he may lose an eye or a finger. . . . But all these mean nothing to him. One has to kill, and life goes on. In this he is like a soldier). Noel similarly connects this construction of masculinity with wars: "Un hombre ha de oler a pólvora, y si no huele a esto no es hombre. Tampoco será hombre si no vuelve a la Patria con las tripas en la mano y haciendo chistes. Tampoco será hombre si deplora ir al sitio de donde casi siempre no se vuelve." (1968, 341) (A man needs to smell of gun powder, otherwise he is not a man. Nor is he a man if he does not come back from war holding his intestines in his hand and joking about it. Nor is he a man if he does not want to go where most do not come back). Heroic masculinity is thus defined by the subject's willingness, indeed a spiritual need, to suffer and to die in order to produce suffering and destruction in others as disposed by the state.

In *Pan y toros* (1913), Noel criticizes Spanish stoicism that insists on seeking suffering and enduring it rather than attempting to remove its causes. This stoic attitude toward danger and harm would be later similarly ridiculed in Goytisolo's *Reivindicación del conde don Julián* (2003) in the pose of a bullfighter. Noel writes in *Pan y toros*: "Enamorados de las causas finales y firmes en nuestro propósito de no ver degeneración alguna, atribuimos los pretendidos errores y equivocaciones . . . o a premeditación y la complicidad de la Providencia" (59) (In love with final causes and strong in our determination, we attribute errors and mistakes to others or to Destiny).

Goytisolo, in reference to Manolete, sums up sarcastically: "Genio y figura hasta la sepultura" (25) (With heroic attitude to the grave). For Larra, Noel, and Goytisolo, the rituals of bullfighting and hunting represent examples of a most conservative attitude, which is a constant fight against the adversary and consequent subservience to power in the capacity of its victim or its hangmen. In this way, the system propagates itself through infinite cycles of violence. According to Noel, as a result Spaniards are reluctant to get excited by constructive projects involving calmer emotions. They complain of boredom and wait for the next bullfighting season while dozens of everyday practical issues await solutions.

Noel follows Costa's teaching that the structure of masculinity has economic consequences. En "Miscelánea taurina" (1914), he contrasts the enormous sums of money spent on bullfighting with the number of children who do not go to school, for there are no schools in certain districts of Madrid. He compares money collected at the entrance to bullfighting plaza just for one event, 90,000 pesetas (183), to that produced by a Goya exposition admired by the whole world, 800 pesetas (165). He reminds his readers that Galdós "si no muere de hambre es porque no come" (165) (if he needed to eat, would have already died of hunger) because, since the writer declared himself as Republican, his aristocratic benefactors abandoned him. He sums up all these comparisons by declaring that bullfighting consumes so much energy and financial resources that nothing is left for remedying the real problems plaguing the people or for rewarding real artists.

For Noel, almost all features of Spanish culture were connected to bullfighting.[12] In "Crónica antitaurina," published in *El Liberal* on 18 August 1918, Noel writes,

> No hay en España nada que de lejos o de cerca tenga la repercusión en el alma del pueblo como la fiesta taurina; . . . el ir a los toros presta al alma no sé qué enormemente macho. Es como una angustia que me conmueve el corazón preparándole a grandes cosas. Es como la ilusión de que los lidiadores no son otra cosa que uno mismo, que se necesita el mismo valor para actuar en esa fiesta que para verla. (n. pag.)

---

12. His contemporary, Ramón Pérez de Ayala (1880–1962), agrees. In *Política y toros* (1918), he writes, "Es un hecho de profunda significación en la vida española y de raíces tan hondas y extensas que no hay actividad social o artística en que no se encuentren sus huellas, desde el lenguaje hasta la industria o el comercio" (n. pag.) (It is a very important fact in Spanish life, whose roots are so deep and extensive that there is hardly a social or artistic activity, language or industry, where its marks are not found).

There is nothing in Spain that shapes people's souls like bullfighting; . . . going to the feast feels enormously masculine. It is a sort of sadness that grasps the heart, preparing it for great things. It is a sort of illusion that one is like a bullfighter, that one needs the same kind of courage to fight the bull and to watch the fight.

In other words, through a mimetic desire and by identifying themselves with bullfighters, people are eager to have spectacular violent confrontations, conquests, and triumphs. In this way, bullfighters become models of masculine behavior and national heroes.

Sensitivity to Animals' Pain

Larra opposes heroic masculinity, defined by the capacity to kill and conquer, to his own sensibility as someone who is already "fuera de combate" (Larra, 1828, 252) (disqualified), because he would only duel to the first blood. With a self-irony characteristic of his most famous articles, he makes the point that those who do not go to bullfights have lost, in the eyes of Spaniards, their pretense to masculinity. He classifies them as outcasts, marginal examples of maleness who cannot compete with traditional Spanish machos. The choice of words, "fuera de combate" (literally, out of fight) is not accidental, as it points to the fact that this marginal maleness that Larra exemplifies goes together with a dislike of violence, in contrast to its attractiveness in the Spanish culture of his time. Larra promotes a new masculinity, sensitive to self and to other's pains, wary of violence, critical of the state's power. As a part of this new sensibility, he speaks against hurting animals. In his article on bullfighting, he expresses an amazement at a complete lack of compassion for animals among the public: "Allí parece que todos acuden orgullosos de manifestar que no tienen entrañas, y que su recreo es pasear sus ojos en sangre y ríen y aplauden al ver los destrozos de la corrida" (251) (There it seems like everybody goes, proud to show that they have no heart, and that they are entertained by staring at blood, and they laugh and applaud when they see torn flesh). In his article on hunting, Larra points with more precision to the causes of human cruelty toward animals, formulating what Richard Ryder (1971) would a century later conceptualize as speciesism. Larra reflects: "Precisamente no puedo hallar otro origen de la diferencia que el hombre establece entre matar hombres y animales que su infinito amor propio; sin embargo hay animales que valen más que hombres y hombres que deberían darse la enhorabuena si no fueran más que animales" (Larra, 1835, 439–40) (In fact, I cannot find any other origin of the difference that humans establish between killing men and killing animals than man's own self-love. Notwithstanding, there are animals that are more worthy than men and men who should congratulate themselves if they were only animals). Some men (hunters for example) behave in such a way that they would be better off if their ethical stances were measured according to the criteria we use for animals. Although as Larra confesses, he did not do much to feel guilty; toward the end of the article, he displays remorse over having even accompanied his friend in the hunt because "aquellos animales ni nos hacían daño, ni nos estorbaban ni podían oponernos resistencia" (those animals neither hurt us nor bothered us and could not defend themselves in any way) and he firmly pledges not to ever again try this "ejercicio para el cual sin duda no debo de haber nacido" (441) (exercise for which obviously I was not born). This last sentence, which could obviously be taken as a personal confession, may also suggest a conviction that men, all men, are not born for killing animals, in spite of the hunting culture's and Spanish culture's beliefs to the contrary.

The sensitivity to animals' pain and various rhetoric devices used to express it in Noel's writings are similar to Larra's. Both Larra and Noel take the perspective of a bull when they describe a bullfight. Larra writes: "Van a ver a un animal tan bueno como hostigado, que lidia con dos docenas de fieras disfrazadas de hombres, unas a pie y otras a caballo, que se van a disputar el honor de ver volar sus tripas por el viento" (1835, 251) (You will see a good animal harassed, fighting against two dozen beasts dressed up like men, some on foot, some riding horses, that will all compete for the honor of seeing his intestines fly up in the air). Among various articles, in Noel's *Pan y toros*, there are two imagined interviews with animals: one with a bull, Canito, and another one with a horse, el Cid, used by bullfighters. The interviewed horse recalls the moment when his eyes were covered, he was led to a frightening, noisy place and suddenly felt cold and pain in his side, which was later repeated many times. Bulls gored this horse at every bullfight, and his wounds were sewn and then reopened time and again. The horse expresses hatred toward humans dragging him to the bullfighting ring, saying that he would kill them with kicks if he could. Noel sums up his exchanges with the horse by stating that "La fuerza es la razón de los hombres" (97) (Might is the reason of man). Agamben explains that the strategies of the "anthropological machine" (33) consist of constructing artificial differences between human and nonhuman animals that somehow authorize animals' mistreatment. In his interview with the horse, Noel mentions one such difference, namely the presence or absence of a soul. The horse reflects on his lack of soul, but Noel asks: "Y el que martiriza o se goza en el sufrimiento la tiene? . . . qué otra cosa debe ser [alma] que el respeto y veneración de lo que en torno de ella existe?" (100) (And the one who tortures or enjoys suffering has one? What is a soul if not respect and veneration of what exists around it?). This is a deeply ecological approach, where soul is understood as a connection to reality rather than a separation from it.

The "anthropological machine" produces literary representations where animals appear as devoid of feelings and awareness and as lifeless moving objects. By imagining the experience from the point of view of the animal, both Larra and Noel exemplify an opposite approach, very common in the activism songs and pamphlets of today's animal rights, whose aim is to mitigate human cruelty. A contemporary critic, John Simons (2002), calls this figure "strong anthropomorphism" (120) and considers it as an important strategy for animal studies.[13] For Simons, the purpose of this movement of imagination toward the other is multifold. It is ethical as the imagined other gains a place in the reader's heart, but it is also a quasi-mystical venture toward otherness. Similarly, Jorge Riechmann in his *Todos los animales somos hermanos* (All Animals Are Brothers and Sisters, 2005) envisions an encounter with an animal as a spiritual moment and a chance for a better understanding of the world, given that it takes men out of their anthropocentric perspective where their point of view blinds them to everything else.

Another strategy used by both Larra and Noel that sabotages the functioning of the anthropological machine is the deconstruction of the discursive strategies that represent violence against animals as attractive. They negate the artfulness of bullfighting. Noel goes to bullfights and writes *Crónicas* that compete with the official ones, unmasking the artificially constructed grandiosity of the aesthetics of bullfighting. He focuses his gaze on what *really* happens, sees the wounded flesh of the victims and opens his ears to their moaning, howling, and death rattle, imagining their physical pain: "El toro, herido, muge horriblemente. Le trae cerca de mi barrera y oigo gruñir a los dos, al toro y al torero. . . . El toro muge cada vez más, trota; el sol destaca sobre la piel negra el húmedo grosella de su sangre. . . . Delante de la espada, de ese morrillo sangriento, las magras chichas. . . . De esa masa horrenda salen voces estentóreas de cuando en cuando." (n. pag.) (The bull, wounded, roars horribly. They bring him close to my seat, and I hear both the bull and the bullfighter grunt. . . . The bull roars more and more, the sun highlights the wet dots of blood on his black skin. . . . In front of the sword, the

---

13. Simons distinguishes "strong anthropocentrism" from the fable-like treatment of animals and other uses of anthropocentrism that do not lead the imagination toward the animals' perspective. I develop commentary on Simons's terminology and analyses of rhetorical strategies that bring readers closer to the nonhuman realm in chapter 4.

lean meat of the wounded hump . . . Out of this horrendous mass a death rattle is heard from time to time). To conclude, Noel makes judgements of value: "¡Qué triste es todo ello, qué primitivo, qué estúpido! . . . Nada más chabacano, insulso y memo. . . . Pero ¿qué aplauden? . . . Ni pasión, ni arte, ni tragedia. . . . Lamentable, todo muy lamentable. . . . Todo sucede horriblemente vulgar, dando la impresión de un espectáculo lelo, memo, repugnantemente absurdo" (n. pag.) (How sad, how primitive and how stupid! . . . Nothing more tasteless, dull and idiotic . . . Regretful, all these, very regretful . . . Everything that happens is horribly vulgar, giving an impression of a stunning, silly, and an absurdly foul spectacle). Noel's descriptions of bullfighting are more than just expressions of distaste. He attempts to change the way people see things, freeing them from the culturally constructed lens that provides cloth for the naked king. He tries to see bullfights without the cultural discourses that elevate them to the realm of "art."

In his recent *Historia de la literatura española (1900–1930)*, José Carlos Mainer (2010) reviews in the same section works by various writers who were passionate about bullfighting and who were passionate about anti-bullfighting (including Noel), bringing all these works to the common denominator of "oscura convivencia con lo violento" (72) (dark intimacy with violence). In a similar, way Mitchell (1991) quotes Azorín claiming that Noel's *Crónicas*, depicting torn flesh and agonies are in fact enjoyable reading for bullfighting fans. They suggest that anti-bullfighting activists persist in representing the most drastic moments of violence due to some conscious or subconscious enjoyment that they draw from depicting it. Even if these observations are accurate (and I am not sure that they are), the fact that Noel's possible subliminal attraction to violence merits more attention from the critic than his conscious engagement against violence—that is, the way he acted in life shows an indifference toward ethical stands. Mainer's approach to anti-bullfighting is yet one more example of Hispanism's affiliation with the bullfighting culture, a projection of its worldview onto the anti-bullfighting literature, where he refuses to see sincere rejection of the national spectacle, suggesting indirectly that in fact they loved it. Mainer makes no comment on the anti-bullfighting writers' activism, thus dismissing it as a superfluous circumstance accompanying writing. In Noel's case, this was, however, the conditioning factor of writing. The fact that anti-bullfighting aesthetics may be enjoyable for someone who consciously indulges in violence, as in the case of Azorín, is yet a different matter. It raises a question about how to criticize violence without representing it or, in other words, what are the most efficient strategies of criticizing violence that would contribute to its demise rather than propagate it.

Noel was an activist and a public intellectual who lived according to his ideas and intended that his articles and conferences would have concrete effects on reality, even if they could bring about only small improvements. For example, in various articles, including the interview with a horse, El Cid, he fights for protection of the horses that are habitually disemboweled by the bulls' horns during the bullfights. His later article on this matter, "El caballo de picador" (The Picador's Horse, 1916) was republished in the veterinary press, fomenting deep criticism of bullfighting among the vets employed in the national feast. A doctoral dissertation by María Begoña Flores Ocejo (2009) recounts a great number of critical articles that followed. As a result, the *Reglamento Taurino* of 1927 introduced protection for horses and removed the designation of bullfighting as "a national feast." Even though in most anti-bullfighting literature (De Lora, 2003; Mosterín, 2010) this first change was attributed to a certain French lady who had accompanied Primo de Rivera to a bullfight and was shocked when sprinkled by a disemboweled horse's blood, Noel's articles and conferences conditioned to a great degree a part of Spanish public opinion about this reform.

The connections that Noel establishes between bullfighting and militarism, machismo, social injustice, as well as lack of education, indifference to animals and the destruction of nature are largely the same issues that today's animal rights philosophers discuss. Noel's ideas and arguments are repeated and rephrased, and they are gaining strength, although his merits are seldom recognized. Although it is from Noel that Mitchell (1991) takes most of his arguments about the social and political role of bullfighting, Mitchell never fully acknowledges it, portraying Noel as a "quixotic" activist and stressing that his views need to be taken with "a grain of salt" (54). This distance allows him to appropriate Noel's ideas, giving them the "coolness" of academic objectivity while putting down Noel's missionary zeal. Yet, it is interesting that Mitchell's vision of bullfighting is, in spite of his anthropological cool, deeply Noelian. After presenting the ideas of Gil Calvo (1989) and Tierno Galván (1988), who attempt to integrate democratic values with bullfighting, Mitchell forgets his anthropological distance for a moment and comments angrily: "Does watching the agony and death of a large bovid really teach Spaniards to take charge of their lives as free, independent, civic-minded adults?" (125).[14]

Noel's life was summed up as fruitless by several of those who wrote about him (Gómez de la Serna, García Mecadal, and others), but that could not be further from the truth. At least, the judgment depends on the criteria. I

---

14. Tierno Galván's article was published after his death.

am convinced that few of those who see Noel as a failure themselves saved the lives of six men and of thousands of horses condemned to death. Additionally, Noel's work laid foundations, supplied ideas, and was an example of personal attitude for the movement against cruelty to animals that lays foundations for an alternative biopolitics. Even if he has not become as famous as several of his contemporaries, his reflections, his way of looking at things, and even his metaphors appear in works of the most important writers and intellectuals who follow, such as Martín-Santos and Goytisolo. Even though he did care about the recognition that he never fully achieved, as an activist he was also aware that faithfulness to his commitment was an obstacle for this kind of success. In his *Diario íntimo*, when most depressed by his material situation and marginality, he stops to reflect on this obstacle at various times, yet every time, he decides to keep fighting against what he believes to be wrong rather than compromising his fight for the sake of personal success.

CONCLUSION: NOEL VERSUS ORTEGA

The fact that Noel is studied by very few and almost never taught must have been in part due to the fact that his criticism of bullfighting and "flamenquismo" was culturally annulled by the writings of the Generation 27, most of whose members were fascinated by these features of Spanish culture and promoted them in their works. This promotion of the culture of bullfighting and flamenco was especially prominent in Lorca's poetry and theater, which attained the greatest popularity. Even though flamenco and bullfighting have since turned into cultural clichés and grown old even in tourist propaganda, they still provide master metaphors to Hispanism. Due to the great literary caliber of the Generation 27 works, and due to their allegiance and in various cases active fight on the side of the Spanish Republic during the Spanish Civil War, it is their vision of Spanish culture as equipped with a spirituality emanating from bullfighting-like tragedies that the first postwar Hispanists in exile were interested in promoting. On the other hand, Ortega, a near contemporary of Noel and himself a very well-accommodated intellectual whose voice is also prominently present in Hispanism, similarly displayed an enthusiastic attitude toward bullfighting and hunting. In *La caza y los toros* (1944) Ortega refers to the ecological crisis, already then perceptible, as a problem of "escasez de piezas" (4) (scarcity of prey) for hunting. Ortega analyzes hunting as a form of entertainment and bullfighting as form of art, only once mentioning the ethical dilemma that they pose and dismissing it immediately. For Ortega, who stated that all Spain's problems began when the lower classes rebelled against the aristocracy and demanded more equality, the Society for the Protection of Animals is laughable. While he admits that avoiding suffering is an ethical norm, he argues that it is not the only norm at work. Without explaining any further, the philosopher abandons the issue. Ortega's lack of attention to this question could not be caused by a lack of awareness of its deep significance for Spanish culture. Ortega read Noel's work, praised him, sent him to war, gave him advice to "make himself a man," and even invited him once to publish in *España*, although he withdrew this invitation right away when Noel would not compromise his positions. If Ortega considered some of Noel's writing important enough to say it was the best of his generation and to solicit Noel's articles, it is puzzling that in his writings on bullfighting and hunting he never mentions Noel's name or his ideas. It is not totally absurd to suppose that Ortega, who had quickly appreciated Noel's talent and intellect, was wary that popularization of Noel's ideas might threaten his own intellectual capital and position. Perhaps, however, this was not so and Ortega was simply deeply conservative as an academic who analyzed the culture in which he lived but did not challenge its order and prejudices, instead reacting to all attempts at social change with skepticism.

Rosario Cambria (1974), who has researched the attitudes of intellectuals toward bullfighting, writes that "Ortega approached the phenomenon of bullfighting as one would expect of a strong Spaniard who had received his philosophical training in Germany—that is, with controlled passion, a favorable attitude, and the open-mindedness and serene objectivity" (211). This tone of admiration is characteristic for most Hispanists commenting on Ortega in any context, including that of bullfighting, even though his arguments in fact are often shallow, mystifying, and not at all convincing. For example, for Ortega, bullfighting's main social role consists of giving people something to talk about. In an article titled "Crítica de un discurso 'en yo menor'" (Critique of a Discourse in "I Minor," 1924), Noel criticizes Ortega's elitism and lack of activism for the Republican cause, implicitly accusing him of banking on his greatness instead of working for the change. Perhaps, however, this is how greatness in academia is built, by excelling at compliance with the rules of the game and distancing oneself from risky adventures. Noel himself was more interested in contributing to what he believed was the common good; he opted for a more organic kind of intellectual work where personal capital is shared among many. Also in this, his attitude was a precursor for new progressive ways. The comparison of Ortega's and Noel's life trajectories and their ideas invites us to question the notions of "success" and of "greatness," in the political context and in our respective academic fields.

CHAPTER 3

SCIENCE, POLITICS, AND ANIMALS

TIEMPO DE SILENCIO AS AN ANTI-BULLFIGHTING NOVEL

> Like all neuroses, mine is rooted in the problem of metaphor, that is, the problem of the relation of bodies and language.
> —Donna Haraway

ROBERTO ESPOSITO writes in "Biopolitics" (2013a) that for years the body has been "the most influential metaphor used in the political discourse to represent life in the society" (318). In "Enimgma of Biopolitics" (2013b), he explains how in modern history parts of the government were compared to bodily organs, political problems were equated with malaise or illnesses, and regeneration was often postulated through an extirpation of sick parts that in reality may have corresponded to a particular person or a group of people, sometimes ethnically defined. This conflation of medical and political discourses based on a metaphor activated particular strategies, such as repelling human groups that were perceived as threatening to the health of the nation, for example, in Nazi Germany. *Tiempo de silencio* (1962) by Luis Martín-Santos reflects this dynamic by constructing an image of Spanish society whose poorest people are treated as sick organs that need to be isolated from the healthy ones, presumably the middle and higher classes, which hope to sustain their alleged "health" by enforcing this separation.

In this context, the novel compares man to a city: "Un hombre es la imagen de una ciudad y una ciudad las vísceras puestas al revés de un hombre (13) (A man is an image of a city and a city is a man's entrails turned inside out). The city appears as a super organism of which a man constitutes one cell, a super brain of which man is just one thought. This metaphor of a society that is a large body constituted of small bodies reminds us of the figure of the bull formed by the activists in front of the Guggenheim museum. In Martín-Santos's vision, however, this is a much less voluntary grouping of bodies, administered from above rather than by the community itself. It is administered by the economy and the rhetoric of the postwar Spain, which is structured around bullfighting metaphors. The bullfighting rhetoric takes over the thinking of men, and the body of the city smothers them. If all the humans in the city are connected by the same "will to power,"[1] the vision of this power is deeply influenced by the bullfighting culture. The circulation of the power in the capillary vessels of the organism of the city is shown to repeat the same pattern that is retraceable to the structure of a bullfight. Sex, science, and an intellectual argument are compared to matador's maneuvers.

Human/animal comparisons appear in various contexts of the novel and in particular in the scientific laboratory. It is because of the human's alleged difference from animals that experiments on them can be conducted, but it is due to the deep similarity between human and animal bodies that these experiments can make sense. The novel not only discusses the human treatment of animals in the scientific laboratories but also human animality (in Luis Buñuel's understanding of the term as *vulnerability*) as well as the process of animalization of lower class humans. In these two contexts of animality and animalization, *Tiempo de silencio* reveals that the crucial concept of biological sciences, the gene, is yet one more metaphor serving the biopolitical status quo because it justifies the unavoidable character of hierarchies, destitution, and harm. Human and animal bodies transcend into each other in the novel, blurring the divide that, as Martín-Santos shows, is to a degree artificial. What emerges instead is vulnerable flesh, common for all living beings, which, in the words of Esposito, can be defined as "the unitary weave of differences between bodies" (325).

---

1. "Will to power" is a concept derived from Friedrich Nietzsche's philosophy. It was never systematically defined by the philosopher, but its significance consists in the assumption that all great humans are driven to achieve the highest possible position in life, defined by an amount of power over others. *The Will to Power* is a book of notes drawn from the literary remains of Neitzsche. *The Will to Power* is also the title of a work that Nietzsche himself had considered writing.

Pedro's expulsion from the city at the end of the novel can be viewed in light of Esposito's theory as an action of the city's immune system that expulses the sick cell outside of its walls. This medieval immune procedure used in cases of leprosy coexists in the novel with the modern system of special distribution of the bodies of people belonging to different social classes that Esposito calls "quadrillage" (342). The network of boundaries between high- and low-class neighborhoods that Martín-Santos describes when Pedro and Amador walk toward the slums reflect a special distribution of the population, a geographical separation of classes that went far beyond sanitary needs and serves the status quo of the political. While this separation is realized through the urban development of Madrid, the social discourses condemn crossing the borders to certain neighborhoods. The inhabitants of those forbidden places are thus themselves banished from the city as possible contaminating agents, below other humans, and like animals, their lives are experimented upon by the power curious about the limits of resilience of living flesh. As the bodies of the poor find their way to the "healthy" spaces of the city, their fight for survival becomes a menace for others' safety. Martín-Santos compares the biopolitical experiment with poverty that sickens the society like a cancer, to an autoimmune disease that consists of an error of in the immune system: Society is like the body that is obsessed with expelling the elements that threaten to destroy it, not noticing that it attacks parts of its own flesh.

BULLFIGHTING BIOPOLITICS AND ITS METAPHORS

Bulls' heads, real and imaginary, accompany Pedro, a young biomedical researcher, in his nighttime adventures. As he goes out on Saturday night, "Andando con paso rápido, pasó junto a una taberna con cabeza de toro" (He walked along quickly, passed a tavern adorned with a bull's head, 57/60).[2] After drinks with his close friend Matías and a newly acquainted German painter, taxi lights transform into bull's horns: "Algunos taxis con su cuerno verde amenazante haciendo sonar una bocina aunque nocturna penetrante" (A number of menacing bull-like taxis charged near their shining horns and uttering bellows that pierced the night, 71/76). When the young intellectuals are getting drunk with the cheapest cognac at the next bar, they hear "*cante hondo* degradado y de rasguear una aguja vieja" (a degenerate *cante hondo*

---

2. Translations by George Leeson. The first page number is of the Spanish text, and the second is the page number in the English edition.

made worse by a needle scratching on the record, 73/77). When Pedro returns from a night of drinking to the pension and secretly enters the owner's daughter's room and makes love to her, he sees himself as a bullfighter killing a defenseless bull. The comparison of lovemaking to killing connects to the next scene where rape leads to abortion and death. While Dorita loses her virginity, Florita, a shantytown girl, bleeds to death as if she had been gored by a bull's horn. Through these metonymic connections of parallel actions, Martín-Santos shows that the bullfighting master metaphor has power over real life. Bull's horns as phallic symbols condition gender relations and discourses where sex is represented as goring and where it often becomes as destructive and harmful as a real goring. Horns are also symbolic of conservative politics and of intellectual discourses structured with bullfighting metaphors and allied to Franco's power. Bull's horns merge with the horns of Beelzebub, evoked by the owner of the pension (75/80) and with the horns of Goya's "Great Bouc" that is projected onto the bullfight-loving Ortega. Even the science proceeds by killing animals in the lab, "como si fuera una lidia" (8/4) (like a bullfight), in order to receive proof in experiments that constitutes a measure of a scientist's triumph.

The novel tells a story of the failure of a young intellectual, Pedro, who came to Madrid dreaming about finding a cure for cancer. He receives a fellowship so he can lead laboratory experiments on rats in order to prove his hypothesis, which is that cancer is not caused by genetic inheritance but rather by a virus. If a virus was responsible for the cancer, it would be possible to prevent it by vaccination. Pedro's research continuity is threatened by a high rate of mortality in the laboratory mice. When all the mice in the lab have died, his assistant, Amador, informs him that he had given some to his cousin, Muecas, who breeds them now in a shantytown in the outskirts of the city. In pursuit of the mice but also moved by curiosity, Pedro goes to the shantytown where he meets Muecas, his wife, and his two daughters who carry the mice between their breasts to warm them up and make them horny. The same night Muecas shows up in Pedro's pension, imploring him to save his older daughter Florita, a victim of a badly performed abortion. Pedro arrives too late to save the girl, but as a result of his attempt to help her he is accused of having performed an illegal abortion and imprisoned. He is freed after the declaration of his innocence by Florita's mother, Ricarda, but due to the scandal his scientific career is finished. His mentor advises him to start a private practice as a provincial doctor.

The bullfighting biopolitics in the novel is deeply connected to language. The narration is structured by a series of intertwined metaphors that reflect on various levels of symbolic, structural, and real violence during the postwar period in Spain. The structural violence (Galtung, 1969) is one that power exercises over its people through the maintenance of social hierarchies, which produce slums. The symbolic violence (Bourdieu, 1977, 1991) is inherent in the intellectual discourses that justify this inequality. The symbolic violence resulting from language, the structural violence resulting from the social hierarchies, and the real violence that hurts the flesh are also portrayed as interchangeable and connected through metaphors. Bullfighting master metaphors—of which the most visible are social life as bullfighting, sex as a bullfight, and intellectual argument as a bullfight—are complicit in the oppression and contribute to real violence. Martín-Santos makes them visible in order to confront them because, as Pedro reflects, "la mejor máquina eficaz es la que no hace ruido" (the best and most efficient machine is the one that makes no noise, 216/244). Consequently the most dangerous are the hidden metaphors (Lyotard, 1988). Using an avant-garde strategy, the author turns metaphors into real events to make readers understand how they work. It could even be argued that all that happens in the novel results from the powers of the figurative language.

Lakoff and Johnson's *Metaphors We Live By* (2003) shows how metaphors can be turned by the human brain into scripts that direct their behavior and how they are constitutive of culture. As various bullfighting aficionados announce with enthusiasm, Spanish is impregnated with bullfighting metaphors used especially to comment on love, politics, sports, and other domains that conceptualize power as domination. Ramón Pérez de Ayala in *Política y toros* (1918), Eugenio Noel in most of his anti-bullfighting writings, Timothy Mitchell in his *Blood Sport* (1991), Manuel Vicent in his *Antitauromaquia* (2001), as well as many others on both sides of the political divide have continuously seen the image of Spanish life in the bullfighting ring. For example, José María de Cossío, author of a monumental bullfighting encyclopedia (1995), writes that in Spain there is no activity, whether social, artistic, or economic, devoid of marks of bullfighting. Carlos Abella, contemporary bullfighting journalist and writer, still in 1996 suggested that the bullfighter was a model figure for Spaniards: "Se produce coincidencia e identidad con el héroe (el torero) y su esfuerzo" (23) (Identification with the hero and his effort is produced). From the anti-bullfighting perspective, Vicent (2001) arrives at similar conclusions, stating succinctly that "En España todo es tauromaquia" (40) (In Spain everything is bullfighting) because bullfighting is "el espejo donde se mira la raza hispana" (86) (the mirror in which Hispanic culture sees itself). Vicent finds it deplorable that Spanish identity is structured through bullfighting metaphors because they naturalize violence as a governing principle of relations between living beings. In his *Antitauromaquia*, he obsessively compares bullfights to the Spanish Civil War, which he calls "una buena lidia que duró tres años y en la que se cortaron unos a otros infinitas orejas" (27) (a good fight that lasted three years and in which an infinite number of ears were cut). He argues that those who appreciate bullfighting transfer its rules to social life and that these metaphorical transferences have real consequences. In this sense, for those concerned with learning how to function in the political status quo, bullfighting is educational because, as Vicent argues, through the extension of the metaphor of bullfighting to the political realm, political violence and persecution may be accepted as rules of the game. Similarly, from an antibullfighting perspective, Paula Casal writes in a personal communication:

> Bullfighting has not been an isolated part of Spanish culture. Spaniards have no other way of saying "brave it" or "be brave and go for it" than "courage and to the bull" or "take the bull by the horns." They refer to "crowd pleasing exaggerations or lies" as "toasts to the sun" (the sun being where low-class aficionados seat in a bull ring). They refer to "the final blow," or "what broke the camel's back" as "receiving the *puntilla*" (like a bull), and they call malicious attacks "puyas" like the ones that the bull receives. They also negotiate objections [by] "bullfighting the problems," and describe as "of fallen cape" something that is down, sad or not well attended. (n. pag.)[3]

There is a long passage in *Tiempo de silencio* that appears to be a fragment of an earlier anti-bullfighting essay containing a historical and social analysis of bullfighting by the author in which Spanish national culture is similarly represented as built around bullfighting:

> Si el visitante ilustre se obstina en que le sean mostrados majas y toreros, si el pintor genial pinta con los milagrosos pinceles majas y toreros, si efectivamente a lo largo y a lo ancho de este territorio tan antiguo hay más anillos redondos que catedrales góticas, esto debe significar algo. . . . Acerquémonos un poco más al fenómeno e intentemos sentir en nuestra propia carne—que es igual que la de él—lo que este hombre siente cuando (desde dentro del apretado traje reluciente) adivina que su cuerpo va a ser penetrado por el cuerno y que la gran masa de sus semejantes, igualmente morenos y doliococéfalos, exige que el cuerno entre y que él quede, ante sus ojos, convertido en lo que desean ardientemente que sea: un pelele relleno de trapos rojos. Si este odio ha podido ser institucionalizado de un modo tan

---

3. See also Luque Duran and Juan de Dios (2002).

perfecto, coincidiendo históricamente con el momento en que vueltos de espaldas al mundo exterior. . . . Que el acontecimiento más importante de los años que siguieron a la gran catástrofe fue esa polarización del odio contra un solo hombre y que en ese odio y divinización ambivalentes se conjuraron cuantos revanchismos irredentos anidaban en el corazón de unos y de otros no parece dudoso. Llamaremos, pues, hostia emisaria del odio popular a ese sujeto que con un bicornio antiestético pasea por la arena con andares deliberadamente desgarbados y que con rostro serio y contraído muerto de miedo traza su caligrafía estrambótica ante el animal de torva condición? Tal vez sí, tal vez sea eso, tal vez puesto que la fuerza pública, la prensa periódica, la banda del regimiento, los asilados de la Casa Misericordia y hasta un representante del Señor Gobernador Civil colaboran tan interesadamente en el misterio. (166–67)

If the illustrious foreign visitor insists on being shown *majas* and bullfighters, and if the great artist paints *majas* and bullfighters with his wonderful brushes, if there are more bullrings than Gothic cathedrals throughout the length and breadth of this ancient land, then there must be a reason for this. . . . Let us approach the phenomenon a little closer and try to feel in our own flesh, which is the same as his, what this man in the tight-fitting spangled costume feels when he is about to be pierced by the bull's horn while the great mass of his fellows, also dark and dolichocephalic, clamor for him to be gored and yearn to see him reduced before their very eyes to a dummy stuffed with red rags. It must be significant that this hatred should have been institutionalized so perfectly, coinciding historically with the time when the country turned its back on the outside world . . . . There seems to be no doubt that the most important factor of the years that followed the great catastrophe was a polarization of hatred against one single man, and that this ambivalent hatred and deification exorcised all the unredeemed feelings of revenge in many hearts. Shall we then call this man the chosen victim of popular hatred, this man who goes in this strange bicorn hat and walks across the arena with a deliberately ungainly step and serious and constricted face, half dead with fright, to make his ridiculous postures in front of a wild animal? Perhaps that is it, since the forces of public order, the daily press, the regimental band and even a representative of His Honor the Civil Governor are so much involved in the mystery. (184–85)

Martín-Santos's interpretation of *tauromaquia* as religious-like activity, absorbing negative emotions of the crowd but also draining all potentially positive social energy coincides closely with Eugenio Noel's. For both, the hypnotic arrest of social life through *corridas de toros* occurs through an identification of the people with the bullfighter. This identification results from a complex of inferiority through the sarcastic reference to dolichocephalic skulls, characteristic of people with mental retardation. Both Noel and Martín-Santos coincide in retracing these feelings of inferiority to 1898, "el año del desastre" (the year of the catastrophe), when the Spanish military failure in the war with the United States was compensated for with bullfighting spectacles. Thus frustration and anger caused by this national failure is portrayed as vented through hatred and aggression against the bull and the bullfighter instead of bringing national reforms. While in Noel's essay "Los toros de Carabanchel," analyzed in the previous chapter, 1898 is the only reference; Martín-Santos's passage alludes to a more immediate and painful present, namely the loss of the Second Republic in the Spanish Civil War. The cultural mechanism of compensation for this loss is, however, the same as in 1898. The defeated people vent their rage on the bullfighter who transfers all their hatred into the bull. For conservative commentators on culture for whom the goal is to maintain the status quo, it was a wise way to deal with social discontent and prevent potential unrest. For social reformers such as Noel and Martín-Santos, this way of reacting to the frustration through venting anger at innocent victims is the kernel of all other social evils because it takes people's interest away from useful political action and it deactivates anger through harm. The pattern of venting frustration on the weaker is present in other social contexts, such as family interactions where women, children, and animals are victimized for failures of the men (Labanyi, 2011). Muecas systematically beats up his wife, Cartucho abuses all the women that he encounters on his way, and the professor of the institute humiliates his postdocs. In this way, power flows through the whole superorganism of the nation, establishing a hierarchy of bodies based on widely understood strength.

According to Jo Labanyi, the death of the bull for Martín-Santos is not the symbolic sacrifice of a God who dies to redeem the nation (as it was for Giménez Caballero and so many others) but rather a result of deceit, of a misdirected violence that should have been employed in a fight against the source of oppression. Were the energy directed toward implementation of a change or a political fight, the evils causing popular anger could possibly be remedied. The social anger that does not know its causes ends up destroying the most defenseless bodies. In this way, through the bullfighting spectacle, exhausting negative energies, social unrest caused by injustice becomes neutralized, pacified, and deactivated at the expense of the weakest. Just as Noel, in order to deconstruct the fascination of bullfighting, Martín-Santos focuses on the pain of the bullfighter's flesh, torn by the bull's horns, hidden by the spectacle. This focus on a suffering body is present in various passages of the novel, such as descriptions of life in the shantytown where we see rats' bites on the breasts of Muecas's daughters and then Floritas's deadly abortion, Ricarda's bruises or even the physical suffering of the police inspector who arrests Pedro. Sensitivity to pain becomes a point of departure in the search for an alternative biopolitics.

SEX AS BULLFIGHTING

Pedro's nightmare after making love to Dorita subverts the metaphoric connection of bullfighting to sex, rendering this connection as monstrous. Pedro feels imprisoned in the role of a bullfighter:

> Como el matador con el estoque que ha clavado una vez, pero que ha de seguir clavando en una pesadilla una vez y otra vez, toda la vida, aunque haya avisos . . . allá el torero ha de seguir clavando su estoque en el toro que no muere, que crece, crece, crece y que revienta y lo envuelve en toda su materia negra como un pulpo amoroso ya sin cuernos, amor mío, amor mío, mientras la gente ríe y pide que se les devuelva el importe de sus localidades. (91)

> Like a bullfighter who has struck once with his sword but has to keep striking again and again. In a nightmare, again and again all his life in spite of the protests of the crowd . . . the bullfighter has to stand there thrusting his sword into the bull, which does not die but grows and grows until it bursts and covers everything with its black mass, clinging pulp, without horns now, my love, my love, while the crowd laughs and ask for its money back. (98)

Pedro's perception of having decapitated Dorita results from the bullfighting metaphor often applied by Spaniards to the sexual act, where as Abella (1996) explains, "La suerte de matar se presta a todas las insinuaciones . . . por la sugernecia fálica del estoque y la penetración del mismo" (40) (The moment of killing can be compared to certain other activity . . . because of the phallic appearance of the sword and its penetration in the flesh) and where failure is summed up in brief by "se fue viva" (40) (she left alive). Martín-Santos subverts the prevalent cultural metaphor that bullfighting is like sex, which manifests itself in Castilian through numerous comparisons of different stages of sexual conquest to different parts of the bullfight, including the comparison of the actual sexual act to the final act of killing, by showing that this structure of love can have literally murderous results for women. Cartucho, the shantytown alpha male, who can only fuck or kill and who ends up killing Dorita, is a caricature of the bullfighting masculinity. He waves and thrusts his knife into all living creatures that get in his way. His pride is based on a bullfighting menace: "Y nadie se atrevía a darle la cara" (Nobody else has guts enough to face up to him, 43/43). Yet in his fear-driven courage, desperate and confused, he himself is a victim of an unjust social hierarchy, and lowest in it, he takes revenge on defenseless women.

Pedro's dream after sex ends with a phantasmagoric logic; the bull shows its deceptive façade both as lover and as enemy. In fact, the aggression comes from the public. It is the public perception of Pedro's acts that will decide from that moment on how his life will change. It is the public's opinion that conceives of *chabolas* as contaminating, that condemns Pedro's involvement there, and that decides the end of his career as a researcher. It is also the public's opinion that decides his love life. If on one level, the decapitation of the lover in the dream may be symbolic of the loss of virginity, on a different level, it indicates a different kind of mutilation, namely male castration and his loss of freedom. According to the social mores, Pedro and Dorita are now condemned to each other. Pedro perceives his lovemaking to Dorita as entrapment and a fall because he always dreamt of a love that would be based on intellectual partnership and not a purely sexual, animalistic attraction. The society portrayed in *Tiempo de silencio* is like the one described by Foucault in "The Right of Death and Power over Life" (2013) in that it excites sexuality through entertainments and spectacles at the same time as it represses it. Such society controls the freedom of its members through sexual traps like the one into which Pedro falls.

ARGUMENT AS BULLFIGHTING

The interpretation of Goya's painting *Le Grand Bouc,* whose reproduction Matías shows to Pedro right after his failed rescue of Florita, again summons analogies to bullfighting. Bouc's meaning hesitates between that of a "buco expiatorio" (scapegoat, 119/127) and a "buco gozador" (pleasure-loving goat, 119/127), corresponding to an accommodated intellectual. Both the scapegoat and the pleasure-loving goat participate in the national spectacle, but their predicaments obviously contrast since the first one dies to entertain the second. The narrator's fixation on Bouc's horns as well as women's adoration of the figure of the professor who gives "terebrofilia del futuro" (a piercing revelation of the future, 118–19/129) also brings to mind bullfighting through a synergy of "horns," (119) "piercing," (129) and female excitement, which are all parts of the *tauromaquia* semantic context. The bullfighting caste, that is, the Spanish lower classes, imprisoned by bullfighting rituals, repeat infinitely their "*cante hondo, media veronica*" (*cante hondo* song, an elegant swinging movement of the cape) with "fidelidades de viejo mozo de estoques" (the faithful eyes of the bullfighter's old sword bearer, 119/130).

While the Professor gives his lecture about how an apple can look from different perspectives, Pedro hallucinates, seeing Florita's cadaver floating in the midst of thick carpets in Matías's house. This hallucination is described in a passage directly following the description of the Great Bouc's horns in Goya's painting, "signo de glorioso dominio fálico" (signs of glorious phallic domination, 117/127), lifted high to gore (118/129), threaten, seduce, and rape all those women and undernourished children. Through the similitude of the arms lifted to bless and preach by both the Bouc in Goya's painting and by the Professor, who gives a lecture in Matías's house and who is a caricature of the bullfight-loving Ortega, Goya's Bouc, the bull, and the assimilated bullfight-loving intellectual all blend into one. The rhetoric of the lecture in the salons of high-class society is characterized by "veronicas intelectuales" (intellectual *veronicas*). A *veronica* (119)—the swinging movement of the cape meant to avoid the bull—as a figure of an intellectual argument makes it appear elegant but empty. In light of the novel, intellectuals like Ortega, producing discourses that entertain elites, distracting them from reality, are responsible for structural violence-producing deaths. They reinforce necropolitics not only through their propagation of bullfighting culture but also through their emphasis on aesthetics, which becomes a strategy for the segregation of social classes. Intellectuals such as the Professor/Ortega are to be blamed for the suffering of all the victims of Cartucho, who as his name suggests, is like a bullet shot out by the system of social injustice, blindly revenging his own deprivation by thrusting himself around.

Martín-Santos points out those of Ortega's ideas that he considers most harmful, which are the feeling of superiority and the disdain for the people: "O buco, a todos nos desprecias. Sí, realmente sí, qué bien lo has visto: Todos somos tontos. Y este ser tontos no tiene remedio." (O Goat! You despise all of us. Yes indeed, how clearly you have seen that we are all stupid. And it is a stupidity that cannot be cured, 119/131). In *España invertebrada* (1921), Ortega connects Spain's underdevelopment to its people's racial inferiority. The Western Goths, who populated Spain, were, according to him, less well-equipped genetically than those who settled in France. Ortega mentions the divided caste system of India as an example of a wisely organized society. Martín-Santos ridicules Ortega's fascination with Indian hierarchies by calling Andalusian cultural forms "reflections of Elephanta" and ironically comments on how "en Bhuvaneshwara la infancia inmisericordiamente de hambre perecía, pero fue en tales templos grande la adoración a los ritos . . . sin que el óbice de la mortandad habrienta y los otros perecimientos irritara como posible masa fermentativa al pueblo—en que tales procesos ocurrían—habilidosamente segmentador (en sectas)" (in Bhuvaneshwar children died unwept from hunger, [yet] death by starvation and other killers did not ferment the revolt among the people, skillfully segmented into sects and distracted by the greatness of the religious rites, 118/129). The bullfighting ritual and the Professor's talk are both compared to the religious rites of Bhuvaneshwar, which distract from the fight for justice and contribute to the acceptance of high rates of child mortality. Again here, the academicians' engagement in deliberations on art for art's sake contributes to the status quo of social power. Ortega as a bullfight-loving intellectual lecturing on an apple for high-class ladies is turned by Martín-Santos into a patron of the child-killing culture whose discourses imprison Spanish life in cycles of the infertile repetitions of cape movements.

HUMAN AND ANIMAL

The concept of the *human animal* is significant in descriptions of different spheres of social life. Human/animal comparisons do not classify certain kinds of humanity as inferior, but they suggest that recognition of the human traits shared with animals is needed for a deeper understanding of a human/animal nature. While the professor giving a lecture in the rich salons is compared to a goat, the women listening to him seem like exotic birds, Burdelmama is a queen ant, and the poor people are compared to mice. The search for Pedro undertaken by the inspector, Similiano, who follows Matías and is then followed by Amador with Cartucho meandering at the end, is compared to the pine-moth caterpillar procession (144/158). Ricarda, Muecas's spouse, mourns her daughter like a faithful dog, and Pedro leaves Madrid, going to the province as if he were a werewolf abandoning the court for the forest. Bullfighting displays the animality of the humans, both the matador and the cheering public, rather than their differences from the animal:

> Pero, qué toro llevamos dentro que presta su poder y su fuerza al animal de cuello robustísimo que recorre los bordes de la circunferencia? Qué toro llevamos dentro que nos hace desear el roce, el aire, el tacto rápido, la

sutil precisión milimétrica según la que el entendido mide, no ya el peligro, sino—según él—la categoría artística de la faena? Qué toro es ése señor? (168)

But what animal is there inside us that lends power and strength to the animal with the mighty muscular neck which charges around the arena? What animal is there inside us that makes us clamor for the close-shave pass, the gust of wind, the rapid touch, the subtle hairs-breadth precision according to which the experts claim to measure, not the extent of the danger, but the artistic standard of the kill? What animal, sir? (186)

No animal has been able to develop an art of killing with so much sophistication except the human animal, which in light of the passage, is a specialist in the art of cruelty. This temptation to regress to the primal instincts and primitive behavior is taken advantage of by the state institutions because it is easier to administer animal-like humans than highly rational ones. The vision of basic equality among all animals as creatures of flesh who suffer pain does not preclude the need for human improvement through the selective revision of human animalities.

While the novel insists on ethical revisions in the human domain, it simultaneously questions the concept of *human* as opposed to *animal*. Man is for Martín-Santos an "hombre-lobo" (werewolf, 212/238) that moves between humanity and animality as day becomes night. Thus the borders between humanity and animality are blurred, and it is not for sure, as the author states somewhat cynically, that "el hombre es la medida de todas las cosas" (man is the measure of all things, 212/239). He finishes the passage with an enigmatic remark that a wolf's mouth is four times larger, which perhaps signals the strength of the animal instincts. In light of these comments, culture appears as an elaborate camouflage for human animal nature. Females of the higher classes are compared to colorful birds, seductive women are sirens, peasants are compared to partridges, and they are all "taurínamente perseguidas" (pursued like in a bullfight, 202).[4] Males are compared to beasts with horns or swords, creatures in between bulls, goats, and bullfighters. The "distinction of taste" (Bourdieu, 1984) in the salons of Matías's mother, which is revealed by dress code, hair, and nail styles as well as the gestures and language of her guests, is presented by Martín-Santos as an elaborate cover for their human animality. The writer, however, compares this "distinction" to distinctions present in the animal kingdom; the colorful dresses of rich ladies

---

4. My translation. The translator omits the reference to bullfighting.

are as birds' plumage. (This vision will reappear later in Juan Goytosolo's *Las virtudes del pájaro solitario* [2013]). The salon culture that insists so much on camouflage rather than understanding produces nausea in Pedro who sees through it, noting that the distinguished people's denial of their own animal traits, otherwise so visible, makes some of those traits grow uncontrolled, leading them away from an ethical life. The concept of humanity as separate and opposed to animality, which these higher classes cultivate, justifies their inhuman treatment of the poor, whom they see as not human enough.

One of those poor is Ricarda, whose most striking human-animal feature is honesty. Her honesty becomes an obstacle for symbolic power as it consists of an uncalculated interruption of the hegemonic discourses. Due to Ricarda's linguistic limitations, she may be a victim of power, but in certain contexts, she is also impervious to manipulation. The social discourses, indifferent to her victimization and pain, unpeople her since, as the narrator ironically suggests, only a person can feel pain. In his brilliant analysis of the significance of Ricarda's character, Joseph Patteson points out that the narrative irony ubiquitous in the novel dies out when Ricarda's life is presented through a series of visual memories and that Ricarda is the only mother in the novel who appears rebellious and resistant to the social order, and as various other critics notice, her acts redeem Pedro and punctually overturn injustice.

Ricarda's human animality is thus presented in the novel as a space of the resilience of human morality. Resilience is a concept conceived by a forest ecologist C. S. Hollings (1986), who defines it as adaptive capacity, "characterized by constant movement between periods of stability and processes of transformation" (O'Brien, 2013, 5) whose pivotal moments are defined by tensions between "mobilization" and "retention," (5) or in the terms of Carl Folke (2006), between "revolt" and "remembrance" (259). Ricarda can be read as an ecosystem in which human and nonhuman elements connect for an ethical alliance. She is an extension of the metaphor of man as an image of the city, city as an image of the man, to the nonhuman environment. Compared to earth, to an onion, and to a dog, Ricarda is a complex system of life that fights for survival holding on to the most basic principles: care, sacrifice, honesty, work, subordination, but also if needed, rebellion, and aggression. Resilience could be a problematic concept to represent the poor as it coincides with the higher classes' discourses that dehumanize the lower classes. In the novel, however, Ricarda's toughness appears contrasted with the dramatic rendition of her pain caused by the fragility of life to which she is connected. The readers can't stop hearing her howling after the death of her daughter, Florita, killed by a rape and a badly performed abortion.

In *Tiempo de silencio* (1962), humans are animals not only because they possess animal-like traits but also because they share the animal predicament in various ways. Mice with tumors, next to dogs and cats, objects of laboratory experiments undertaken by science later become connected to the poor living in slums, although it is also hinted that even higher class ladies may also be victimized by cancer. This connection is established metonymically because the mice are carried by Pedro's assistant, Amador, to Muecas's family, who live in slums and whose daughters will breed them. The metonymy formed by physical proximity further strengthens the symbolic standing of the mice for poor people, who die like mice and are treated like pests when they invade the "human" space of the higher classes. Here the metaphor extends and points to other similarities. As the mice's multiplication is needed for science's business, certain levels of poverty help to maintain the status quo and the market economy. Mice-like people, unaware of the *doxa*, cannot contribute to politics through language but only through the body, as Ricarda does. Although these are significant contributions, they are also ridden by pain.

GENES AND SCIENCE

> Like a pair of Founding Fathers, acting in concert to promote one and the same innovation in political theory: the representation of nonhumans belongs to science, but science is not allowed to appeal to politics; the representation of citizens belongs to politics, but politics is not allowed to have any relation to the nonhumans produced and mobilized by science and technology.
> —Bruno Latour

The description of the experiments on animals makes it obvious that they have no chance of success and serve only as a justification for the maintenance of the scientific structures.

> Perro aullador que no orina. Al no orinar, víctima de su violenta carga afectiva, el perro elimina sus esencias por el sudor. Al no sudar más que por la planta de los pies, el perro elimina su aroma también por el aliento, con la lengua fuera así colocada a los fines de la transpiración. Cuando el perro ha sido operado y se le ha colocado un fémur de poliestilbeno o polivinilo, sufre tanto que demos gracias a que—aquí—. . . anglosajonas no existen para proyectar el renco insatisfecho sobre la Sociedad Protectora. . . . Porque, a quién le importan los perros? A quién molesta el dolor de un perro

cuando ni siquiera a su propia madre le importa lo más mínimo? Bien es verdad, que de esa investigación del polivinilo nada puede resultar puesto que ya sabios, en laboratorios transparentes de todos los países cultos del mundo han demostrado que el polivinilo no es tolerado por los tejidos vitales del perro. Pero quién sabe lo que puede aguantar un perro de aquí? (11)

These dogs, which cannot urinate, can only transpire completely by emitting a noxious vapor through their mouths their tongues hanging out. Also when the dog is operated on in order to be grafted with polyvinyl femur, it suffers so much that we are lucky there ... [is] no Society for Prevention of Cruelty to Animals.... Who cares about the dogs? Who is troubled by the dog's pain, if its own mother does not care? It is true that this research in polyvinyl can produce no results since scientists in the civilized countries of the world have demonstrated that the dog's vital tissues do not tolerate the introduction of this substance. But how can we tell what one of these dogs will endure? (6–7)

It is significant that Martín-Santos's novel begins with the passages criticizing scientific experimentation on animals, which later in the novel reflect on social injustice. As science justifies the need for torturing and killing animals with the need for progress and an increase in knowledge, politicians argue that the suffering of the lower classes is needed for the prosperity of the nation. Both arguments exclude from among the beneficiaries of improvements those who pay the price for it. In the last passages of the novel, when Pedro leaves Madrid on the train, he reflects on war and poverty with irony, ending with yet one more comparison of the poor to the lab animals: "When there is so much hunger that they turn into mummies, throw them out" (247). Like the animal that suffers and dies, "preparado para que la ciencia medre a expensas de su sangre" (prepared to offer its blood to the advancement of science, 193/216), in necro-economics, the poor feed the market.

The comparison of the laboratory experiments to bullfighting appears on the very first page of the novel when a researcher is characterized as "el terebrante husmeador de la realidad viva con ceñido escalpelo que penetra en lo que se agita y descubre allí algo que nunca vieron ojos no ibéricos. Como si fuera una lidia. Como si de cobaya a toro nada hubiera" (a piercing explorer of live reality with sword-like scalpel that penetrates what moves and discovers what no Iberian eyes have ever seen. As in a bullfight. As if there were no difference between the animal experimented on and the bull, 9).[5]

---

5. My translation.

Similar to the deprivation of the lower classes, these scientific experiments seem to be designed to check the limits of its objects' endurance. Parallel to bullfighting and political torture, experiments on animals put on stage the superiority of human over nonhuman life, perpetuating the hierarchy by administering pain. Martín-Santos hints at the habitual argument justifying bullfighting through a mystery that is only available to true Spaniards. This Iberian capacity to learn special lessons from pain provide for particularly cruel science. At the same time, the comparison between suffering animals and humans continues. As we read in a further passage, the cries of suffering dogs are "gritos casi humanos" (almost human cries, 14/9) close to Bentham's famous reflection, that all suffering animals are equal in pain. Martín-Santos notes that mice have livers and a pancreas like human beings (25/23), and it is this similarity to humans that makes them suitable as objects of scientific experimentation.

En *Vidas y muertes de Luis Martín-Santos* (Lives and Deaths of Luis Martín-Santos, 2009), José Lázaro views *Tiempo de silencio* as precursory to the field of medical humanities. The author, himself a medical humanities scholar, reflects that Martín-Santos connects the discourse of humanities and that of medicine in a way that shows that they should have never been separated. *Tiempo de silencio* can be also considered more broadly as a precursor to the field of science studies. It shows that science is not simply objective but rather coconstructed by political and social discourses that limit what is researched and in what ways. Not only science administration but also its hypotheses and the very design of its experiments appear in the novel as conditioned by the discourse of the groups in power and their interests. At the most basic level, the construction of humanity as different from animality is what makes it ethically admissible for the medical sciences to experiment on animals. This distinction leads also to the construction of humanities and biological/medical sciences as separate, with humanities concerned with the human and sciences primarily with nonhuman life. Although science rests on the assumption that humans and animals are sufficiently similar to justify the research of human cancers on mice, this knowledge is not consequential in cultural discourses. It brings no recognition for mice either in the humanities or in politics. The distinguished professor of humanities talks to upper-class ladies about aesthetics only. After being used in experiments, animals are either thrown in the garbage or resold for new experiments: "Los perros olvidados de los de las tesis, que en cuanto han hecho la cosa . . . se olvidan de que tienen un gato con los alambrillos dentro o un perro con su goma colorada dentro de la tripas hala, hala a ganar dinero." (The students with the stray dogs who write a thesis . . . forget that they have had a cat with wires left inside it, or a dog with colored rubbers still in its entrails, go off and earn themselves a living, 33/32). As one of the results of the separation between sciences and the humanities, there is no regard for animals. Human animality, as well as animals' likeness to humans, remain a lab insider's knowledge, which as Latour (1993) states, cannot appeal to politics. For Matías's mother, discussing bodily issues researched by Pedro verges on bad taste. Humanities, religion, and literature construct culture in terms of the spiritual and symbolic, and thus, separating them from reality, they incur error but also avoid political engagement.

When during his visit in the slums, Pedro reflects on the need to study "this Animal," which in its culture produces "supremo mal" (supreme evil, 23), he implies that there is a need to study human as animal in this culturally constructed environment, which would mean a collaborative effort of the humanities and sciences. He sees the slums as an extension of his laboratory, indeed better fitted to do research as it contains the real conditions in which both humans and mice live, multiply, and fall sick. These real conditions are opposed to the lab, since in the slums, life processes develop in a different way. Pedro reflects on sickness as culturally, contextually, and environmentally conditioned rather than only genetically constructed, which is what the laboratory approach assumes. Seeing slums as a perfect laboratory for research, Pedro challenges the concepts of genes and of heredity, which in medical sciences of his moment, are considered to be the main factors of cancer. In this moment, indeed, a metaphor proves helpful as a research tool. Pedro reflects on poverty as metaphorical cancer and realizes that an environment ridden by poverty could be in fact an even more important cause of cancer.

"Las chabolas" (Spanish slums), marked by deprivation, contamination, violence, and stress, make up that environmental evil produced by humans that Pedro connects to the autoimmune sickness he researches. The metaphor, which identifies poverty with cancer, guides him to see that illness is indeed a sociopolitical problem and cannot be cured without political changes. Sick humanity, as he observes, results from its material context—the lack of hygiene, bad nutrition, gender and family relations impregnated by violence—rather than being conditioned only by genes. Paradoxically, however, these ideas cannot be proven by any laboratory experiments precisely because of the laboratory's alienation from life and its separation from culture. On the other hand, what can be proven in the lab does not inform on the real mechanisms of life because the lab isolates life from reality to in order to experiment.

Scientists are buried in their laboratories and immersed in their experiments on organisms that have been removed from their habitual context. This arrangement leads to a very limited vision, for scientists cannot see the whole picture where a variety of factors synergize. Seeing life in a lab as biological only, separated from its sociopolitical and ecological contexts, they incorrectly perceive the mechanisms governing their functioning. But this separation from the real environment makes them unthreatening for the political status quo because anything they conclude from their experiments has nothing to do with the context of real life. Thus they make claims of truth, but do not speak truth to power. On the other hand, humanities, such as practiced by the professor (the caricature of Ortega) who gives lectures at the salons of the rich, are also so far away from reality that they turn into a snobbish society of admiration for art for art's sake. One of the results of the separation between biological/medical sciences and the humanities is their incapacity to understand life fully and use this understanding to challenge the power that subjugates life for its own interests. In a sense, Machiavelli's maxim "divide and govern" applies also in this context. Modern separation between nature examined by scientists and culture glossed over by humanists is a crucial element of biopolitics that facilitates the power to administer life, encountering very little opposition from academia.

The story of Pedro points to the autonomy of science as a domain where, as in no other social realm, new formulae of knowledge can be hypothesized and tested while it contains deep criticisms of the functioning of this domain. Initially, Pedro can in fact obtain certain financial support that makes it possible for him to pursue experiments which may prove that cancer is caused by viruses. This autonomy is, however, incomplete or perhaps even illusive. Pedro fails because what he proposes to prove cannot be proven in a laboratory context. As the increased rate of death of the mice held in the laboratory shows, live organisms do not function in the laboratory context in the way they do in their habitual settings. The freedom that science gives to pursue new knowledge is limited by the very construction of its experimental space, which isolates life from its natural conditions. Additionally, what counts as knowledge is rigidly determined by the definition of what constitutes proof. Even if autonomous from the field of production, science is also measured by its productivity, accomplished through providing proof. Pedro's mentor, initially interested in his young researcher's idea, at the end reproaches him for not having managed to prove anything. Admittedly, Pedro chose to work on a task too complex to succeed, but as the novel shows, the questions that he asks are the ones that need to be answered for science to have real impact on human life. It can be concluded that the organization of the scientific domain privileges the production of knowledge that has no significant impact on the sociopolitical arrangements of life.

For example, the way that the concept of the gene had been incorrectly constructed is subservient to racism and to the discourses defending nationalism, social hierarchy, and class separation. Martín-Santos was likely familiar with the idea of *epigenetics*, coined by C. H. Waddington in 1942 to describe the interactions between genes and the environment to form a phenotype. The concept of the phenotype points to the fact that it is not the genes but rather the shape they take in given circumstances that determines the health and behavior of an organism. Epigenetics, as a field that has only developed recently, discovered that functionally relevant modifications to the genome that do not involve a change in the nucleotide sequence can be introduced by changes in behavior and environment. These findings, documented after the novel was written, confirm the intuitions of Waddington and of Martín-Santos. The image of life that emerges from the epigenetic vision is much more plastic, both malleable and corrigible, and thus it calls for politicized environmental concern. It can be opposed to the scientific discourses that argue it is the genetic inheritance that is responsible for sickness, poverty, family violence, and other social ills. The fact that they are called ills, rather than problems, points to the deterministic character of the discourse based on the metaphor of society as an organism structured by genetic inheritance. Martín-Santos deconstructs this way of thinking in various passages of the novel, for example, when he ironically mentions dolicocephalic skulls as characteristic of Hispanic "raza" (race/culture). The concept of race/culture was built through emphasis on certain genetic traits that recent findings show are not really there and that were thought to not only condition but even condemn people to certain forms of life. It is in the intersection of these two scientific myths, based on metaphorically constructed concepts of gene and race, where much of social evil occurs. In the paragraphs describing the Spanish "chabolas" (slums), Martín-Santos ridicules this discourse, exposing its affirmative use by Franco's propaganda that praised the inventiveness of the Spanish race and confronted it with the deeply grim reality of life. Here again the resilience of life as an argument for a lack of care voiced by the regime is counterposed by the descriptions of its fragility in a later passage when Florita dies. The call for studying "this animal" in its political, social, and ecological context should be viewed as a positive and important proposition. It is no less than a postulate connecting the humanities and sciences for new more complex research on human and nonhuman life as they interact in everyday life. The role of both the humanities and interpretive social sciences would be to revise the myths and metaphors sustaining oppressive politics and politically subservient science.

Interrogated in prison, Pedro explains that he dreams about discovering that cancer is caused by a virus and not genes because, if it were so, there would be a possibility of preventing it with a vaccine. Parallel reasoning holds for poverty. If it is thought of as an inevitable part of any well-functioning economy, the only justifiable action is charity. But if it is defined as changeable, political action that can transform these conditions should follow. If poverty is a metaphoric cancer in the novel, it is in the interest of the class in power to impart the belief in its unavoidability, its being "in [the] genes" of the Spanish organism, as in Ortega's writings, as a "parte del mismo organismo, de la misma sustancia del ser vivo" (part of the very organism, very part of the living being, 180/200). Martín-Santos argues that discourses that represent poverty are as unavoidable as Ortega's are harmful but also that the discourses that represent sickness as just a biological phenomenon and not as culturally constructed are just as incorrect and misleading. The fact that the medical sciences do not consider the material culture and the environment of life as an important, perhaps decisive, factor of illness is visibly subservient to the status quo since no social change needs to be postulated by the scientists.

CONCLUSION: *TIEMPO DE SILENCIO* AS A HETERODOXY

In *Tiempo de silencio,* reflections on life, animals, politics, science, and love are structured by extended and interrelated metaphors, which provide schemata for thinking and behavior. What happens in the bullfighting ring becomes what Juan Marsé (2003) calls "paisaje moral" (202), which unfolds in the life stories of individuals who are actors in social and political life. Bullfighting interiorized leads to life according to social expectations, where the public is in power to condemn and expel an individual who does not comply. In his nightmare, Pedro is a bullfighter, whistled at with disapproval by the public. It is the same public that later in real life condemns him to give up research and retire to a provincial practice. Cartucho is another caricature of a bullfighter, whose determination to win leads him to thoughtless aggression toward anything that moves, as well as, consequently to a thoughtless acceptance of aggression from others. The bullfighting metaphor of love leads to an eroticization of violence and brutalization of love. Finally, the professor/Ortega is also compared to a bullfighter, who aims to impress high society with cool talk without making any attempt to reflect on the evils of reality surrounding him. *Tiempo de silencio* implies that these bullfighting metaphors that model public and individual behavior, as well as various others that construct humanity and animality in scientific and cultural discourses, need to be rationally revised for language to be conducive to transformation.

The obsessive contrasting of humanity against animality in bullfighting, which is thought of as a justification for inflicting pain and killing the animal, is translated into the *modus operandi* of biopolitics in the human realm. Humans holding to power separate themselves from the poor through a special distance and aesthetic distinction and thus do not feel responsible for the suffering of the poor. The suffering of the animals and the poor come to be regarded as natural and disappear from the ethical awareness of the classes that are financially comfortable. People belonging to these classes deny their own animality in order to repress any possibility that their behavior may be inhuman. The more they deny it, however, the more inhumanly they are able to behave, as trusting in the superior status of their humanity assists them in their callousness and cruelty. Also science benefits from the distinction between humans and animals, which justifies killing animals in the lab for the sake of a "proof."

In the frame of Bourdieu's theory of social change, *Tiempo de silencio* can be read as a *heterodoxy* that is as a discourse making *doxa* visible in order to deconstruct them. According to Alfonso Rey's essay on Martín-Santos, this focus has become even more central in *Tiempo de destrucción* (Time of Destruction, 1985), which undertakes the task of deconstructing myths and metaphors needed to reform life. *Tiempo de silencio* not only questions the discourses dominating the life of Spanish society but also argues against passivity and in favor of political activism. In the last passages of the novel, the passivity of accepting misery and pain is metaphorically equated with castration, desiccation, and mummification (215/245), as well as compared to the martyrdom of St. Lawrence who, when he was roasted alive on the fire, only asked to be turned to the other side (217/ 247). Early in the novel, Pedro's long interior monologue on Cervantes and *Don Quijote* is a reflection on the possibility of activism. It takes a form of six spirals of "racionalismo mórbido" (morbid rationalism, 58/61) which begin with the supposition that "en la locura reposa el ser moral del hombre" (madness is the essence of man's moral being, 58/61). The thought circles around relations between the real and the symbolic, which are significant for Don Quijote's attempts to change the world. The first spiral consists of the observation that morality exists in a fictional or symbolic realm that is not real. The second spiral is an affirmation that the symbolic realm becomes real through its powers of producing evil, "El mal—que sólo era virtual—se hace real con ese hombre" (Evil which was only virtual, becomes real for this man, 59/61). This is indeed the power of metaphors. The third spiral states that those who fight against evil are believed to be good. The fourth is a note that one person cannot improve the world and that such an attempt reveals madness and can be laughable. Fifth is a warning that the madness of taking morality literally and acting upon it (which is good) is dangerous for the order of the world, which rests on hypocrisy, as there is always a gap between words and deeds. Therefore, an activist who insists on ethical honesty may need to be executed. The sixth spiral modifies this last statement: "No hay que exagerar" (We must not exaggerate, 59/62), all talk about the impossibility of establishing goodness on Earth is only about the obstacles encountered by those who attempt to fight evil. At the end, "todas las puertas quedan abiertas" (all doors are open, 59/62). The six spirals end with an optimistic statement of faith in the possibility of change and meaning of activism, after all. The story of Pedro's attempt to bring change through science ends with a failure, and for this reason, *Tiempo de silencio* has been systematically interpreted as a deeply pessimistic novel.[6] The above passages, however, point in the other direction and allow the interpretation of the novel as an expression of faith in the need for activism in the midst of doubts about its social effects and contextualized by a story of failure.

Various problems described by *Tiempo de silencio* were specific to Franco's Spain, but some still persist and are not uniquely Spanish. Although substantial improvements occurred in terms of the care for human life and some in the treatment of animals, the biopolitics has not essentially changed since the times of Martín-Santos. This was made visible by the crisis of 2008 when poverty was exacerbated, returning in some areas to the levels of postwar Spain. The analyses of the discourses contained by *Tiempo de silencio* are pertinent for reflection on today's societies, both Spanish and otherwise. They allow us to conclude that an alternative biopolitics requires a new conceptual structure, where all life is considered to be interrelated and constitutive of the environment, as Pedro suggests during his visit to the shantytown. This alternative biopolitics needs different rhetoric and consequently it requires a partnership of different disciplines, such as the sciences, social sciences, and humanities, which traditionally have worked in separation. The first fundamental concepts that need to be revisited and redefined in different disciplinary contexts are "human," "animal," and their mutual relationships. This is the direction in which the next chapter takes us.

---

6. See, for example, Jesús Pérez Magallón's."El proyecto acosado: El fracaso en *Tiempo de silencio* de Luis Martín-Santos" (1994) Carole A. Holdsworth's "Two Failed Scientists: *Tiempo de silencio*'s Pedro and Pynchon's *Pointsman*" (1993), as well as Gustavo Faverón Patriau's "La máquina de hacer muertos: Enfermedad y nación en *Tiempo de silencio* de Luis Martín-Santos" (2003).

CHAPTER 4

MEANINGS OF ANIMALIZATION AND HUMANITY AS ANTHROPOMORPHISM

THE PHRASE "human is an animal" both is and is not a metaphor. Darwin showed that we evolved from apes, and Donna Haraway (2008) reminds us that we share up to ninety-eight percent of our genes with various other species. Comparing biological data and rhetorical figures, it strikes one that language provides for more differences between human and nonhuman animals than bodies themselves. In common understanding, the concept of "human" implies a set of essential differences from animality, while in fact they are not there, except and debatably for language. Animals have been animalized and the concept of "human" is a result of a long process of "humanization." According to Giorgio Agamben (2004), there were a number of reasons—political, economical, and psychological—to animalize animals and to humanize humans. The difference between human and nonhuman animals is constructed so that it justifies an administration of life for the benefit of humans at the expense of nonhumans or not-quite-human humans. This difference is a basic concept of the exclusive biopolitics that justifies harming the "lower" forms of life and that provides arguments for selective care and selective destruction of some living beings. Various thinkers and artists have not been content with this structure of privilege and have undertaken searches for alternative rhetoric, restructuring relations between humanity and animality. These searches for alternative rhetoric explore the figures of animalization and anthropomorphism.

Animalization is a crucial rhetorical figure of biopolitics that, according to Foucault (2003, 2013), is concerned with administering purely biological aspects of life, reducing humans to their animal bodies. This implies treating the human population as if it were an animal one except for the life of the mind, whose manipulation requires more sophisticated mechanisms than in other species. This exception is not there as far as the killing is concerned because human beings can be killed exactly like animals, regardless of their mental life. As a reaction to the rhetoric coined by Haeckel and Nietzsche that deconstructed the difference between human and nonhuman animals, the most disastrous biopolitics of Nazism emerged. In *Quelques réflexions sur le hitlerisme,* Immanuel Levinas (1997) blames the loss of human exception on the Holocaust and other disasters of the war. Laura Bossi (2008), who seems to hesitantly agree, notices, however, that this was not the right reaction to the blurring of the human/animal divide and that as a result of the discovery that humans and animals are one big family, it would have been much more logical to respect more animals than to lose respect for humans. Jorge Riechmann (2012c) suggests in one of his poems: "*El ser humano es un animal enfermo* decía Nietzsche / La respuesta nunca debiera haber sido / Buscar una sobrehumanidad sana / . . . / Sino encogerse de hombros: / . . . / Sin ningún desapego, sólo robusteciendo / Nuestra voluntad de cuidarnos / Unos a otros" (n. pag.) (Human is a sick animal, said Nietzsche / We should have never tried to find a healthy superhumanity, but rather to shrug our shoulders: / Without any hesitation, strengthening / Our will to take care / Of each other). In this kind of more life-friendly biopolitical discourse, animalization appears as an argument for more care for the needs of the human species and justifies the human incapacity to overcome certain basic biologically conditioned problems of life. It leads to a more tolerant and more caring outlook on life, as in Buñuel's films. In this vision, poor humans who fight for survival or rebel against oppression are innocent even if they hurt others because a fight for survival is a necessity of life. Belief that humans are animals implies a need to build a society where the fight for survival and rebellion against oppression are not necessary so that an ethical improvement can be made. Therefore human/animal comparisons are not to be viewed as degrading for the human, but rather as an attempt to look afresh at life in order to overcome the works of the "anthropological machine" that built an abysmal and unjustifiable difference between the human and animal realms. This fresh look allows one to appreciate the facts of life that are to a degree common for both humans and insects. In those welfare-oriented discourses, the figure of animalization becomes close to what could be considered its opposite: anthropomorphism, or in other words, humanization of the nonhuman.

Humanizing the nonhuman is, according to Timothy Morton (2007b), not a negligibly naïve strategy, but rather a bridging device essential for our understanding of the outside world. Morton argues that anthropomorphized elements of the environment are dialectical images in the sense that Walter Benjamin gave to the term. Dialectical images can be interpreted as complicit with the dominating ideology but can be also read against the grain, as a way out of the ideological schemata. Anthropomorphism as a literary device has been criticized for projecting the structures of human subjectivity toward "otherness" and thus being complicit with anthropocentrism. Morton notes, however, that this projection does not destroy the object's otherness altogether, but rather sets it halfway between the known and the unknown, thus opening the door for environmental awareness. For John Simons (2002), anthropomorphism is a strategy necessary to understand animals and, as a result, to deepen and revise the understanding of humans as similar and different from animals. Anthropomorphism is, according to him, necessary to compare, regardless if the stress is placed on the properties of the comparing subjectivity or on the compared otherness. Morton, from the perspective of environmental studies, and Simons, from the perspective of animal studies, both appreciate anthropomorphism as a means of empathy, which stretches beyond subjective projection aiming at an inclusive politics of life.

This chapter searches for correspondences between rhetorical figures of animalization and anthropomorphism and biopolitical discourses. The analyses of literary texts and films are focused on depictions and relationships between human and nonhuman life in their historical and political contexts. The analyses feature four important moments when the thought on human/animal relations evolved as a result of new theories and of sociopolitical transformations. The popularization of Darwinism and especially Social-Darwinism in Spain led to a series of attempts to redefine humanity as opposed to animality, as in Emilia Pardo Bazán's writings or in Clarín's (Lepoldo Alas's) stories. Nietzsche's theories prompted projects of building superhumanity inspired by animal violence, as in Pío Baroja's novels, which inspired the Falangist cult of violence, represented in Carlos Saura's *Caza* (Hunt, 1965). On the other hand, Freud's theory and surrealism pointed to a need for a society that better accommodates all of life's needs, thinking of human and nonhumans as deeply interconnected, just as in Buñuel. The animal rights movement brings about a realization that the ethical treatment of animality may give us an opportunity to save the essence of humanity that has been partially lost in our civilization.

EMILIA PARDO BAZÁN'S BEASTLY PEASANTS AND CLARÍN'S THINKING COW

According to Giorgio Agamben (2004), allied discourses of humanism, which he calls the "anthropological machine," through the centuries have established a caesura between humans and nonhumans that until today justifies the subjugation of animals and the superiority of men. Therefore this subjugation is based on arbitrary conceptual differences. What Agamben considers most interesting, however, is that the separation between human and animal would in fact run within man, who for centuries, even before Darwinism, has been viewed as "a conjunction of a body and a soul, a living thing and *a logos,* of a natural (or animal) element and a supernatural or social or divine element" (16). In fact, according to Giovanni Pico della Mirandola's *Oration on the Dignity of Man* (1996), which Agamben calls "a manifesto of humanism" (29), man's nature is not established, but rather can realize itself in two ways: it can degenerate into brutes and regenerate into higher soulful forms. Agamben suggests that as a result of this kind of duality in the perception of the human, the anthropological machine began to exclude the nonhuman from the human and the subhuman within humanity itself. That strategy of exclusion intensified after Darwin's theory was popularized when humans had to be "saved" from the ape, and as a result a few decades later, the Jew was separated as "the non-man within the man" (37). When popularization of Darwin's theory provided for a simplified conclusion that humans are in fact animals, logically there were two possible discursive outcomes: animalization of humans and humanization of animals. These amounted to a number of complex representational strategies and slightly less complex political discourses, ranging from coining new definitions of humanity and animality, where insufficiently human individuals or groups became animalized, as in racism, to the visions where common features of humans and nonhumans provided for more understanding and care for both. Two classical examples of such responses from Spanish literature can be appreciated in some of Emilia Pardo Bazán's and Leopoldo Alás's (Clarín's) short stories.

Emilia Pardo Bazán (1851–1921) was one of those intellectuals who because of her religious worldview turned impulsively against Darwinism and at the same time became very influenced by it. In some of her short stories and novels, the lower classes are represented as animal-like brutes, and their inferiority becomes deeply troubling for human morality.[1] Pardo Bazán's short story "La ganadera" (1910), is one of the most striking examples of animalization, and it puts on stage Agamben's thesis that the separation of the human from the animal within human nature is an essential first step in the process of exclusion of certain human groups from rightful "humanity." Lower-class individuals are portrayed as incapable of meeting the criteria for morality due to their animal-like nature and thus unable to fulfill the demands of true humanity that Christianity poses to them.

In "La ganadera," Penalouca lies next to a rocky bay where ships are often wrecked. On stormy nights, the villagers come out to the beach with torches and attract ships to the rocks so as to make them wreck and then they murder the survivors who manage to come ashore in order to take their property. This "hunt" is their way of subsistence. The only individual able to rise above this wolf-natured society is the priest. It is his Christian spirituality that allows him to loathe this animalistic nature of his parishioners and that leads him to oppose the bestiality of shipwrecking. He tries to stop the evil and pays with his life for his daring. One stormy night when the bloodthirsty crowd rushes to the beach with sharp tools to kill and rob, the priest follows them, positioning himself between the aggressors and the shipwrecked victims. With his arms opened in a shape of the cross, he calls on Christ in order to stop them. The people, however, excited by their greed, push and step over him like a pack of wolves. The next day the priest's corpse is thrown by waves onto the beach, together with those of the shipwrecked travelers. In its simplistic reading, "La ganadera" excludes from humanity the criminal, beast-like peasants and saves the ideal of Christian humanity, marked by a sacrifice, for those who can live up to it, as did the heroic priest.

The simple story of "La ganadera" increases in complexity once placed in the context of the debates on Darwinian and Haekelian ideas, as well as in the context of social tensions caused by the unrestricted capitalism of the nineteenth century. Pardo Bazán's negative take on Darwinism remained consistent from 1877 when she wrote "Reflexiones científicas sobre el darwinismo" (Scientific Reflection on Darwinism) published in a Catholic journal *Ciencia Cristiana* (Christian Science). In this article, she discredits evolutionism as an unproven theory that should not threaten the religious vision of the world, as so many were afraid it would. The writer, however, appears alarmed by the application of Darwinism to social theories such as Haeckel's, stating that evolutionism has taken Germany in "una dirección trascendental,

---

1. For example *Los pazos del Ulloa* (1886), "Un destripador de antaño" (1998), "La ganadera" (1910), and others.

belicosa y utópica" (40) (a transcendental, bellicose and utopic direction). Thomas Glick (1982) explains that, in Spain, Darwinism, and especially Darwinian ideas recycled by Haeckel and applied to the social sciences, influenced more thought about society, economy, and culture than about the sciences, since this was the order of interest among Spanish positivists. At the end of the decade of 1970, *Revista Contemporánea* and *Revista Europea* printed a number of articles commenting on Haeckel's theories. Their common idea was that Haeckel discovered a global system that explained the evolution of the universe. In brief, in all its aspects, the development of the universe was to be governed by the principle of natural selection, by which stronger and better-equipped individuals would win over the weak ones, thus improving the species, societies, and political systems. In this way, the strategy of the exclusion of the "weak" became naturalized and transformed into a matter-of-fact cosmic principle. This went hand in hand with the popularization of market theories where "the invisible hand"[2] guaranteed some sort of economical justice for all (Montag, 2013). Written as an anti-Haeckelian exemplary narrative, "La ganadera" shows that allowing a free play of the laws of nature in society neither brings moral results nor leads to general improvement, but rather it makes society work like a jungle, where people hunt for each other like animals. The story, which intends to show a negative take on Haeckel's ideas, can also be read, however, against its grain as an affirmation.

The paradox of the story is that its criticism of Haeckel's theory is only viable from the position guaranteed by the laws of the social and economic development condoned by the social evolutionism Haeckel inspired. The narrator of the story, the priest, and its author, the Countess Pardo Bazán, owe their superiority to the hierarchy maintained by a jungle-like economic law that privileges the rich. The (im)morality of early capitalism had been criticized by the emerging Marxist left and anarchists as it became a reminder of the Darwinian, animal-like fight for survival. Indeed from a Marxist point of view, the poor people from "La ganadera" are not the fittest, but to the contrary, they are precisely those who were not fit enough to secure themselves places on the top or even in the middle of the social hierarchy, and so they await an opportunity to get back at those who pushed them down. In the descriptions of the poor people's attacks on the shipwreck victims, whose

---

2. The "invisible hand" is a metaphor used by Adam Smith (1759, 1776) to describe unintended social benefits resulting from individual greed, the idea being that the riches accumulated by higher classes will spread down to the lower ones, which as we know has not really occurred.

pockets are full of gold, resounds a fear of social revolution. In this, "La ganadera" reflects on the social tensions produced by the escalation of inequalities resulting from the developments of nineteenth-century capitalism; various followers of Haeckel justified these disasters (and some still do) as necessary for overall economic progress. In this context, exclusion of the lower classes from "humanity" was a means for propagating the ideal of economic expansion that resulted in further pauperization of the lower classes. Animalization was a rhetorical strategy of necro-economics (Montag, 2013) that let some die so that others can triumph. Since the Catholic Church justified the existing social hierarchy, the sacrifice of the priest in the story could be read as his expression of ideological alliance with the rich under attack.

It is hard to know to what extent Pardo Bazán did or did not foresee the possibility of this alternative reading of "La ganadera," which would put into doubt the ethics of the existing social system. The ambiguity of the story may indeed be intended as a warning for Pardo Bazán's own social class. They may see that in a situation of a shipwreck, the positions of "the strong" and "the weak" suddenly switch, the well-adapted becoming a prey of the poorly adapted. If as social evolutionists imply, might is right, then the crowd of the local poor would be indeed in its right to take over the possessions of shipwrecked, rich travelers. But if the social evolutionists are wrong and might is not right, it also follows that the rich are not morally authorized to live in luxury at the expense of the poor. Pardo Bazán's middle- and upper-class readers, who may be tempted to believe that they naturally deserve their position in society because they originate from a superior lineage, should be shaken by the prospect of the situation described in this story. The story does not do enough, however, to make the rich readers morally question their social status, and the ending, which represents the sacrifice of the priest, makes us forget all considerations about the ethics of the social order since it solves all doubts with the figure of sacrifice, that is, an homage to Christian morality. The dramatic scenes of the killing of the rich sailors by the poor as depicted in the story further characterize the rich as innocent and noble, and the poor as beasts.

While Pardo Bazán's stories, such as "La ganadera," suggest an exclusion of the beastly poor from humanity, her contemporary, Leopoldo Alas (Clarín) humanizes both the poor and animals. As Sarah Brenneis (2007) notes, in Clarín's four stories on animals ("El Quin," "¡Adiós Cordera!," "El Gallo de Oro," "La mosca"), the author deliberates on the value of life, both human and nonhuman, and he asks whether this value changes in light of the Darwinian theory of evolution. The answers are surprising. The perception of all life connected through evolution, rather than provoking attempts to get rid of animality in the human, leads Clarín to spread the sacredness of life, and as with Eugenio Noel, to consider the presence of the soul in animals. The motives that prompted Clarín to write these stories shed a new light on ethics. Clarín writes about animals to prove that the strategies of psychological realism are superior to those of modernist writings. He proposes that close observation, empathy, and imagination applied to writing about animals will render them as interesting as humans. As he applies realism to writing about animals, anthropomorphism becomes a natural consequence of close observation. As a result, animals appear to share various characteristics with humans, and the stories reveal injustice in the relationship between humans and other species. A literary exercise *nolens volens* turns into an ethical manifesto. In "¡Adiós Cordera!" (1892), an application of psychological realism to a cow takes the writer even further into a deep ethical questioning of human civilization, based on the narrative of technological progress.

In this story the indirect speech provides an ideal narrative strategy to identify with humans and nonhumans victimized by the transformations of life caused by nineteenth-century industrialization. The humanization of Cordera is not a superfluous measure but a deeply significant literary strategy connected to close observation and resulting from the relationship between the cow and the children. As the narration reproduces the consciousness of the cow, the readers are led to think that Cordera's sensitivity and awareness are not that different from the children's. In a situation of fear and suffering, the boy, Pinín, and the cow, Cordera, are portrayed as having a similar degree of awareness, limited to an immediate state of consciousness and unable to locate themselves in the larger framework of meaning. In some ways, in her thinking Cordera knows more than the people. She is also humanized through her role as a substitute mother for the children through their bond and, metonymically, through the last gaze of their real mother who set her eyes on the cow at the moment of her death. Cordera is also human-like in the sense that she has subjective interests and is capable of reciprocity;[3] for example when she has calves, it is obvious to the children that she prefers to give her milk to them rather than to the farmer, and they bring her calves to her so that she can feed them. The children take her to the meadows where she can enjoy better grass since they know that she cherishes it. In exchange, Cordera is cooperative and helps the children take care of her and even takes care of them. Empathizing with the children, the

---

3. As Donaldson and Kymlicka argue in *Zoopolis* (2011), according to Rawls, this is a quality that makes citizenship possible.

reader deplores the cow's death and her transformation into cutlets as much as or more (since she is the heart of the story) than he/she grieves for Pinín, who is on his way to become "cannon fodder." The corresponding Spanish metaphor, "carne de cañón" (literally "cannon meat"), found in the text establishes an even closer similarity between the situation of the cow and the boy, who is forced to fight in the Carlist War for benefits he will never receive. Cordera's death in the slaughterhouse appears in the story as an absolute evil, similar to war, and suggests that the system has been transformed into a killing machine that feeds on the least aggressive, most loving, and thus indeed most humane human and nonhuman forms of life. Clarín's brilliant analogy incorporates a reflection on technology's role in the growth of the abyss between the higher classes and those excluded from the space of privilege. The train, a symbol of technological advancement, first takes away the cow to the slaughterhouse and then the boy to the war. The slaughterhouse in this story functions as a precedent of the war, and it the analogy is stretched somewhat, even as a cause of war. The parallelism is enforced by the repetition of the farewell scene in which the despair and tears are the same, only the name of the soon-to-be-slaughtered changes. The two farewells that structure the story represent the act of separation between the human and the nonhuman or the not-quite-human, like the poor.

In contrast to Pardo Bazán, Clarín sympathizes with the excluded ones, whom he portrays as genuinely loving and deeply suffering, in fact more humane than the rich gluttons to whose excessive appetites they fall prey. If psychological realism applied to animals leads Clarín to an appreciation of animality, this in turn appears as connected to his sympathies and his solidarity with the lowest classes. The slaughterhouse cattle trader who purchases Cordera and the rich people who would eat her later—just as the train, the telegraph, and those who start and lead wars—are nameless and devoid of subjectivity, reduced to the gears of the capitalist system of production and consumption. While those who eat the Corderas of this world and send the Pinins to war are "humanity," they are not humane. The disjuncture between the human and the humane reveals the conceptual split within the humanist worldview, which exposes the injustice of the capitalist biopolitics.

BAROJA'S LION AND BUÑUEL'S SCORPIONS

Nietzsche famously questioned human superiority over animals in *On the Genealogy of Morality* (1887) where he envisioned the human as a "sick animal" (2006, 88) and suggested overcoming this sickness through overcoming the ethics of the weak, but his writings, often fragmentary and enigmatic, have received a number of different interpretations. Matthew Calarco (2008) argues that Nietzsche was misunderstood by those who attributed to him an apology for animal-like violence transposed into human society and a eulogy of the proverbial jungle law that would allow the strong to win. While the notion of the superman, the idea of devaluing moral values and criticism of reason, led some like Pío Baroja to view his philosophy as an invitation to animalization as violence, Calarco believes that Nietzsche's main goal was to contest metaphysical anthropocentrism and urge "possibilities of thinking from other-than-human perspectives and modes of existence" (40). Notwithstanding the debate, in *Camino de la perfección* (2002), Baroja attributed his vision of humans, human society, and politics to Darwin and to Nietzsche.

In his talk given at the Real Academia Española in 1935, Baroja described his intellectual trajectory in the following terms: "Me sentí darwinista, consideré, como espontáneamente consideraba en la infancia, que la lucha, la guera y la aventura eran la sal de la vida" (qtd. in Close, 2000, 65) (I became a Darwinist, believing, like I used to in my childhood, that struggle, war and adventure were the salt of life). Baroja's naïve social evolutionism results in a fascination with violence as a fight for survival, which is evident in the title of one of his early cycles, the trilogy *La lucha por la vida* (Fight for Life, 1904–5). Anarchism is represented there as a failed attempt to avoid the real brutal fight for survival, which is in fact unavoidable and whose results, as the novels show, are determined by might and not justice (Close, 2000). Nietzsche's philosophy is commented upon enthusiastically in *Camino de perfección* (1902), where Baroja attributes to the philosopher an affirmation of egoism as a healthy animal-like attitude and a disdain for altruism viewed as a sort of a moral decadence imposed by the Christian vision of humanity. Visibly inspired by these thoughts, Baroja measures the success of Fernando Ossorio, the main protagonist of this novel, employing the criteria of "health" understood as a willing adoption of animal instincts in his life. Ossorio gives up his search for spirituality because it leads to melancholia, and he ends up marrying a healthy peasant woman, not an intellectual at all, who gives birth to a strong baby. Looking at his newborn child, Fernando promises himself that

> dejaría a su hijo libre con sus instintos: si era león no le arrancaría las uñas; si era águila, no le cortaría las alas. Que fueran sus pasiones impetuosas como el huracán . . . , libres como leones y panteras . . . , y si la naturaleza había creado en su hijo un monstruo . . . que lo fuese abiertamente, y por encima de la ley entrase a saco en la vida, con el gesto gallardo del antiguo jefe de una devastadora horda. (272)

he would leave his son with free instincts; if he were a lion, he would not cut his nails short; if he were an eagle, he would not cut his wings off. He wished that his passions be impetuous like a hurricane . . . , free like lions and panthers . . . , and if nature had created his son a monster . . . that he openly be one, above the law, that he go on living and plundering, with a brave gesture of the ancient chief of a devastating horde.

The attributes of a different species, projected onto a newborn human express simplified, fairy-tale ideas of animal nature as viewed from a human perspective. Ultimately, it is a human fantasy of power that naturalizes violence, turning it into the defining feature of beasts. This motivation is visible in the ending of the paragraph which shifts suddenly to the human domain as "an ancient chief of a devastating horde."

Baroja's animalization of his characters follows parameters much more complicated than Pardo Bazán's and is often inconsistent. He is aware that even "the highest" forms of human culture, looked at carefully, display an animality that he welcomes. There is no worthy humanity for Baroja, and there is no moral worth: "Lo justo en el fondo es lo que nos conviene" (95) (justice is a matter of convenience), and "la vida es una cacería horrible" (98) (Life is a terrible hunt). As a result, Baroja deconstructs humanity and morality altogether, and he calls it "dislocación del terreno moral" (140) (movement away from morality), which follows the pattern of Nietzsche's depreciation of moral values. As Hurtado's uncle, Iturrioz, explains in *El árbol de la ciencia* (1911), "Se puede decir que la psicología humana no es más que una síntesis de psicología animal. Así se encuentra en el hombre todas las formas de la explotación y de la lucha: la del microbio, la del insecto, la de la fiera" (97) (One can say that human psychology is simply a synthesis of animal psychology. In human society one can find all forms of exploitation and of struggle of a microbe, of an insect, of a beast). Iturrioz concludes that "la justicia es una ilusión humana" (98) (justice is just a human illusion). These observations lead Iturrioz from "is" to "ought." He ends up stating that since violence and exploitation dominate in both human and animal domains, this is the way things should be, and our morality should accommodate them since their actions are "natural." Even if taking Iturrioz for the author's spokesman might be a simplification, since all the voices in the dialogue need to be recognized, the story shows that he is right, and Andrés's objections to his uncle's ideas fade out as Andrés fails in life. Andrés objects to Iturrioz's eulogy of the "fight for life," arguing that this very concept is "anthropomorphic" (94). It is debatable to what extent the principle "might is right" really governs the animal world as opposed to reflecting human perceptions of animal relations. But Iturrioz considers this irrelevant since, as he replies, things have to be viewed from the relative human perspective rather than the cosmic one since it is unavailable. We can notice an inconsistency in Iturrioz's philosophy as he rejects justice as a "human illusion" but is not willing to question the illusory character of "fight" and "strength." Based on Baroja's anti-Semitic essays from the thirties, when while writing about the Spanish Civil War he openly takes the side of the stronger, the voice of Iturrioz ends up becoming his own.[4] While Baroja's narrators recommend becoming animal-like for the purpose of health, happiness, and triumph, Buñuel's early films similarly portray humans of both higher and lower classes with a number of animal-like features, and he compares them to animals, but Buñuel's take on them is different.

The rhetorical animalization in Buñuel's films can be also opposed to Pardo Bazán's in that it focuses critically on higher classes while justifying the behavior of the lower ones. In the most memorable scene of "La ganadera," the Penaloucans, armed with whatever they have and gathered in a wolf-like pack, move toward the beach, which is ironically named La playa del Socorro (The Beach of the First-Aid). There they will attract ships to the rocks with light signals in order to cause them to wreck. Waves will throw the shipwreck victims and their possessions to the beach where Penaloucans will take over their goods and kill the survivors. This can be compared to the second sequence of Buñuel's film, *L'age d'or* (1930), which represents the march of the bandits through the cliffs over the sea to fight against the Mayorcans, whose ships are approaching to take over the land. In both cases, the wretched of the earth are about to attack the rich and civilized in a fight for survival. The representations of this situation could not, however, be more contrasting. In Pardo Bazán's story, these lower-class humans are like dangerous animals, an armed mob, while the rich are noble and sympathetic victims. In "La ganadera" a helpless woman carried by the waves to the beach in vain implores that her life be spared. The raging beasts know no pity. By contrast, Buñuel's bandits are represented as so weak that they can hardly walk; one after another, they fall from exhaustion before they manage to do any damage, while the rich and civilized sailors who come from the sea are dynamic and efficient as they take possession of the place (which used to belong to the bandits) and inaugurate their new order. In Buñuel's vision, it is the civilization of the Christian religion that, as we see in the following sequence, grows into a militarized Empire of Rome that takes away everything from people who live on its margins. These people, whose fight for survival takes over their lives as if they were animals, are in fact represented as handicapped and disabled.

---

4. See chapter 2 of my *Del infierno al cuerpo* (2007).

Similar to Pardo Bazán, Buñuel also compares his pitiful bandits to animals, but instead of suggesting the need for their exclusion from humanity, he exposes the unfairness of this exclusion. The first scenes of the film adopt a nature documentary style, showing the life habits of scorpions with an accompanying voice-over reading a fragment from the scientific work of French entomologist Jean-Henri Fabre, *The Life of Scorpion* (Begin, 2007). The scorpion's innate violence is the most memorable feature mentioned in this description. The beginning commentary on scorpions is interrupted as Buñuel switches to the bandits whose behavior, as Begin claims, illustrates those missing passages of the description of a scorpion's life in Fabre's work. As Linda Williams (1981) notes, both sequences are connected by the subtitle, "A few hours later." When the sequence on scorpions ends and the bandits take their place, the camera continues documenting human life in the same way as it recorded the movements of the animals. Williams observes, "The similarities between the bandits and scorpions are readily apparent. Both are indigenous creatures living in a rocky, barren terrain. Both have organs of warfare and information—the scorpions have pincers, and the robbers have guns, knives, pitchforks, ropes and a sentinel (an organ of information) who keeps watch" (114). The comparison of the bandits to the scorpions brings a reflection that the fight for survival is a basic fact of life for all species, including humans, but also that it is the most tragic and the most respectable of all fights and this surrounds violence perpetrated on the way with an aura of an innocence.

The same technique can be observed in various scenes of Buñuel's earlier film, *Tierra sin pan* (Land without Bread, 1927) where sequences from the everyday life of inhabitants of Las Hurdes are filmed in the same way as those depicting a donkey in agony or a goat falling from a rock. According to Mendelson, quoted by Begin, this "anti-artistic approach" (429) is a new facet of surrealism, and it seeks to equate scientific exploration with humanism. Lotar, Buñuel's cameraman, had previously worked in Painlevé's nature films, which had been viewed in Residencia, where various members of the Generation 27 lived and met, and he also had contributed to a pseudo-ethnographic journal edited by George Bataille. The surrealist project of that journal consisted in bringing humans from different societies together in order to "humanize across the board" (Begin, 2007, 431). According to Begin, this is precisely what Buñuel intends to achieve in *L' age d'or* and in *Tierra sin pan*, comparing humans and animals. This comparison should not be interpreted then as dehumanization, but rather as an attempt to equate all living beings by showing what we all have in common. In the first two sequences of *L' age d'or* and in *Tierra sin pan*, Buñuel challenges the established hierarchy of the natural world, where human beings are placed above all other natural forms of life. This, however, does not amount to constructing the portrayed humans' inferiority, as in Pardo Bazán's story, but rather it constitutes an effort to demystify the artificially constructed superiority of the human as a species. In this, according to Begin, Buñuel abandons the framework of humanism, approximating his perspective to that of the posthumanist debate.

In representing human behavior, habits, culture, and society as if they were scorpions, Buñuel undertakes the project that he himself calls "comparative biology" (1984, 113). The surrealists' theory of culture consists of challenging established sublimated values, but instead of endorsing Haeckel's belief that human law should be transformed to be more "natural" and more animal-like, surrealists show that human mores are already animal-like enough, even in the aspects that are never suspected of being so (which is also what Martín-Santos does in *Tiempo de silencio*). In the third sequence of *L' age d'or*, where the Empire is represented as "the phase of the scorpion" in the development of human civilization, sophisticated forms of human life appear exaggeratedly animal-like. For example, the bourgeois way of hanging out in their living room is compared to a cow's hanging out in a pasture. A huge cow is slowly moving from one room to another as we first see the interior of a house in the city of Rome. She is shooed away by humans unwilling to recognize that they are just like her. At the same time, however, Buñuel is skeptical about the possibility of happiness portrayed as liberation from the cultural mores of everyday life. Unlike Baroja, the film director does not propose happiness as animalization, but rather, similar to Martín-Santos, he suggests that animal instincts in human life should be revealed, analyzed, and consciously accepted or rejected.

According to Williams (1981), the film passage from the sequence on the bandits to the following one, intertitled "la fundación de Roma," marks the end of an objective, scientific-like commentary on the life of species, both animal and human, and the start of the subsequent criticism of society that is the main theme of the rest of the movie. Even if the imperial state's violence can be retraced to the behavior of scorpions as an outgrowth of their death instinct, the state's violence is evil; scorpions cannot be accused of immorality. While violence is as natural as the fight for survival in the human/animal kingdom, once it is institutionalized, for Buñuel it becomes an ethically questionable form of oppression. *L' age d'or*, in contrast to "La ganadera," shows that the ethical status of life depends on its place in the social hierarchy, with those least powerful and the least complex forms of life deserving the least reproach. Their innocence, however, does not amount to their inferiority. This humanization of simpler forms of life, ranging from scorpions to bandits, is parallel to a critical deconstruction of the sublime cultural forms and to harsh criticism of the imperial violence that is shown to have grown out from the simplest animals' instincts.

AN ANGLE OF ANIMALIZATION IN CARLOS SAURA'S *LA CAZA* (THE HUNT)

In his recently published *Spanish Holocaust*, Paul Preston (2012) suggests that the national rebellion in Spain was characterized by "nearly racist" loathing that national commanders had for the Spanish lower classes, which were considered "sub-human" (1). He writes that "Mola's secret instructions . . . [for purging the south of Spain] echoed the practice of the Africanists against the Rif tribesmen" (132) and that executions were justified by the need for "racial cleanliness" (514). Preston entitles his book *Spanish Holocaust* to point out that the national leaders aimed at an annihilation of people who threatened the status quo of power and who, in a great number of cases due to their lower class origin, were looked upon as a different race or different species. Sebastiaan Faber (2012) synthesizes Preston's ideas in the following way: "The military Nationalist leadership saw Spain's workers and peasants not just as dangerous subversives or fellow citizens gone astray, but actually as subhuman, mentally and morally deranged creatures whose physical extermination was necessary for the good of the country" (1). In Carlos Saura's *La caza* (1965), this is precisely how Paco regards the crippled Juan. The mixture of disgust and supernatural fear that he feels toward the peasant leads to animalization as Paco observes that "el cojo tiene cara de hurón" (the crippled has the face of a ferret). The analogy between the extermination of the poor during the war and the biological war on rabbits is implied by Paco's statement that the weak and crippled have no place in life. Saura's film pointedly reaches back to the roots of genocide, as structured by the treatment of animals and, in particular, by the ecological control of animal populations as well as masculinity, shown through hunting.

Similar to bullfighting, the hunt as entertainment, as Ortega (1944) defined it, turns the modern attitude toward the domination of nature that Adorno and Horkheimer (1972) blame for the resurgence of fascism, into a purposeless display of passion for killing animals. Saura's film stages this passion in relation to the Spanish intimacy with death in a parabolic fable about three old friends who got together to hunt rabbits and ended up hunting each other. José, Paco, and Luis, presumably companions from the Spanish Falange who fought together during the Spanish Civil War against the

Spanish Republic, meet for a hunting expedition that is officially designed to renew and celebrate their friendship, but the plan goes awry because there is no more friendship to celebrate. Falangist friendships based on mutual admiration of flawless and fearless youthful masculinity cannot pass the test of time. During the years that had passed since the war, the three men lost their fearlessness and invulnerability. They became bitter as their physical transformation made them betray their ideals of masculinity. Instead of the mutual admiration that they once felt toward each other's strength and that constituted the basis of their friendship, they now hate each other because they became weak and old. Unable to accept this, the men constantly reproach each other for the weakness they perceive.

Paco's fear of weakness, sickness, and handicap is especially obsessive. He would prefer to die rather than become crippled, and he develops a superstitious fear of the limping Juan, (José's' servant): "Los cojos traen mala suerte" (The crippled bring bad luck). Paco has reasons to be afraid of a handicap since he believes in the laws of the pitiless world where, as he explains, "el pez grande se como el chico" (the big fish eats the small one) and where, obviously, he wants to be a big fish. He loves to hunt because it gives him the illusion of being powerful. The scene in which José, Paco, Luis, and a young Enrique walk, shooting at all the rabbits they can see, raising clouds of dust on the mountain slope, has a cartoonish character, providing them with a primitive illusion of power. This illusion is fed by the destruction of the weaker, but rabbits soon stop satisfying the hunters and, needing bigger animals to destroy, they begin to look at each other as possible prey. Luis announces this in a statement that functions as a prolepsis: "La mejor caza es la caza del hombre" (The best hunt is the hunt of man). The hunt of man is a war.

The parallelisms between the hunt and the war are constant in the film, and the early criticisms by John Hopewell (1986), Virginia Higginbotham (1988), Marsha Kinder (1993), and others pointed this out, concluding that the mutual destruction of the three hunters at the end announced the self-destruction of Franco's regime. The remark, "Oye, esto no es una operación military" (Listen, this is not a military operation), voiced by one of the hunters as the three are climbing the slope and Luis whistles a war song melody, hints at this interpretation. However, Sally Faulkner (2005) suggests that the hunt sequences are not purely allegorical. She argues that images of "rabbit slaughter are so direct that they must be seen as what they are rather than as indirect metaphors" (460). In her view, these scenes function as interruptions from the narrative of the film and "shake us out of our viewing complacency" (461). In an article entitled "Animals Were Harmed during the Making of this Film: A Cruel Reality of Hispanic Cinema" (2004), Rob Stone similarly claims that in *La caza* the scenes of hunting constitute "irruptions of reality" (n. pag.). Faulkner quotes Saura's reflections on how his idea of the film slowly moved away from the hunt/war allegory toward "a broader meaning" (461). While both metaphoric and realistic meanings are simultaneously active, the metaphoric meaning is also not limited to war. As in *Tiempo de silencio,* the depiction of harm toward animals provides for the humanization of the animal victims and consequent animalization of the humans killing them.

In his psychoanalysis of fascism written on the basis of a study of narratives of Wehrmacht soldiers, Klaus Theweleit et al. (1987) explains that most of them experienced a fear of women and hatred toward the feminine (perceived as weakness, softness, or a handicap) within the masculine self. In *La caza,* this fear is experienced by the three friends and especially by Paco, and it is also a force behind the social and political structures that they represent. But, women, the poor, and the disabled are animalized in the film, and it is the rhetoric of animalization that justifies the violence perpetrated against them. The parallelisms between animals and humans as well as between violence against animals and violence against humans in the film build beginning with the first scenes. As the three friends first arrive and look at the farm, the subjective long camera shot follows the shape of an arid, dry mountain slope, spotted with the black stains of holes in the ground, whose size from a distance is hard to estimate. They look like rabbit holes, but in fact these are the caves where Republican soldiers were hiding from the National army during the last days of the Spanish Civil War. The camera constantly plays with their visual similarity. Thus rabbits hiding in their holes bring the memory of those besieged Republican soldiers who were similarly hiding and were shot right after coming out to the light. In a real "memory sight" (Nora, 1989, 7), inside of one of the caves on the slope, there is an unburied skeleton of a soldier who died there during the war, still dressed in a ragged uniform that reveals his bones. When the camera takes us inside the cave's dark space, the scene evokes another very dark space when a ferret enters a burrow and attacks a rabbit, scaring the others out to the open space where the hunters quickly shoot them. In various moments of dialogue between the hunters, the hunted rabbits are not only implicitly compared to Republican soldiers but also to women and to the poor, who are a menace due to rapid multiplying. In Luis's interior monologue, the hunters are compared to sick rabbits suffering from myxomatosis.

The epidemics of myxomatosis provide a metaphor for the state of masculinities of Paco, José, and Luis, which as Felipe Aparicio Nevado (2011) notes, points to the "franquismo agónico" (271) (agonic francoism) of the mid-1960s and which Faulkner (2005) brilliantly considers as a problem of aging, but it is also a biopolitical measure. The myxo virus was purposefully released in order to decimate the rabbit population in Australia and France in the 1950s because as a result of their tremendous fertility, they were considered too numerous. The fear of the fast growth of rabbits—"Los conejos nos invadirán" (rabbits will invade us), Luis apocalyptically predicts—brings to mind the fear of the rapid growth of human masses expressed by Ortega's *La rebellion de las masas* (1937). According to Ortega, the masses' arrival in public spaces upset the status quo, threatening the natural equilibrium of Spanish society. The fear of the masses and the disdain for the lower classes were also essential elements of the National campaign discourses before, during, and after the Spanish Civil War.

The analogies between the purposeful spread of the virus myxomatosis to kill off rabbits and the thought of population control that appeared in Europe in the 1930s and led to the Holocaust are among the most interesting in the film. In *The Birth of Biopolitics* (1978), Foucault points to the origins of modern population control in the Nazi state of the 1930s when the control was administered through the systematic killing of unwanted humans. The analogies between the human treatment of animals and subaltern humans do not end with rabbits. As the beginning credits of the film run, we see ferrets moving anxiously in their cages, almost touching the camera with their noses, awaiting liberation. Faulkner notes that throughout the conversation between the three old hunters with the crippled Juan's sick mother, powerful similarities are established between them and the ferrets. The hunters and the ferrets are tense, aggressive, and anxious to bite each other. Ferrets are kept in separate cages, just as Saura systematically separates the hunters, placed in separate frames. The ferrets' anxiety and their imprisonment in these first sequences may also evoke the image of the Spanish people under Franco, following the orders of the powerful. As the film progresses, however, the ferrets' real presence in the film overrides their symbolic meaning. Ultimately, the three friends kill each other because of a ferret.

José invites Paco to hunt in order to ask him for money so he can continue his relationship with the young Maribel. In order to remind him of their bond, he takes him to the cave and shows him the skeleton of the dead Republican soldier, now turned into a ghost that haunts the guilty victors of the Civil War. Paco, however, rejects the guilt. Faced with the past, he moves his arms as if scaring an insect, leaves the cave immediately, and apparently regains equilibrium. The cave continues to haunt José. During his nap, a voice-in-off echoes his dreams: "quemen esta cueva, quemen esta cueva" (burn this cave, burn this cave). When after the nap the three friends begin to hunt for rabbits again, this time with ferrets, Paco, irritated by José's obsessive remorse and his interest in friendship, kills the ferret that belongs to José's servant, Juan. He pretends that it was an accident, but Juan tells José that he saw Paco aiming at the ferret. It is then that José decides to kill Paco, not because he refused to lend him money, but rather because, after rejecting the bond of their common guilt, he subsequently kills the ferret as if affirming his determination to do all that he had done wrong once again. On another level, however, the immediate reason for José's shooting at Paco is *an angle of animalization*. Only a few minutes later, standing at the same angle in respect to Paco, just as Paco stood in respect to the ferret that he killed, José suddenly discovers the angle and feels compelled by it to shoot at his old friend. The angle expresses the situational similarity that amounts to the identical predicament of the animal and the man, both of whom have a rifle pointed at them.

    Posing with the camera held as a rifle in a poster announcing *La caza*, Saura shows that he is aware of the significance of the homonymy in the two kinds of shooting. Killing is established as the main theme of the film when, in a very long close-up, the hunters' fingers load their rifles and try out the triggers by aiming at indefinite targets that slowly concretize in the scenes that follow in the film. The comparison between rifle angle and camera angle appears when Enrique, looking through the lens of Luis's rifle focuses for a moment on the body of Juan going down the mountain slope at a distance. It is the zoom that allows him to realize that he is pointing at a human being. Enrique brought his camera to the hunt and takes pictures of Juan's niece, Carmen, but Paco rejects this modern variant of search for trophies, insisting that "los mejores recuerdos son las piezas cazadas" (the best memories are hunted animals). The story told by the film announces this complicity of the camera with the hunt through the picture of the hunters with their trophies. Shooting by a hunter and the shooting of a film director do not only coincide in the angle of killing, but in this particular film (as in various others) they similarly resulted in death. As Stone notes, in many Spanish films animals die to express a director's criticism of cruel treatment of animals. For rabbits and ferrets, the shooting angle means both filming and killing at the same time.

    The distinction between animalization through a shooting angle of a rifle vise or camera sight is connected to the way they relate to the skin, the protective coat of the body which the rifle destroys and the camera analyzes but cannot cut through. This distinction amounts to the difference between life and death. As we see Juan pulling the skin off the killed rabbits, revealing the red, still-lifelike flesh underneath, Luis and Enrique are driving to town to get bread. They enter a dark chamber and cross into a courtyard where a sheep is hanging upside down, and here also people pull its skin down. A close-up of the skinned head of the sheep and of the butcher's hands cutting pieces of the animal's flesh is accompanied by the voice-over revealing Enrique's thoughts: "Con ese cuchillo parece un criminal" (With this knife he seems a criminal). At this moment, an alternate montage takes us to the cave where with a horror-like flash of a match, José shows to Paco a human body, skinned by time. In the next sequence, an exaggerated close-up of Paco's naked arm exposes a scar on his skin precisely at the moment when a jeep with a mannequin stops in front of them. During the siesta, as we hear the men's dream voices, the camera offers us the microscopic close-up of the skin on Paco's torso, where we can distinguish another visible bullet scar and an additional close-up that allows us to inspect the porosity of José's face. The irregularity of the human skin's surface resembles the irregularity of the surface of the dry land that Enrique admires through binoculars. Faulkner explains that the two minutes of skin analysis on the two napping men function as an "idle period," a term used by Gilles Deleuze (1986) in reference to "pauses in the film which are not justified in narrative terms" (472) and which are aimed at an intellectual reflection on time, space, and matter. Saura compares human bodies with the mountain slopes. This comparison between human skin and the earth's surface—on which a long-angle camera shot makes human shapes appear minuscule, like insects—brings, one more time, an ecological awareness and highlights the similarity of all life. Similar to that found in *Tiempo de silencio*, the metaphor of the human as animal turns to reality when the wider context of the environment in which they are both embedded is considered. Thus two discourses of animalization can be distinguished in this film. One, voiced by the three friends, announces that life is a jungle and that it is better to be in it as a strong animal with a rifle. The other one belongs to Saura's film crew and is deeply critical of the first one, conscious of the disastrous consequences of hunting as it escalates. The visual rhetoric of Saura's crew, however, is still unable or unwilling to construct its argument without reproducing, through the killing angle, the worst procedures of the biopolitics that it criticizes.

MAYORGA'S APES AND DOGS

The animal rights movement that made itself heard by a considerable amount of the Western population at the end of the twentieth and the beginning of the twenty-first century, demanding consideration of animals' pain and suffering, prompted as much change in the thinking about humanity in relation to animality as Darwin's or Nietzsche's ideas. It asked for the inclusion of animals in the ethical realm, and in some cases, also legal realm, which was until then only available to humans. Animal rights movements humanized animals by focusing on their capacities, their behavior, and even their facial expressions. Paula Casal, arguing on behalf of the Great Ape Project, asked people to imagine that great apes were a newly discovered human race. Granting the human-like status of a person to the great apes, it was argued, is necessary for them to be liberated by the courts from imprisonment and torture. This rhetoric constitutes a step toward an alternative biopolitics that is focused not on the mechanisms of exclusion in an all-embracing fight for survival, but rather in search of a functional system of inclusion that protects as many living beings as possible.

The question of the ethical treatment of animals has been widely discussed in Spain during the first years of the twenty-first century, not only in the context of bullfighting but also as the result of the initiative to grant limited human rights to baboons, orangutans, chimpanzees, and gorillas. The bill, introduced by Francisco Garrido (Spanish United Left and the Green Party), was approved by the Spanish Parliament in 2008, but it was never ratified because ETA, the Basque Terrorist Organization, took center stage. The economic and political crisis and the change of government in 2011 pushed the rights of apes even further away from the political agenda. The significance of the project has been to initiate a national debate, questioning the privileged space of the humans, the treatment of animals, as well as reflection on the meaning of rights.

The Spanish chapter of the international organization the Great Ape Project was founded in 1997 and called in Spanish "Proyecto Gran Simio." It has been recently led by Paula Casal and Jesús Mosterín. As Casal declares in an interview for *La Voz de Galicia,* people engaged in this NGO (Non-Governmental Organization) connect the ideals of environmental activism and animal rights. They believe in the need to transform human attitudes toward nonhuman life. The Great Ape Project's manifesto reminds the public of the similarity between humans and hominids in terms of common capacities of empathy, self-awareness, foreseeing the future, and the production of tools, in addition to the vulnerability to pain that connects all animals. On their website, Proyecto Gran Simio activists argue that the simians' human-like characteristics qualify them for rights. They recommend "the slow but constant stretching of the golden rule 'treat others as you want to be treated yourself'" (n. pag.) and remind us that rights have been granted to more and more groups as "the line between 'us' and 'them' has dissolved past the boundary of tribe, nation, or race to reach the human/animal divide." Great apes deserve to be protected first, but the species that should follow are dolphins and elephants, who display a human-like, complex social and emotional life, memory, and self-awareness. Ultimately the platform of rights should be extended to all animals. The difficulty in establishing the endpoint of this transformation often turns into an argument against beginning them.

In an article entitled "La pequeña simia" (A Small Ape), published in *El País* in 2006, Adela Cortina, a well-known Spanish philosopher, comments on the Great Ape Project in the following sentence: "Si you fuera una pequeña simia, estaría francamente molesta" (n. pag.) (If I were a small simian, I would be very disappointed). She accuses the Great Ape Project team of working through an anthropocentric framework and excluding from legal protection animals that are not sufficiently human-like. She argues convincingly that a similarity to humans should not be the only reason for privileging the treatment of some groups of animals over others. Thus she rejects anthropomorphism as an unfair rhetoric that extends rights to kin only. In her article entitled "Los derechos homínidos" (Hominids' Rights, 2012), Casal argues that the genetic proximity of apes does not have any moral relevance, but the question of capacities is much more significant because they provide for experiences that make suffering in captivity human-like. Hominids, for example, make plans for the future, have complex social relations, and are consciously aware and afraid of death. Their long-term memory and cognitive continuity contribute to an additional human-like suffering when they are subjected to torture. In sum, certain capacities intensify hominids' experiences in such a way (similar to humans) that they make death and torture for their subjects harder than for other animals. As a reaction to the debate, some voices protested that attention should not be given to the pain and suffering of the apes while so many humans are suffering. These protests made it apparent that attempts to liberate animals from suffering bring more (and not less) attention to the suffering humans. As a result, it could be argued that the activists fighting for animal rights are constructing discursive bases of biopolitics that decrease suffering and pain among all living beings, including humans.

Juan Mayorga's theater brings out the effects that the animal rights debate has on biopolitics. As many others, Mayorga connects the inhumane treatment of animals in contemporary societies to the treatment given to marginalized social groups, but his theater takes strong anthropomorphism to new levels of complexity through the structure of his characters. Morton claims that anthropomorphism "enacts a non-essentialist awareness of the interdependence of subject and object, perceiver and perceived" (2007b, 8) and that this opens up not only different possibilities of interpretation but also of interaction with anthropomorphized objects. Characters in Mayorga's theater are constructed precisely in this way. A 200-year-old lady turtle that evolved into an old woman, a white gorilla named Copito de Nieve (who reads Montaigne), and especially various human-dog characters exist on the threshold between humanity and animality that is constantly blurred. In Mayorga's plays, the blurring of the human/animal divide, as in Clarín, brings out a number of ethical problems found in human civilization that are not otherwise visible: a loss of the sense of freedom, the incapacity to think independently, and the hypocrisy of hegemonic social discourses. As the characters invoke and negate boundaries between humanity and animality, moving between barking on all fours and sophisticated metaphors, the public has to negotiate different conceptual frameworks. Constant fluctuation between human and animal beings results in a reframing of both.

Mayorga's plays stage the very process of animalization responsible for the harm that occurs to those who are animalized. The protagonists, marginalized humans and animals, show and tell how the process of "othering" took place in their stories. As a result, they (and the animals they were turned into) become tentatively rehumanized. For example, in *Las últimas palabras de Copito de Nieve* (The Last Words of Snow Flake, 2004), staged in Nuevo Teatro Alcalá, a human-like white gorilla, who is naturally inclined to philosophical meditation, tells the story of how he has assimilated to the expectations of the public, pretending he likes bananas to make children happy. Copito tells us from the stage that he is not really the ape as we imagine it, but rather he has qualities that we do not suspect in him and that he may be much more human-like that we would like to think. In *Palabra de perro* (Dog's Word, 2004), two main characters are enclosed in a space surrounded by an electric fence and watched by guards. It is not clear throughout the whole play if they are illegal, dark-skinned immigrants animalized by how they are treated by the society or fantastic dogs who talk as in Cervantes's short novel *El coloquio de los perros*. The indeterminacy of their specie status provides for doubling of meaning in all the exchanges that brings out the parallels in the predicament of dogs and illegal humans, as well as dogs and humans in general. While the multiple layers of humanity and animality that mesh in all these characters make it impossible to clearly distinguish between human and nonhuman, humane and inhumane remain clear. In particular, all the arrangements justified by a hierarchic perception of species strike us as inhumane, such as putting homeless dogs to sleep, pushing illegal immigrants back to the sea, or torture applied to human or nonhuman bodies. In *Animales nocturnos* (2003) immigrants working night shifts are compared to animals that only come out at night.

Mayorga's play *Las últimas palabras de Copito de Nieve* (The Last Words of Snow Flake, 2004), that received the Premio Telón Chivas in 2005, has been inspired by the real story of a favorite albino gorilla in Barcelona Zoo and by the debates on the great apes' rights. The real Copito baby (1964–2003) was cared for by the zoo's veterinarian, Dr. Román Luera, and his wife, Maria Gracia, and became so attached to Mrs. Luera that in order to transfer the ape to Barcelona Zoo, the woman had to move to the cage with the ape child. A documentary on Copito shows Mrs. Luera sitting next to little Copito on the animal side of the zoo's bars.[5] The image of a human, next to an animal, inside of a zoo's cage is one with a *punctum* in the sense that Roland Barthes (1981) describes it, something that wounds or pricks the sensibility of the viewer. The warm, patient facial expression of Mrs. Luera signals her readiness to share the cage with Copito as well as an awareness of the temporary character of this arrangement. Copito's expression is enigmatic. He seems comfortable, although there remains a doubt as to what extent he may in fact be (un)able to intuit the future separation from his adoptive mother and his loneliness inside the bars. Years later the mayor of Barcelona called the same Copito "an exemplary citizen of Barcelona," (Mayorga, 2004, 9) and when he fell incurably sick in 2003, two thousand children came to say "good-bye" to him. Mayorga imagines the stage as the picture of Copito with Mrs. Luera, except that instead of a warm and caring motherly figure, the White Gorilla is accompanied inside of the cage by an indifferent guard and by another gorilla who is black. Copito's monologue takes place in front of the children who came to say good-bye to him, and the Barcelona mayor's famous words serve as a reflection on human and animal citizenship.

Mayorga's play does not question that the great apes are capable of thinking and feeling almost like humans, as the authors of the Great Ape Project argue. Even if the Copito in Mayorga's play falls short of a perfect memorization of Montaigne, forgetting some words, he is beyond doubt more intellectual than his guard, whose mediocrity is living proof that humans are not a superior species. Mayorga's play can be taken as an argument in favor of the great apes' rights as it disposes the public against the imprisonment of the gorillas and especially against the torture used to tame rebellious animals, both human and nonhuman. (The taming manual in the zoo is called the "Guantanamo Bible"). The questions that the play raises, however, are whether rights like those that most of the public that comes to the zoo have will really make the apes free and whether similarity to humans is the correct justification for anybody's rights. The play expresses a doubt regarding the ultimate gains that the apes would obtain from the rights, pointing out that all these rights have not made Spanish people free. Toward the end of the play, Copito mentions that he has only recently learned of Franco's death and that

---

5. See http://www.rtve.es/noticias.

he could not tell that anything has changed in the people who were coming to see him.

Questioning the distinctions between humans and animals in the play brings to light other perhaps more important ones between some humans and some animals, namely the capacity or incapacity to rebel, which is a significant component of freedom. Rebelliousness and conformity can be found among both humans and animals, but in this play, it is only the black gorilla who is capable of rebelling. Thus the play invites the public to reflect that freedom and the capacity to fight for it, which is one of the greatest human ideals, is in fact more an animal than a human attribute. Freedom appears in the play as an interior quality, which cannot be instituted by the law.

In Teatro Buero Vallejo, where the performance is directed by Adrés Lima, Copito's guard stands most of the time right next to the White Gorilla; however, he sometimes moves in and out of the cage while Copito and his companion black gorilla, pacing anxiously in the back, always remain inside. This design of the space points out that the human and nonhuman animals may similarly suffer from the lack of freedom, and they may be similarly conforming. Most humans, however, have a way out of the cage, while animals are condemned to imprisonment. The cage around Copito shows that even if freedom cannot be given, indeed, it can always be taken away, by cages and walls that are erected to discriminate between life with and without rights. This vision is as much a call for the liberation of animals as it is a complaint about various kinds of prisons and walls among humans, including the self-inflicted loss of freedom through conformism. In light of the play, all these liberations can only occur together.

Copito's love of literature and especially his search for help in the philosophy of dying in Montaigne points to his moral suffering. In the monologue preceding his death, Copito remembers various reflections on dying that he found in reading Montaigne. The most important one, never mentioned directly by the gorilla but put into practice in the play, is that the proximity of death gives one freedom not to conform. Remembering how the mayor of Barcelona called him an "exemplary citizen," Copito deplores his past life assimilation and suddenly turns against the people who watch him, crying out: "Pero ¿qué idea de ciudadanía tiene ese hombre? ¿Cuál sería su ciudad ideal, un zoológico?" (9) (What idea of citizenship does this man have? What would be his ideal city, a zoo?). This comparison between the zoo and the city turns the mirror toward the spectators (citizens), suggesting that they should think of themselves as animals in the zoo.

Twenty-first century zoos now have ravines instead of cages so that the visitors can be oblivious to the loss of freedom suffered by animals. Similarly, the contemporary city, as a zoo, builds invisible barriers and invents strategies that prevent people from realizing that they have lost their freedom. While believing that they pursue their dreams, they are in fact conforming to the rules that society imposes on them. As Copito rebels and begins to yell truth to those in power, his human guard, himself painfully lacking freedom even if he does not realize it due to his mediocrity, kills the gorilla with an injection, showing what happens to those who transgress and how the ethics of conformism is implemented. The city as compared to the zoo, which is indeed a prison, is a space that cannot beget any freedom-giving rights. While Clarín's anthropomorphism brings out criticism of early capitalism, Mayorga focuses more on those features of biopolitics that survived fascism, becoming an integral part of twenty- and twenty-first century politics. The question that emerges as a result is how to transform the human to avoid future political tragedies.

In *Palabra de perro* (2004), anthropomorphism serves as a tool to stage the animalization of illegal immigrants. *Palabra de perro* is a rewrite of Miguel de Cervantes' *El coloquio de los perros,* contextualized in contemporary Spain. According to Giovanni Previtali-Morrow (n.d.) and Julio Rodríguez-Luis (1997), the deep meaning of the Cervantine novel emerges from the implicit comparison of the author's own life to that of a dog. Mayorga applies an avant-garde turn to this metaphor and puts it on stage as if it were a story. In other words, the dog, who metaphorically stands for a human in Cervantes's work, *is* an illegal, dark-skinned immigrant in Mayorga's play who was metaphorically turned into a dog by society's treatment.

In the recent staging of the play at the Sociedad Cervantina (2010) by the director Sonia Sebastián, human and animal elements constantly interchange. The actors are human with naked torsos, but below the waist, they are covered with dog-like fur and have tails. They talk in dog-like poses, make dog-like facial expressions, bark, but also get up to walk and talk like humans. Their species status changes back and forth. Early in the play, we realize that Berganza is the one for whom the police patrol searches when they say: "Buscamos a un moreno. A cien metros de la costa se les hundió la patera. Vimos a uno alcanzar la orilla" (32) (We are searching for a dark-skinned one. Their raft sank a hundred meters from the coast. We saw that one managed to get to the shore). The police want to find the survivor in order to remove him from Spain, if he is still alive. In the last scene of the play, two guards approach Berganza and Cipión with syringes to put them to sleep. This powerful ending allows for various interpretations: it comments on the practice of putting unwanted animals to death, and simultaneously, it denounces the political discourses that are antagonistic toward illegal immigrants, establishing similarities between their situations. The camp where we find them surrounded by an electric fence reminds us of Giorgio Agamben's (1998) camp, built by modern states to dispose of those deprived of citizenship and rights, reduced by politics to "bare life" (9) that can be manipulated according to the needs of the privileged inhabitants of the city.[6] According to Agamben, "the camp is merely the place in which the most absolute *conditio inhumana* that has ever existed on earth was realized" (166, emphasis in original), and the "inhumanity" of this condition opens up the metaphorical space of "the animal." Mayorga's construction of the characters of Berganza and Cipión, whose status remains uncertain during the whole play as they slip between humanity and animality, leads us to reflect on the common space demarcated for animals and "illegal" humans in society. It seems to be meaningful that Berganza is metaphorically reduced to an animal by dramatis personae called "ciudadano uno" and "ciudadano dos" because he is lacking in citizenship. The electric fence functions as a symbol of the condition shared by animals and people rejected by the polis, both groups devoid of the right to move freely (Netz, 2004).

The initial exchanges between the dogs in Mayorga's *Palabra de perro* make the public realize how difficult it is for humans and animals without legal status to find a safe space within the city and how they have no voice. The words of Cipión: "Más vale cerrar el pico, que esto de que los perros hablemos es salirse de naturaleza" (1) (It is better to shut up, it is not natural for the dogs to speak), and "un perro que habla está en peligro" (2) (a dog who speaks is a dog in danger) can be interpreted as a complaint about numerous social practices that patronize illegal immigrants, who are conditioned by their willingness to remain in positions of subalterity and silence. The old woman, Canizares, offers protection to our protagonist in exchange for sexual services, and the two "citizens" who find him at her house offer to arrange papers for him in a few years as long as he works for them, hides, and shuts up every time it is needed.

In *Palabra de perro*, the anthropomorphism of the dogs allows the public to hear how they speak. The difference between animal and human speech in Mayorga's theater leads us to question the functioning of human speech, as reflected by the following episode. The guards of the camp where the dogs are enclosed hear them talk and one wants to inspect the cages, but the other one calms him down by saying that these are the voices of people who brought vodka. Then one of the dogs summons the other: "Habla como hablan las personas" (19) (Let's talk like people), and they begin to denigrate Moors, Gypsies, and Jews. The guards hear that and it satisfies them. One says to the

---

6. In Agamben (1998) "bare life" is the unprotected life exposed to death.

other: "Tenías razón. No es dentro de la jaula. Podemos dormir tranquilos" (20) (You were right. It is not coming from the cages. We can sleep quietly). And the guards go to sleep. If "let's talk like people" sounds surprising, since the dogs were talking in a language spoken by people from the very beginning of the play, we realize that they were not exactly talking like people. The playwright suggests that human talk's most distinctive feature is the denigration (often animalization) of others, which later justifies their mistreatment. These are strategies for the stratification of life. Just as in *La tortuga de Darwin* (2008), this play also constructs an argument that due to their dishonest and manipulative usage of language—speaking against others, justifying cruelty and death—humans are in fact losing their humanity, whose residue can be found in animals not capable of a manipulative use of speech. Mayorga's animals are genuine in their speech, and they attempt not to denigrate anyone. They even abstain from generalizing statements about humans, in spite of their bad experiences with so many of them. In *Palabra de perro*, Berganza is supposed to bite his tongue every time he makes a negative generalization while he talks about his life among people.

The play suggests that language and its metaphors have the power to provoke behavior and are responsible for the construction of human and animal status, as well as their mutual interactions. Cipión explains the meaning of "metaphor" to Berganza: "Por ejemplo, cuando dices, 'Fue la gente la que me convirtió en perro,' lo que quieres decir es que a veces se te olvidaba tu ser animal, pero la gente se encargaba de recordártelo" (38) (For example, when you say, 'People turned me into a dog' what you are trying to say is that sometimes you were forgetting that you are an animal, but people would always remind you). An anagnorisis of *Palabra de perro*, somewhat alike in Cervantes, reveals that Berganza is really a human being turned into a dog by the society, but this realization is one more time put into doubt when Cipión says: "Grandísimo disparate sería creer que alguna vez fuiste hombre ... eso que a tí te parecen recuerdos quizás no sean más que sueños" (38) (It is a fantastic idea that you were once a man ... what you think are memories are just dreams). This perplexing remark may suggest that the whole play, as in Pedro Calderón de la Barca's *La vida es sueño* (1969) may have been a dream of a dog who wanted to be human. If so, however, while Calderón de la Barca's public is invited to doubt the reality of their own lives, Mayorga's public shall put into doubt their own humanity and consider for a moment that, in fact, we are all animals, who make all sorts of efforts to forget about it and to feel above our animality. Cipión says that Berganza's humanity (and we should think, as everybody else's) should be viewed as a metaphor, that is anthropomorphism. Among animals, the most powerful of us, equipped with the capacity to manipulate with language, construct ourselves as the normative subjectivity: "humans," imposing on others a lower "animal" status. Both constructions, according to Mayorga, need to be revisited.

When Berganza wonders if he had really been a human in the past and if it would mean that he could get that status back sometime, Cipión suggests that he better forget about it and conform to being a dog. Berganza, however, clings to the hope of his humanity. In the morning, when he sees the two guards ominously approach them, evoking Rafael Alberti's poem of Spanish Civil War ("Un perro rabioso" [A Furious Dog, 2003]), he summons his friend to defend "como hombre rabioso" (as a furious man) and run together to "un lugar mejor. A un lugar donde ser hombres" (39) (to a better place. To a place where we could be human). These last lines can be read as a call for a revised version of "humanity," defined by the courage to rebel. The human and the animal are confused again in a metaphor when we realize that it is precisely when man fights like an angry dog that he seems most human. Mayorga's play suggests that the capacity to rebel has in fact been a part of animal nature. It is humans who have been gradually losing it. This implies that a revised version of humanity requires that in some ways we become more animal-like, act honestly, and be spontaneously rebellious. Maurizio Lazzarato (n.d.) writes that a state's strategy, which is aimed at extracting a surplus of energy from living beings that lack citizenship (both animals and humans) for the benefit of powerful citizens, is forced to transform when it encounters resistance. Lazzarato reminds us that, according to Foucault, this resistance is a key factor of political change, and in Mayorga's play, it is also a key factor of ethical change. In other words, according to Mayorga, there is a duty to rebel and fight for a better humanity.

CONCLUSION

Animalization serves as a foundational figure of politics because it justifies the mistreatment, abuse, and even killing of the animalized. As analysis of Pardo Bazán's short stories shows, selective animalization of the lower classes justifies social hierarchies. Baroja's protagonists' search for superhumanity through purposeful self-animalization goes together with an admiration of the strong and disdain of the weak. This is the principle of support for authoritarian governments and military solutions such as in Fascism. In Fascist regimes as in Saura's film, not only war but life in general is conceived as a hunt of the stronger for the weaker. As *La caza* shows, animalization can be compared to an angle with an invitation to turn a living being into prey. Animalization is a prelude to persecution and killing, except if it is an animalization across the board—one that, as in Buñuel's films leads to a consciousness of similarities that connect all humans and nonhuman animals. In this last case, however, rather than animalization, we should see it as an argument for equality in ethical consideration. Humanizing animals allows us to see not only the similarity of their predicament with humans but also to compare and contrast human and animal responses to imprisonment, taming, and torture. The results of these comparisons are often surprising as they lead us to see that in order to be more fully human, oftentimes we shall be more like animals. In other words, thanks to these comparisons we can discover the essence of humanity, some of which we have lost but remains preserved in animals. The seemingly naïve perception of some animal lovers that "animals are better than humans" coincides with highly sophisticated stories by Clarín and plays by Mayorga that point out that it is in the animal bodies where some of the greatest human virtues reside, such as a capacity for rebellion, unconditional love, honesty, and freedom. In addition to genes, humans and animals are connected by these abilities to be good or to be cruel to others. Humans, however, have developed an awareness of ethics, which makes it logical for them to adhere to ethical standards.

While the habitual way of seeing nonhuman life envisions it as so different from human that it is often not treated as life at all, anthropomorphism proves to be a very useful strategy to challenge these perceptions of animals, showing that the similarities between humans and animals are often not just metaphorical, but real. They were made invisible by the discourses that shape modern culture, which conceive of animals as livestock or beasts or represent them as machines. Anthropomorphism appears in this sense as an efficient deconstructive strategy that returns life and personalities to animals. While it can be viewed as not so different from the "anthropological machine" in that it assimilates the environment to the subjective perspective of humans, stripping them of their otherness and equipping nonhuman life with human properties, it may be just a first, imperfect step in a complex process of building a new conceptual framework for a more humane administration of life.

CHAPTER 5

ANIMAL RIGHTS MOVEMENT FOR AN ALTERNATIVE BIOPOLITICS

In my native Spain, many traditions involving the painful death of animals enjoy state protection, and animals are sacrificed to honor local virgins or saints in most provinces. For example, to mark the beginning of San Vicente's festival in Manganeses de la Polovorsa, Zamora, a terrified goat, is thrown from a bell tower, while by the chapel of San Vicente del Palacio, Valladolid, blindfolded young women strike and stab rows of hanging roosters. In other fiestas in Ávila, Cáceres, Madrid, Biscaya, Rioja, and Zamora, rows of roosters, ducks, or geese are hung upside down so that riders or swimmers can tear down their greased heads.

A wide variety of animals are used throughout the country. Donkeys are beat up and dragged to the point of collapse in Cáceres, greased pigs are crushed to death by the crowds in a churchyard in Cantabria, while in Torrevieja, Alicante, they drown at sea while people watching play "greased pig fishing." Even worse fates befall goats or young cows in much of Cáceres, Guadalajara, Salamanca, Toledo, Zamora, and Madrid; they are chased and bound; forced to drink liquor and then raped or sodomized with various objects while men beat them, tear off their horns, and torture them to death. The finale varies. In Las Rozas, the goat's throat is cut, in Illana the cow is pushed off the cliff, in Consuegra, the young cow is collectively stabbed (and the video is shown at the local disco).

The three most common fiestas, however, involve cows and bulls. They may be tied and dragged around town all day and night while locals stone them, drop heavy weights on them, cut and beat their heads and genitals, and stab them hundreds of times until they die.

—Paula Casal, "Is Multiculuralism Good for Animals?"

IN A CARTOON published in *El País*, Forges draws a bull whose bloody back is decorated with banderillas, swords, and other ornamental and sharp objects, and signs it, "se/añas de identidad" (signs of identity). Forges, one of the most popular *El País*'s cartoonists, publishes this vignette alluding to the title of Juan Goytisolo's novel *Señas de identidad* (Marks of Identity, 1966) and showing how a change of one letter turns it into "sañas de identidad"—that is, "cruelties of identity." Given the figure of the bloody bull, the cartoon seems to be criticizing bullfighting in the context where cruelty is one of the marks of Spanish identity.

Forges, one of many anti-bullfighting artists in Madrid, suggests that a critical attitude toward bullfighting amounts to a critical attitude toward certain glorified features of the national identity. He sees cruelty and suffering where splendor and glory are officially celebrated, and consequently, he feels the need to intervene and stop the harm rather than to aestheticize and eroticize the violence that is part of the national spectacle. The fundamental character of the animal for the national biopolitics—indeed a necropolitics in Mbembe's understanding of this term—and for the cultural sensitivity that is subservient to it is established by the phonetic similarity between "saña" (cruelty) and "seña" (mark). Cruelty toward animals becomes a modeling attitude in the treatment of humans in Spanish culture.

This chapter analyzes the discursive strategies and artistic means with which Spanish activists transform the sensitivity that lies at the foundation of the national necropolitics, opening people's eyes to cruelty in its most basic form—when it is committed against animals. The ideas of human/animal relations are connected to other debates occurring simultaneously in the public realm on the environment, on traditions, on the relations between men and women, and on class. For many an alternative biopolitics has to emerge in all these domains at once.

The first section of the chapter provides a brief history of the animal defense movement as it developed after 1975 and summarizes its most important campaigns and achievements. The following sections analyze activist performances and other visual art that subverts "the national archive" (Taylor, 2003) suggesting a new way of looking at suffering, transforming masculinity and social relations. The chapter argues that even though politically, in terms of the votes that PACMA (Partido Animalista Contra el Maltrato Animal [Party against the Mistreatment of Animals]) received during the last elections, the movement for the defense of animals does not seem significant, it has been culturally influential. Little by little, the activists and artists have forged an opposition against cruelty of all sorts, including a respect for non-

FIGURE 5.1
Cartoon from *El País*, by Forges: "Señas/Sañas de identidad"

human life and an understanding of their significance in politics. Even if most people did not vote for PACMA, most people are no longer fond of bullfighting and other festivities where animals are tortured or killed.

While previous chapters invoke the animal studies and biopolitics theory, this one focuses on the rhetorical strategies of the animal defense movement in dialogue with frame theory (Snow et al., 1986), as well as Diana Taylor (2003) and Tony Perucci's (2009) theories of performance activism. The dialogues creatively established by animal activists with many other important issues are interpreted as "frame bridging," which is the "linkage of two or more ideologically congruent but structurally unconnected frames regarding a particular issue or problem" (Snow et al., 1986, 467). As Snow explains, "frame" is a "schemata of interpretation" that helps individuals make sense of their experience within the world, therefore, also guiding their actions. In order to participate in a social movement, various individuals need to transform their private frames of experience and then align them (1986, 464). According to Taylor (2003), performance activism works through the subversion of the archive, as a new repertoire interprets anew symbols and histories. Perucci's (2009) concept of "ruptural performance" is helpful for understanding the dynamics of the reception of activist performances by a captive audience that is surprised by them in the midst of everyday activities in the city.

The research for this chapter, apart from the existing written sources, has also included personal interviews with activists and artists who belong to the movement, as well as on-site ethnography. In December 2011, I attended some of the performances that I analyze. I talked to Jorge Riechmann, Paula Casal, Oscar Horta, Manel and Carmen Cases, and Fernando Turró, and I exchanged e-mails with Albert Riera. The previous chapters have argued for the need of a "situated knowledge" in academic writing (Haraway, 1991) rather than an "objective" one that distances itself from harm in its attempt to be "balanced." Situated knowledge "from below," (191) as Haraway defines it, takes the side of the victimized and turns against those who hurt them. As I analyze the movement and its artworks, my perspective is very close to that of the activists, although in some cases my views about strategies that they adopt are somewhat critical.

SPANISH MOVEMENT FOR DEFENSE OF ANIMALS: DROP BY DROP

While the intellectuals with the most symbolic capital during the period of transition to democracy, often called simply Transition, attempted to inscribe democratic values onto bullfighting or alternatively promoted a Spanish democracy with bullfighting values, the animal rights movement was formed at the grassroots level by the lower classes, those who did not care for symbolic capital or decided to form it in alternative spaces. The first nongovernmental organization (NGO) to be founded after Franco's death in 1975 was ADDA (Asociación Defensa Derechos Animal [Association for Defense of Animal Rights]). Benito de Benito, a railway worker who loved animals and deplored the way they were treated, placed an advertisement in a local Catalan newspaper asking people opposed to the cruel treatment of animals to contact him. Those who attended the first meetings agreed to form an association devoted to the protection of animals, establishing an office in Mataró, later transferred to Barcelona. Manel and Carmen Cases, retired architects who today rotate as president and vice president of ADDA, although not among the founding members, joined the association and volunteered during its first years. ADDA's sites, both the real one on Bailén Street in Barcelona's *L' Eixample* and the virtual ones[1] are treasure troves of current and archival information about campaigns undertaken, changes in the law, debates, and animal emergencies in Spain and in Europe, carefully maintained in Cata-

---

1. See www.addaong.org and on Facebook.

lan, Spanish, and English. Since 1990, ADDA has published a journal, *Revista ADDA*, and since 1996, has published the *Antibullfighting Tribune*, which reprints letters, manifestos, and articles on debates on bullfighting in Catalonia, Spain, and other countries.

Practical, well-organized, and realistic, Manel and Carmen talked to me about what they consider to be the greatest successes of the campaigns in which they participated, which resulted in the prohibition of killing homeless animals in the shelters, the pronouncement of Barcelona as an anti-bullfighting city in 2004, and their favorite event, an organization of the yearly salon, *Animaladda,* where hundreds of homeless animals are adopted. During three days in October, the public assists in educational workshops, talks, debates, and concourses, as well as a display and the sale of pet industry products that help to fund the event. ADDA is funded by the dues of its members, which include Barcelona's sustainable construction company *Contratas y obras,* (Contracts and Works) whose web page proudly announces its sustainable, green, and ethical construction. Among the values of the company is "social responsibility," and to comply with it, the firm donates close to one percent of its budget to "socially disadvantaged environments" (www.contratasyobras.com). As a result, it solicits projects from various NGO's around the world and carries out those that are compatible with the company's vision and its workers' qualifications. For example, it has constructed a school in Nicaragua, an ecological chicken farm in Senegal, and an apartment house for handicapped people in India. The company donates money for organizations such as Bolivian Children with Glaucoma and Marginalized Catalan Women, and sends food to Saharan populations in need. The Web site that visualizes all these projects opens up with a quote from Theresa de Kolkata: "We ourselves feel that what we are doing is just a drop in the ocean. But the ocean would be less without that drop." *Contratas y obras* provides substantial amounts of money for the yearly salon *Animaladda* and other activities of ADDA, such as an international contest for the best article promoting the ethical treatment of animals. In an interview in his office on Freixá Street in Barcelona, the owner of the firm, Fernando Turró, claimed that his love of animals is the foundation of his humanitarian pursuits. The collaboration and investments in small projects between this medium-sized Catalan construction company and the NGO devoted to animal welfare can be classified as "green capitalism," which according to radical environmentalists, does not challenge the system but in fact reinforces it. But these little acts of kindness may have far-reaching consequences. It has been in part due to ADDA's activism, supported by people like Turró, that Spain's international image has changed.

One of the campaigns of ADDA in alliance with the Dutch Society for the Protection of Animals was aimed at changing Spain's tourist publicity in Europe. Through their initiative, Netherlands-based Comité Anti Stierenvechten (CAS) was founded in 1993 to prevent Dutch tourists from visiting Spain for bullfights and for other cruel festivities. As Dutch tourists stopped attending Spanish bullfights, CAS became an international organization, contributing to dimming Spain's international allure as a realm of otherness in many countries. This change, even if not complete, reflects a shift in Spanish identity that depends on outside perceptions.

Festivities featuring cruelty to animals attracted tourists from Europe and the United States who wanted to witness the anthropological "difference." The most emblematic of those festivities was obviously bullfighting. As Dorothy Kelly (2000) argues in her article "Selling Spanish 'Otherness' since the 1960s," Spain insisted on being different not only because the difference united Spain for the dictator's benefit but also because it sold. For tourists willing to travel to small towns, there was quite a diversity of joyful spectacles of torture on all sorts of living creatures, like those described in an epigraph for this chapter by Paula Casal, but the torture of humans remained hidden. Some of these festivities are still celebrated. In the Toro de la Vega Festival in Tordesillas, every year bulls are chased through the streets of the town by hundreds of men on horses, in cars, and on motorcycles with long lances that they stick into the bull's body until he falls down. Often his testicles are cut off when he is still alive to be awarded to the man who thrusts the decisive blow. In the beautiful city of Denia, the chased bull is often forced into the sea to drown. Various Catalan towns celebrate festivities known as "toro embolao" (*embolat* in Catalan), in which balls of flammable material, sometimes also fireworks that would later explode are attached to bulls' horns and set on fire, forcing the animal to run in panic through the town. In the Pero Palo Festival in Villanueva de la Vera, a donkey is violently forced through the streets of the town by rowdy, drunk young men. More troubling descriptions of these kinds of festivities cover many pages of the book by Antonio Lafora, *El trato de animales en España* (2004) as well as Casal's article, "Is Multiculturalism Good for Animals?" (2003), quoted earlier.

According to the Andalusian Organization A. I. D. (Animals in Distress), still today "up to 15,000 towns and villages sacrifice animals as centerpieces of their *fiestas*" (Animals in Distress, n. pag.).[2] Many, however, were either outlawed or transformed under pressure from the animal defense movement over the past twenty years. For example in Lekeitio in the Basque country,

---

2. See animalsindistresspa.org.

similar to the festive celebration portrayed in Buñuel's *Las Hurdes,* geese used to be hung head down on ropes over a harbor and men jumped toward them from boats, competing to stay in the air the longest by holding onto the bird's neck. Until 1992 the geese were alive, but under the pressure of animal rights activists, the festivity was modified, and dead geese are used today. In Manganeses de la Polverosa, one of the most macabre festivities, in which a goat was tossed down from the church bell tower and the awaiting crowd placed bets on if it would break open or not, was outlawed in 1992. In some Catalonian towns, "Toro embolao" is now celebrated as "toro de fuego" (bulls of fire) where the real bull is replaced by a metal frame in flames carried by a running man. The townships that decide to celebrate it in its original way take all sorts of precautions to avoid dealing with animal rights activists and being filmed and denounced by the media. For example, in Los Barrios in 2011, "Toro embolao" was organized on a private farm where access was restricted. Tickets cost thirty-four euros and could only be purchased by those known to the owner (no outsiders), and identification rubber bands on the guests' wrists were checked rigorously at the gate. People whose background had not been checked were not allowed to enter.[3]

In the 1980s, Spanish organizations defending animals were helped by pressure from the European Parliament, which was considering Spain's entrance in the EU. In 1985, the Commission on Youth, Culture, Education, Information, and Sport proposed a resolution to the European Parliament to look into the question of bullfighting. In April of the following year, Cottrell, a member of the European Parliament, proposed that bullfighting be forbidden in Spain, Portugal, and Southern France. In 1988, the European Parliament issued a contradictory resolution that announced legal proceedings against festivities involving cruelty against animals, but at the same time it expressed the will to protect traditional and religious celebrations in agreement with the spirit of multiculturalism.[4] This prompted an intense debate in the Spanish media that allowed the organizations fighting for animal rights to take this opportunity to implement the European law on the mistreatment of animals. As media discussed their demands during the 1990s, animal activists organized protests that passed through the streets of many towns. By then, some intellectuals took the side of the movement. Jorge Riechmann, Jesús Mosterín, Pablo de Lora, Rosa Montero, Pilar Rahola, Rosa Regàs, Elvira

---

3. See http://www.toroembolaolosbarrios.com/leer-noticia.aspx?idNoticia=64.

4. The details about these events can be found in Pablo de Lora's *Justicia para los animales* (2003, 298–303). Paula Casal's article "Is Multiculturalism Good for Animals?" (2003) is an enlightening read that analyzes conflicts of interest between culture and animals from philosophical perspective.

Lindo, Manuel Vicent, Juan José Millás, Antonio Muñoz Molina, Paula Casal, Martha Tafalla, Ruth Toledano, Alaska, and politicians such as Juan López de Uralde, Cristina Narbona, David Hammerstein, Francisco Garrido, and others aligned themselves against bullfighting, which they viewed as incompatible with that of a democratic transformation, the development of new models of masculinity, and more sustainable attitudes toward nature.

Eventually a number of concrete changes in regulations—although slower in social practices—have been achieved. In 2008, fines for the "con ensañamiento" (cruel abuse of animals) were turned into a penalty of imprisonment for up to one year, although they were never enforced. The same year, the Spanish Parliament voted in favor of limited human rights for great apes, a law that has not been ratified.[5] In 2010, the term "ensañamiento" (cruelty) was eliminated from the clause, which increased the efficiency of law enforcement. Anticruelty protocols were introduced for animal experimentation. Broadcasting of bullfights was limited in public media to certain hours, and access to bullrings in many Spanish provinces was once again restricted for children and youngsters (after Corcuera removed these restrictions in 1992). Vegetarianism became fashionable in the cities. For example, Barcelona now has more than twenty vegetarian restaurants.[6]

As the anti-bullfighting movement became strongest in Catalonia, which in 2010 banned bullfighting in part due to the strength and popularity of the grassroots animal defense organizations such as ADDA, it has been argued that it was just an expression of an anti-Spanish attitude (Dopico Black, 2010; Tosko, 2012) because at the same time the Catalonian festivities of "bou embolat" and "bou ensogat" were not banned. These are festivities involving the harassment of bulls that are popular in small towns of the delta of the river Ebro. My analysis of the detailed transcripts of the debates in the Catalonian Parliament shows that the ban of bullfighting in Catalonia was not just an anti-Spanish demonstration, but rather a reform motivated by a growing sensibility toward the mistreatment of animals in this part of the Iberian peninsula. Most of the discussion was in fact focused on the bull's pain. It was, nonetheless, clearly due to the Catalan nationalism of some of the members of the Parliament that "bou embolat" and "bou ensogat" were not banned. The ban on bullfighting could only be given a favorable vote with the support of CiU (Convergencia i Uniò), a party deriving most of its votes precisely from those small towns in the province of Tarragona where

---

5. For further discussion, see the previous chapter.

6. Twenty-five, according to http://www.sincarne.net/barcelona-vegan-restaurants.htm, but I know of some that are not on the list.

these festivities are most popular. This party's members were willing to vote in favor of banning bullfighting only if their festivities were spared, and without their votes the ban would not pass. This group was motivated by a loyalty to their local culture, which is indeed a sui generis form of Catalan culture, rather than by separatist Catalan feelings.[7] Mosterín called it "vergüenza catalana" (Catalan shame; 2010, 104), and most of the other anti-bullfighting activists similarly deplored the way in which politics took over ethics in this ethically inspired law initiative. ADDA presented an allegation against the law that legalizes "bou ensogat" and "bou embolat" in Catalonia, showing that it violates the law for protection of animals. ADDA's journal edition that was devoted to the celebration of the ban on bullfighting was rather sad, as most articles deplored the fact that the prohibition was not extended to "bou embolat" and "bou ensogat."

Even if the anti-Spanish feeling was arguably a part of the movement, it was so in the sense of the resented discourse of Spanishness that bullfighting represented for many. Baltasar Porcel (2004) wrote in *Vanguardia* that the anti-bullfighting movement connected bullfights in imperial Spain with thoughtless traditionalism, the cult of death, honor, Spanish difference, national Catholicism, lack of respect for women, and more, and it was against this kind of Spain that it protested. But this perception was not only in Catalonia. A similar line of associations criticizing the Francoist traditions of "La España negra" (black Spain), appeared, for example, in an article by Antonio Muñoz Molina (2008), who lamented that democratic Spain was not rid of the old rhetoric that he connected to bullfighting:

> Los pasodobles, las monteras, los trajes de luces, la grosera simbología de la sangre, la arena, la cornamenta, la espada. Era la España negra: la de los lugares comunes baratos del turismo, la de la intelectualidad extranjera que fingía apreciar nuestro exotismo y al mismo tiempo nos miraba de arriba abajo, brutos domados por un dictador y tan prisioneros de sus pasiones y sus rituales que no podían entrar seriamente en el mundo moderno.
>
> Creíamos que la libertad, al ventilarnos el país, iría despejando toda esa panoplia de espectros; que el ejemplo de nuestra democracia y la riqueza de nuestra mejor tradición ilustrada disiparían poco a poco en el mundo la fama negra de España. (n. pag.)

---

7. In Xerta, a little town half an hour inland from Tortosa, I found pro-bullfighting and pro-independence signs tacked on the walls of buildings while the loudspeakers emitted prohibitions against topless bathing and announced bullfighting festivities.

> The *paso doble,* the bullfighter's hat and costume, the crude symbolism of blood, sand, horns and the sword was the black Spain: the Spain of cheap commonplace tourist spots, foreign intelligentsia that pretended to appreciate our exoticism and at the same time looked at us from above [and saw] brutes controlled by a dictator and such prisoners of our passions and rituals that we couldn't seriously enter the modern world.
>
> We thought that, after airing the country, freedom would clear up the panoply of phantoms, that the example of our democracy and fine wealth of enlightened traditions would dispel Spain's bad fame little by little.

Both Porcel and Muñoz Molina argue against bullfighting on the grounds that it constituted a symbolic pillar of Francoist Spain, and as a result, in the narrative of national identity it emphasizes values and constructions that the democratic societies should not reproduce.

There has been, however, no consensus among intellectuals. Between 2004, when Barcelona and seventy-two other mostly Catalan municipalities announced themselves as anti-bullfighting cities, and 2011, when the ban was voted on, a number of well-known writers, such as Adela Cortina, Javier Marías, Cristina Fernández Cubas, Almudena Grandes, but especially Fernando Savater, spoke and published in defense of bullfighting. In her presentation "Mi pasión por los toros" (My Passion for Bullfighting, 2009), Cristina Fernández Cubas talked about the "communión con el público" (community with the public), which gives an illusion of forming "una entidad superior" (a superior entity) and which provides an identification with the bullfighter leading to a quasi-mystic experience of self-overcoming, deeply rooted in Spanish history (n. pag.). This group euphoria, indeed very similar to the one that Noel criticizes as responsible for a lack of citizenship, Fernández Cubas believes to be worth preserving. In a similar fashion, Almudena Grandes (2010) praised bullfighting as an art of "miracles," which she compares to a "liturgy" as she describes "600 kilos y dos pitons en punta, un hombre desarmado, una muleta y el arte que le salva de la muerte" (n. pag.). (600 kilograms, two sharp horns, a disarmed man with a cape and the art that saves him from death).

In July 2010, the Catalan Parliament banned bullfighting with 66 votes in favor, 55 against, and 9 abstentions. During the debates in the Parliament, one of the *aficionados* complained that the anti-bullfighting movement was so powerful that it condemned him to a clandestine existence, like "violentos, torturadores, inmorales" (n. pag.) (violent, immoral torturers). Savater (2010) compared the anti-bullfighting lobby with the Inquisition, and Fernández Cubas confessed "me siento en Barcelona como si perteneciera a una secta infernal" (2009, n. pag.) (I feel in Barcelona as if I were a member of an infernal sect). In 1991, Mitchell called anti-bullfighting activists "the Don Quixotes of Spanish culture" (82), and he claimed that they always failed miserably. After the Catalan ban on bullfighting in 2010, however, these kinds of statements could no longer be made. Even if the recent law declaring bullfighting a national heritage may invalidate the Catalan decision to stop them, Barcelona's bullfighting plaza, Las Arenas, has already been turned into a commercial mall while a debate surrounds the future of La Monumental. While the Emirs of Quatar offer to purchase it and turn it into the largest Mosque in Spain was rejected, Catalan architect Xavier Vilalta suggested transforming the bullfighting ring into a public park and sport complex, redesigned with a focus on ecological sustainability and respect for animals.

The Instituto Gallup polls conducted throughout Spain in 2000 and 2006 showed that most Spaniards expressed no (70 percent) or little (20 percent) interest in bullfighting, leaving bullfighting aficionados in the minority (10 percent) (Instituto Gallup, 2002; "Comparativa ICSA-Gallup," 2009; Lafora, 2004, 220–30; "Bullfight Opinion Polls," 2013; "For a Bullfighting Free Europe," 2008).[8] The results of the polls are not only due to the change of sensibility toward animals but also largely due to the fact that other entertainments became more popular. The research by Jordi López Sintas and Ercilia García Álvarez (2002) shows that "young Spaniards' search for consumer products is associated with modern values" (133). Only 6 percent of males and less than 4 percent of females under age 30 attend bullfighting as opposed to 12 percent of men over 65. However, according to the same publication, interest varies also according to the levels of education; for only 3.7 percent of males of all ages and 1.9 percent of females with a university education is bullfighting an entertainment of choice.[9] These results indicate that for most living Spaniards bullfighting is no longer connected to their identity other than as a part of national history whose memory and perception is being constantly reenacted and transformed by the anti-bullfighting activists.

Catalonia is not the only center of anti-bullfighting sympathies. The Canary Islands banned bullfighting in 1991. As the recent film directed by Manuel Gutiérrez Aragón, *La vida que te espera* (2004), shows, Galicians always preferred cows, and the anti-bullfighting attitudes are also very strong in the Andalusian capital of Málaga, which is the center of the Spanish Vegan

---

8. See note 12 in the Introduction for more details on polls.

9. This second set of data suggests that educated professionals distinguish themselves by disliking bullfighting. In this context, it is even more surprising to note that a great majority of Hispanists remain unchangeably devoted to uncritical eulogies of bullfighting in literature and film.

League and since 2007 a place of annual anti-bullfighting manifestations, which managed to attract strong media attention and thousands of "likes" in the social media. Today in Spain, there are more than ten organizations concerned with animal welfare and animal rights: ADDA, ANDA, Libera!, Defensa Animal, Altarriba, Anima Naturalis, Equanimal, Igualdad Animal, Proyecto Gran Simio, and A. I. D., as well as international organizations that often work in Spain, such as a PETA, WSPA, CAS-International, the Baas-Galgo, and hundreds of small charities and animal shelters. There is also PACMA and Equo, a green left party also concerned about animals.

INTERNATIONAL DAY OF ANIMAL RIGHTS

As a way to fight against the discrimination of animals and to stress their ethical stance, two young organizations, Equanimal and Igualdad Animal,[10] since 2008 have prepared a series of events to celebrate the International Day of Animal Rights on 10 December because it is also the International Day of Human Rights. This event is an example of their efforts to extend the frame of human rights to animals by advocating their equality with humans. Anti-speciesist organizations like these aim at a dramatic transformation of human civilization, for they believe that it is unethical to cause pain to any sentient creature, regardless of its level of intelligence. In their manifestos, they do not write about Spanish culture, tradition, and gender roles, as various other organizations do, yet the radical character of the equality between human and sentient nonhuman animals that they put forward is the most consequential for any identity and culture.[11] They demand recognition of equality between humans and sentient nonhuman animals, and as a result, they ask for a total abolition of animal exploitation. For this reason, they are also called "abolitionists." Anti-specisists go further in their demands than the animal rights movement because they not only ask for legal protection but also for the elimination of discriminatory treatment of animals as inferior to humans. In their campaigns, anti-speciesists reject arguments used by other move-

---

10. These two organizations joined forces in 2012.

11. According to Oscar Horta, a professor at the University of Santiago de Compostela and an intellectual leader in these organizations, sentience encompasses all animals equipped with central nervous systems, which would include bees but exclude sponges, for example. The presence of a central nervous system is privileged as essential for transforming information that is acquired in the process of an organism's interaction with its environment into "experience." Experience understood in this way is what guarantees the subjectivity of suffering. Otherwise, damage in living tissues may occur, but there is no "individual" that is experiencing it, and thus ethical violation is not happening.

ments, for example, that it is healthier to be vegetarian or that taking care of an animal prevents depression as well as the argument of sustainability raised by environmental movements. They do not privilege one animal's value over another's because of human interests or, because it belongs to a species close to extinction, or because of its importance for an ecosystem where it lives. In their view, these kinds of arguments detract from the primary issue, which is the suffering of an individual animal as a victim.

Oscar Horta notes that the ecological perspective can often be opposed to that of the animal rights movement. In certain circumstances, such as in areas where one population of animals grows beyond what is considered to be its "natural size," ecologists agree with its culling or other forms of intervention that may be viewed as cruel by animal rights activists. The debate reflects competing claims over a philosophy of life and its ethical consequences and can be read as an instance of boundary work (Gieryn, 1999), wherein movement activists and intellectuals such as Horta distinguish their work from others in the animal rights movement in order to gain authority and influence in internal struggles over credibility.

I observed the International Day of Animal Rights in Madrid's Puerta del Sol on 10 December 2011, a cold and rainy day. Over four hundred activists, predominantly young, came from all over the country and from abroad (see figure 5.2). They stood in rows, two yards from each other, all dressed in the movement's t-shirts pulled over warmer clothes, holding dead animals in their hands. These were animals that had died in factory farms or commercial laboratories and had been retrieved from the garbage to make Madrid's passersby reflect on their deaths. Various activists looked at the animals they held and cried.

None of the dead animals had names, but they were given numbers, and stories of their lives, as they were imagined by the group, were told through the loudspeakers by a dramatic female voice. This is my recording:

> Número XXX nació en una granja de producción de huevos. A los pocos días de nacer le cortaron el pico con una cuchilla. Debido a la herida que le produjeron no pudo comer pienso y acabó muriendo de hambre a las dos semanas de nacer. Hoy un activista sujeta a esta víctima entre sus manos. Hoy es el Día Internacional de los Derechos de los Animales y queremos proponer una profunda reflexión sobre como consideramos y cómo vemos a los animales en nuestra sociedad. Estos animales fueron tirados a contenedores, considerados basura. Y hoy estamos aquí demostrándolos de forma pacífica para que todo el mundo conozca y reconozca a estas víctimas del especismo. Algunos de los ratones que se están sujetando en este

FIGURE 5.2
Activists at the International Day of Animal Rights, Plaza Sol, Madrid

acto son animales utilizados en la experimentación ... Se experimentó con ellos y cuando se acabó se les tiró a la basura. Vidas que no tienen nombre. Individuos que fueron tirados a la basura ...

Todos somos iguales, independientemente de nuestro aspecto, de nuestra raza, de nuestro sexo y también de nuestra especie. La capacidad para sentir nos hace iguales y únicos."

FIGURE 5.3
An activist crying over a dead piglet during the International Day of Animal Rights, Plaza Sol, Madrid

Number XXX was born in an egg factory farm. A few days after she was born, her beak was cut off. The pain prevented her from eating, and she died from hunger just two weeks after she was born. Today an activist holds her in her hands. Today is an International Day of Animal Rights, and we want to propose a deep reflection on how we consider animals in our society. These animals were thrown into the containers, considered garbage. And today we are showing them to you in a peaceful way so that everybody can get to know these victims of speciesism. Some of the mice that we are holding were objects of experimentation, and when the experiment was over they were thrown in the garbage....

We are all equal, independent of our appearance, our race, our gender, and also independent of our species. The capacity to feel makes us equal and unique.

Light rain falling incessantly made passersby reluctant to stop. Most of those who gathered were older lonely people or grandparents with children, and the comments that I overheard indicated limited understanding. "Something about animals" the bystanders explained to each other. For those who stood and watched, in spite of the message announced through the loudspeakers, it was a baffling performance, even though it aimed to construct a precise ethical meaning, encouraging people to see animals' deaths as a tragedy. There was, perhaps, something too direct, didactic, or even forced in this imposed emotional reaction, and this, apart from the rain, may have been the reason that a lot of people did not stop to look for more than a few seconds.

There were an unusual number of friendly media representatives who photographed the demonstration from all possible angles (favoring the bird's eye view) and who interviewed the leaders of the movement. Later in the evening news, in the midst of reporting on celebrations connected to human rights, the spectacle at Puerta del Sol gained in size, clarity, and persuasion. The systematic character of animal exploitation and death contrasted with improvements to the human condition also proudly celebrated on this day. If the real tears in activists' eyes and the dramatic voice coming from speakers defamiliarized the habitual, indifferent way of looking at animal deaths for those who saw the performance on the ground, the camera's high angle gave objectivity to this new way of seeing, making crying over dead rats, piglets, and chickens completely normal.

Although the radical stance of anti-speciesists is far from being accepted widely by any significant percentage of society, and the strategic efficiency of the anti-speciesist movement's unwillingness to connect to other movements is debatable, the frame extension of human rights into animals constitutes perhaps one of the most dramatic cultural changes proposed by any new social movements. The changes in cultural practices that have resulted from the proclamation of equality between humans and animals were discussed in workshops and talks that accompanied the manifestation at Puerta del Sol. They amount to a very significant transformation of lifestyle, patterns of consumption, as well as a new set of emotional identifications. The proselytizing in anti-speciesist activism leads to constant conversations with parents, siblings, and friends, which are often transformed into debates involving accidental visitors, neighbors, and others. Although anti-speciesm is still a radical minority's point of view in Spain, slowly more and more educated young people become familiar with it and some become convinced. Igualdad Animal has over 350,000 "likes" on Facebook, considerably more than any other animal rights or animal welfare organization in Spain.

INTELLECTUALS, ARTISTS, AND "LA BASURA DEL CORAZÓN": ALASKA'S PLEDGE

As a result of the Franco regime's symbolic use of the bullfighter Manolete fifty years after the bullfighter's death, various leftist intellectuals refused to participate in his 1997 memorial celebration. In "Ucronía de Manolete" (1997), Burgos criticized the commemorative bullfight on the fiftieth anniversary of Manolete's death, admonishing the conservative Popular Party (PP) in charge of organizing the celebration for its Francoist origins and accusing its members of feeling "a secret love" for Miura bull, arguing that "si el PSOE estuviera en el poder, el 50-a aniversario sería un homenaje al 'Islero,' el toro que destruyó al torero símbolo del franquismo" (n. pag.) (If the PSOE (socialist party) were still in power, the 50th anniversary would have been an homage to "Islero," the bull who destroyed the bullfighter-symbol of Francoism). Burgos reproached Manolete's admirers for being moved by "la basura del corazón" (n. pag.) (literally: the garbage of the heart). "La basura del corazón" are those aesthetic representations that feed on melodramatic sensitivity and are subservient to the national necropolitics. Necropolitics constructs a spirituality that connects eroticism to violence and sacrifice, which helps to recruit men for war and to convince women that brutality is a part of love. As the analysis in chapter 1 indicates, one of the purposes of fomenting the bullfighting culture for various political regimes has been to build this kind of spirituality because it facilitates maintaining the hierarchies, male domination over women, and in general, the status quo of power. "La basura del corazón" is displayed in melodramas where exaggeration is a key quality, true love is only possible close to death or after, and politics is limited to sacrifice for the nation.

For various intellectuals and artists, the heart of the debate on bullfighting and national identity is in the politics of taste, and they attempt to transform the taste subverting what Taylor (2003) calls an "archive," a set of stories,

FIGURE 5.4
Alaska poses in front of the poster designed by Juan Gatti, "La verdad al desnudo; Tauromaquia es cruel" (The Naked Truth: Bullfighting Is Cruel).
© 2014 Uly Martín (Ediciones EL PAÍS, SL). All rights reserved.

images, and ideas defining the cultural identity. The anti-bullfighting poster by Alaska and Juan Gatti, "La verdad al desnudo: Tauromaquia es cruel" (The Naked Truth: Bullfighting Is Cruel) is yet one more subversive representation of the Spanish national archive: a naked woman with three banderillas sticking out of her flesh.

While the eroticization of violence and comparisons between bulls and women belong to the national culture, Alaska, interpreting them literally, stages "la basura del corazón" (the garbage of the heart). She deconstructs the metaphor by showing its dark side and by exposing its excess. This subversion through staging has been a surrealist strategy whose purpose was summed up by Buñuel as, "to explode the social order, to transform life itself" (1984, 107). Metaphors put on stage explode as their exaggerated character is revealed.

As a reference to the archive, Alaska's naked body with wounds displays the same eroticization of violence against which the artist is protesting, but subversively, this image is placed in a frame of animal defense activism. Thus even if the woman's body may still be causing masculine pleasure, it is used by the woman to express her solidarity with the animal to which she had been compared in various Spanish metaphors.[12] In this way, the pleasure that is involved in the consumption of the image is sabotaged by a meaning that supposes this pleasure's deconstruction, namely an accusation against the kind of masculinity that causes pain through this construction of the erotic. In Alaska's poster, it is through the substitution of the woman for the bull that the frame transformation takes place, where the bull's pain and injury become human-like and in particular woman-like. Men who kill bulls appear guilty of a violent assault on the female body. In sum, the poster implies that violent masculinity hurts, and both women and animals are its most common victims.

The photo has been used in the campaign against bullfighting prepared by Anima Naturalis in cooperation with PETA and has been republished in various widely read Spanish newspapers: *20 minutos, El País, Público,* and others, as well as on numerous online sites. As *El País* reports, when Alaska posed for journalists and photographers next to her poster, she described the rhetoric of the image with the following arguments. First, the "naked truth" of bullfighting is that it is cruel. People who disbelieve this should have the three *banderillas* stuck on their backs. Second, this truth is not easy to see, implying again the need to undress it; Alaska herself grew up in a bullfighting family (her mother's second husband was a bullfighter), and it took her time to realize that it was butchery. This rhetorical strategy of modeling a change of attitude seems particularly appropriate in a country like Spain, where a great number of adults associate bullfighting with childhood and their fathers' passions. The same way of framing the change in one's attitude about cruelty toward animals also appears in articles in *El País* by Rosa Montero, herself a daughter of a bullfighter, and by Manuel Vicent, a frequent contributor to the newspaper and the author of an anti-bullfighting book, *Antitauromaquia* (2001). Similar to Alaska's metaphor of a truth that needs to be unveiled, Montero and Vicent present the national predilection for bullfighting as a cultural blindness that needs to be cured in order to achieve true democratic citizenship.

According to Snow, the catchiest instances of frame bridging are those that achieve greater *resonance*, that is, establish convincing connections with other social movements or debates due to the enduring, significant cultural themes they evoke. Alaska's poster connects the frame of the anti-bullfighting movement to that of feminism and its fight against harmful traditions. As she

---

12. See chapter 3.

turns to the female journalists surrounding her, she jokes that "las costumbres culturales también deben evolucionar. Si no, yo estaría en la hoguera y vosotras fregando cacharros" (n. pag.) (Cultural customs have to evolve. Otherwise I would be burned at the stake and you would all be washing dishes). This remark brings together bullfighting and women's oppression as they coexisted in Franco's Spain and sheds light on the meaning of nakedness on the poster. Alaska suggests that when bullfighting was popular women lacked the freedom to decide about their bodies and could not show naked flesh in public. She implies that a woman who uses her own body to express her ideas is not subdued or objectified, but rather free to do with it what she believes is right. Thus the freedom to undress and even to transform the body is a politically active gesture. It is also a necessary condition to fight for the liberation of other bodies, such as those of bulls. This reflection connects Alaska's poster to Agrado's famous monologue in Almodóvar's *All about My Mother* (2000).

There are similarities and coincidences between Almodóvar's art and this poster of Alaska, for the poster displays a surprisingly Almodóvar-like kind of plasticity of image and colors, and we find that it is designed by Juan Gatti, who used to make posters for Almodóvar's movies. It could be hypothesized that Alaska's choice of artist was not only dictated by her desire to use the great skills of Gatti but also by her attempt to connect the animal question to the ethics developed in Almodóvar's films. As Paul Julian Smith notices in his review of *Los abrazos rotos* (2009), Almodóvar's various films have been obsessed with the fragility of flesh, and the most important turns in his stories often correspond with the protagonists' passages through hospitals. *La flor de mi secreto* (1995) and *Todo sobre mi madre* are arguing for a transcendence of individual life, not through life after death, but rather thanks to organ transplants and a corresponding transfer of love from one's own child to strangers. In *Todo sobre mi madre*, Almodóvar's ethics appear in declarations made by female characters that cruelty is the hardest thing to forgive, that one's freedom goes as far as it does not hurt the other, and that the truth should always be faced. All these are values of an alternative biopolitics, especially if applied broadly not only to humans but also to animals. Alaska's poster, which connects to the subversive language and to the aesthetics of Almodóvar's film posters, surreptitiously links his films' ethics to that of the animal rights movement. The poster feigns belonging to Almodóvar's world, as if it were announcing a film that he has never made but should and in which all ethical norms that he believes apply not only to interhuman relations but also to human-animal relations.[13] Far

---

13. Almodóvar does not sympathize with the movement for the defense of animals, and he has had a conflict with Amnistía Animal-Comunidad de Madrid, which accused him of

from Almodóvar's desire to accommodate national "basura del corazón" (garbage of the heart) into a democratic art in *Matador* (1987), Alaska, as in her famous song, "A quien le importa?" (Who Cares? 1986), has shaped her art according to her own criteria. She achieved her success due to her capacity to resonate with troubled souls and despite the criticism she has always received for her music. Her courage to be different may be connected to her activism for animals. For years she has been collaborating with AnimaNaturalis, the animal rights organization active in various Spanish-speaking countries, as well as BaasGalgo, an international organization dedicated to the rescue of Spanish greyhounds habitually hanged at the end of the hunting season. In her book *Transgresoras* (2003), which features five hundred inspiring women, Alaska mentions some women of different species, such as the female chimps of Jane Goodall.

"EAU DE TOROTURE"

While most anti-bullfighting campaigns compare humans and animals in pain, Albert Riera's poster connects bullfighting to the question of class disparity in contemporary Spain. The image designed by this artist of "Eau de toroture" as a cruel fragrance was published on a full-color page on 17 June 2007 in *La Vanguardia, El Periódico, ABC, El Mundo, El Mundo Deportivo, El Punt,* and *Avui*.

The initiative was part of an anti-bullfighting campaign led by the ADDA and the World Society for the Protection of Animals (WSPA) to protest the return from retirement of the well-known bullfighter José Tomás, and it was financed by Turró. According to Riera, the event was advertised as a sort of social party for well-off people who care about social visibility and attend bullfighting spectacles without ever asking themselves what is actually happening in the ring. Riera thought that this social group's false glamour was feeding on the pain of others (personal communication, April 2012). His project suggests that bullfighting, apart from hurting the animal, is celebrated at the expense of the Spanish lower classes rather than being a lower-classes privilege. According to Riera, even though the popular character of bullfighting is officially proclaimed for publicity purposes, bullfighting is nowadays very much a part of the establishment. It is a business, which apart

---

killing six bulls in shooting *Hable con ella* (2002). Although Deseo S. A, the company that produced most Almodovar's films, formally answered the accusations, explaining that all the legal permits for the use of the bulls were obtained and that the shooting of the film took advantage of the training of a young bullfighter, Almodóvar's informal, arrogant answers to these accusations further distanced his position from that of the animal rights movement.

FIGURE 5.5
"Eau de toroture" by Albert Riera

from its being sponsored by the state, maintains lucrative links with other multimillion-dollar businesses and businessmen from Spain and countries across the world.

The glamorous bottle of perfume filled with blood that is depicted by the ad provokes a visceral reaction. The artist reveals the deceit of the coolness associated with bullfighting. The ad confronts potential customers, suggesting that if they buy into this kind of coolness, it is blood that they will purchase. It is thus a reminder that those who finance the spectacle are responsible for what happens in the ring. The cap of the bottle, which connects the bull's horns to the traditional bullfighter's cap, signals brutal and cruel maleness, ominous in its primitivism, which obviously contrasts with the glossy glamour of the bottle. It is thus an implicit accusation of the elite classes, who are portrayed as a sort of devil-worshipping sect and whose expensive consumption of products is represented as fueled by lives lost. Here the basic framework of the anti-bullfighting movement is extended to include grievances about the growing class divisions in Spanish society. It points to the fact that bullfighting is a business of the rich that deals not only with the bull's blood but, implicitly alluding to discourses of the Left, also with the blood of the poor. In comparing the harm done to animals and to the poor by the globalized, gain-driven economy that takes its managers to bullfights, this image establishes a bridge to the frame of the environmental movement that often contrasts economic progress with damages to the environment, including animals and the poor. In fact, Riera himself is an environmental activist, campaigning against the construction project of yet another high-speed train in Catalonia.

WE ARE THE BULL

The cover of this book contains an image whose symbolism is central to its argument and that connects the ideas of all of its chapters. Here it will be analyzed in depth as a "ruptural performance" (Perucci, 2009) and as a transforming variation on the national archive (Taylor, 2003), that is, as an instrument and symptom of a change in the national culture. Between 2008 and 2011 in numerous towns across the Iberian Peninsula, Barcelona, Pamplona, Bilbao, and others, activists belonging to various organizations for the defense of animals, such as Equanimal, CAS-International, and Anima Naturalis, formed a large shape of a bull with their bodies painted in black and red (see figure I.1 in the Introduction and the cover).

"We Are the Bull" reads the image, as if paraphrasing the famous song "We Are the World," which generated aid sent from the United States to Africa in 1985. But it goes further than that in announcing that humans and nonhumans are together in flesh. The figure seen from above is an impressive example of body art; humans painted in black filled the healthy skin of the represented bull and those painted in red marked the wounds. The resulting illusion of a bull's body as a continent, constituted by small human bodies that can be read as a figurative representation of Spain, which according to Roman historians, was the shape of a bull's skin. Here the animal rights activists' discursive frame most directly connects to the debate on the Spanish national identity. In contrast to Spain's symbolic representation in bullfighting where a Spanish male defines himself as an opponent of the bull, these events by animal rights activists seem to argue for the organic unity of human and animal flesh. Thus this new figure of the wounded bull redefines citizenship and postulates the inclusion of nonhuman animals into the ethical domain on the basis of a capacity for pain that is shared by all animals (human and nonhuman). Notably, the bridging of the frames of animal rights and national identity facilitates the articulation of the central point of the ethics of the animal rights movement—the equality of all sentient creatures.

It is noteworthy that the performance requires a special way of seeing. The image asks for transcending one's limited perspective and seeing "the bigger picture" from above, where individual life appears as only a part of a greater organism whose composition is heterogeneous but whose wounds are shared. This is an ecological perception of the interconnectedness of all life—in this particular image, within the biosphere of the Iberian Peninsula, which is symbolically represented by the shape of the bull. The biosphere that includes humans and animals knows no frontiers. This event presents the anti-bullfighting campaign in connection to the ecological campaign to change human relations with nonhuman life. In doing so, it extends the animal rights activists' framework to include all those worried about climate change and the exhaustion of the Earth's resources. Additionally, this performance is more than an attempt to affect some passerby, but rather it is a proposition about the nation that will be republished in various media and circulated all around the world. It reconceptualizes belonging to a place as an ethical responsibility for all the connections that one's flesh establishes with others.

Perucci's concept of ruptural performance explains how a street happening may cause a frame transformation that can change the mindset of the public. Ruptural performance seeks to challenge the values of the society by creating a shock and suspending "automatism of perception" (Shklovsky, qtd in Perucci, 5) through defamiliarization. Ruptural performance establishes itself not outside of everyday life, like a theater spectacle, but rather is inspired by the surrealist ideal of looking at everyday life with fresh eyes, although it is with a different purpose than the search for the marvelous that André Breton envisioned. Instead of the search for the marvelous, ruptural performances suspend the habitual with a political goal in mind. Perucci's ruptural performances are vague, however, in that they are different from regular political manifestations. For example, the performances of "The Church of Stop Shopping" that Perucci analyzes parody exaggerated shopping. A group of activists enter a supermarket and begin to load their carts with everything they can take from the shelves quite beyond the carts' capacity. Even when all the merchandise drops to the floor, they keep loading more. By that time, the store management has already realized what is happening, and the guards take the activists outside, leaving the regular customers baffled and amused. "Drawing on Brechtian aesthetics and the Artaudian embodiment of 'the poetic state' as well as the (a)logic of Dada and the materialism of Minimal Art, ruptural performance enacts interruption, event, confrontation and bafflement as a form of direct action" (Perucci, 2009, 1).

The performance "We Are the Bull" is more complex because it occurs on two levels. On the ground level it confounds, like the shopping performance. Even the anti-bullfighting message is not obvious at first. Activists undressed and with their bodies painted, lying on the pavement, challenge the idea of humanity connected to a culturally approved dress code and the *habitus* of walking rather than lying on the city streets. At the first moment, an encounter with the performance brings the humanity of a passerby to the animal level of the biological body, which can be of different colors, here black or red, and which lies connected to other bodies. Both male and female participants wear only underwear which equalizes their bodies. Their semi-nakedness is a sign of their freedom to take on the task of political solidarity with other similarly naked bodies of animals for the purpose of their protection. Through the choice of the place, in front of Bilbao's Guggenheim Museum, the activists connect this fresh look on humanity to the ideals of the avant-garde art and in particular to one of the most internationally known structures in contemporary Spain. In this sense, a ruptural performance fits Michel de Certeau's vision of walking through the city as an adventure where space encountered on the ground level opens up with "leaks of meaning" (133), whose totality is only accessible from above. On the ground level, the performance is an announcement, partial and fragmentary, of some important reflection, a mindset change or a rebellion that matures in the city.

The situated character of this first encounter with the performance is then often transformed by fitting it into "a big picture," when the ruptural performance transforms to a revelatory one and when meaning is formed. Various people who have encountered the manifestation on the ground may have not understood it. The banner with the words "Abolición Tauromaquia" (Abolition of Tauromaquia) laid on the ground and was not easily visible. Only when they see the performance's photographs taken by the media from above, do they notice the shape of a bull and connect it to the anti-bullfighting debates and Spain. In this way, the aerial view in the photographic news reporting on activism has the value of God's eye (or a bird's eye). It objectifies, defines the meaning of an alternative spectacle, and is thus essential in its contribution to the activists' mission. By retransmitting these performances in the local news, Spanish media help to hybridize the national spectacle of bullfighting, deconstructing it through the images of anti-bullfighting and other animal defense performances.

The "We Are the Bull" performance, similar to the International Day of Animal Rights analyzed earlier, is constructed with media retransmissions in mind because its meaning can only be completed by a high camera angle. In the media, the performance happens for the second time and "the leaks of meaning" lead into an interpretation. Various local newspapers allowed activists to publish their manifestos next to the photographs of the event. For example, the title for the article accompanying the photograph chosen by *Revista de Opinión de Bilbao* was "Exitoso acto anti-taurino en Bilbao" (Successful Anti-Bullfighting Protest in Bilbao), and El Hada (2013), plausibly an animal rights activist, writes that "la abolición de la taurmoaquia es una lucha por el avance de la ética" (n. pag.) (the abolition of the bullfighting is a fight for an ethical advancement).

It is somewhere between the street encounter with the half-naked bodies of the activists lying on the pavement and the television news or newspaper article that the mindset change could take place. If, as Perucci suggests, the typical reaction to a ruptural performance is a state of confusion, a defamiliarizing surprise during a walk through the city creates curiosity and an openness to new information and ideas, which can be assimilated or rejected. The particular openness of the Spanish public toward the debates on bullfighting is due to their presence in what Taylor (2003) calls the national "archive," a repository of images and stories which historically structure national culture and of which most people are aware. For Taylor, performance activism enacts subversive versions of the inherited cultural scenarios, where the archival motives attract the public while the subversion transforms the archive surreptitiously. If Spaniards see a large shape of a bull on the plaza in front of the Guggenheim Museum, their reaction to it may be tainted by the fact that the bull is a national symbol. This fundamental archival identification does not need to change if the anti-bullfighting message is accepted instead of the bullfighting worldview. This would be a case of a "frame transformation" through a "frame extension," consisting of an inclusion of animals into the ethical domain. In this way, the bull can remain an emblematic animal of Spain.

Frame extension, adding new elements into the "frame" of experience of a particular person is one of the main strategies of movements fighting for the protection of animals, for animal rights, or for the abolition of animal exploitation. For example, if an existing frame is that killing and hurting others is not good, the animal activists' efforts concentrate on including animals in this frame as possible "others." The performance "We Are the Bull" achieves this in a surreptitious fashion through the idea of the continuity between human and nonhuman flesh. This is a basic idea for the construction of an alternative biopolitics, one that would care for an environment understood as a sum of interconnections between humans, animals, and places.

CONCLUSION

If electoral success were the highest measure of success for a social movement, then we would conclude that activism for the defense of animals has failed, since none of the parties representing the new social movements, including the animal rights movements, achieved any significant success in the 2011 elections. According to Jorge Riechmann, these results could have been different if only a common list of all the parties to the left of the Spanish Socialist Party (PSOE) had been prepared. Riechmann laments the historical fragmentation of political movements in opposition to the status quo as a frequent characteristic of the Spanish political landscape in the twentieth century, which constitutes an obstacle for forming any sort of a strong alternative culture.[14] But electoral performance is not the only measure of a movement's success, and the alternative culture may lack political propositions but have a strong transforming influence.

Because of the central location of bullfighting in the archive of national history, myth, and art, the animal rights movement has challenged inherited discourses of national biopolitics and catalyzed changes going deep into the grain of culture. The rhetoric of animal rights movement campaigns synergized with the rhetoric of the debates on masculinity, gender relations, regional autonomies, and the role of traditions, among others, has also challenged the basic political structures of the state: the law regulating relations between the human and nonhuman realms, the state's right to violence, and the economy of life. Because of the radical challenges that the movement postulates, it has been rightly perceived by Fernando Savater (2011) as a threat to

---

14. Personal communication, Madrid, December 2011.

the present civilization, which Savater wants to keep and people caring for animals perceive as cruel and destructive.

In contrast to highly commercialized, institutionalized, and state-subsidized bullfighting, which became an ossified spectacle of false glamour for neoliberal economic management and a spectacle of nostalgia for the oldest generations, the anti-bullfighting movement connects people of all ages at the level of spontaneous protests and cultural practices that emerge as a result of sincere ethical objection. While bullfighting once connected different Spanish provinces on behalf of the unity of a Spain governed from Madrid, anti-bullfighting connects them now as new generations distance themselves from the biopolitics that offers "pan y toros" (bread and bullfights), keeping people politically inactive but well fed and channeling their energy into celebrations. (Since the beginning of the crisis in 2008, bread's availability has not been that obvious). This model, elaborated by absolutist rulers, was in some ways continued after Franco's death by the socialist government, which was connected to its centralism. Instead, the anti-bullfighting movement asks for active and informed citizens' participation in cultural politics that would keep in check institutionalized violence and protect all life from suffering. Even if different organizations belonging to the movement for the defense of animals have disparate discursive strategies—some bridging their frames to the environmental movement, others to feminism, the antiwar movement, or the Catalan autonomy interests, some refraining from all bridging to focus on the issue at stake—all their political and artistic production indiscriminately condemns cruelty. Even if due to their differences, these organizations are not capable of becoming an important political force, they significantly transform the national culture, little by little eliminating cruelty as one of the signs of Spanish identity. As these changes occur, the international image of Spain has also been changing. Even if Spain retains its symbolic connection to the bull, it is now the bull of the national debate on cruelty rather than one that is thoughtlessly killed to affirm the right to kill of those who have the power to do so.

CHAPTER 6

BULLFIGHTING AND THE WAR ON TERROR

DEBATES ON CULTURE AND TORTURE IN SPAIN, 2004–11

ROBERTO ESPOSITO in *Immunitas* (2002) and *Bíos* (2008), and Jacques Derrida in *The Politics of Friendship* (2004) talk about the post-September 11 biopolitical crisis in terms of immunity. Both philosophers are alarmed by the imperative of security that reveals the dark side of democracies as Western states announce the need to "immunize" against Islamic terrorism, invoking the need to risk and to sacrifice some lives to protect others. For Esposito this "crisis" turns immune defense into an autoimmune syndrome. The pictures from Abu Ghraib expose such a dangerous overgrowth of the immune mechanism. Taken for fun by the U.S. guards the photographs show the torture and animalization of Iraqi prisoners. Reprinted in international media in 2004, these images alarmed the world about the reality of the War on Terror.

In his first reaction to the atrocities, Žižek (2004) connects the Abu Ghraib pictures to the fraternity and military rituals in the U.S. culture, but later he relates them to the Holocaust and other contemporary camp and torture practices, as well as to Marquis de Sade and Kant, asking, is there perhaps "a line from the Kantian ethics to the cold-blooded Auschwitz killing machine" or "a legitimate lineage from Sade to Fascist torturing?" (n. pag.). If Kant and Sade are two faces of the same coin, the first one represents the superego of the law and the second its "obscene underside" (Žižek, 2010a).

The relationship between law and sadomasochism that Žižek (2010a) notices points to a trend where the human body becomes an object of conscious economic and political manipulations. Foucault marks this as the beginning of modern times, which is so clearly displayed in the photographs from Abu Ghraib as in Passolini's *Salò* (1975).

In *Birth of Biopolitics* (2006), Foucault identifies two peak moments of such modern biopolitical manipulation: when the Nazi state became obsessed about total control over life in the 1930s and when the economists of the Chicago School in the 1970s treated human lives in society as a resource, extending rationality to the market and minimizing governmental control of the damage to the poorest. These two strategies are based on opposite ideologies, but both pose problems of life in economic terms, aiming at maximizing net gains of chosen populations at the expense of others. According to Agamben, even though Nazi politics has long been condemned, its *modus operandi* survived as camps thrived during the second half of the twentieth century, becoming "a new biopolitical nomos of the planet" (1998, 176). Žižek (2010a), following Agamben, suggests that sadomasochistic practices both in sex and in other realms of life, for example as in torturous initiation practices in student fraternities and some of Mapplethorpe's photographs, reflect a subliminal social consciousness that potentially we are all in a camp and so we need to prepare for this possibility, obsessively practicing possible scenarios. For Tony Perucci, familiarity with the "culture of torture" is not particularly American, but rather characteristic of neoliberal economic culture "free of all constraints," best exemplified by the prisons' rules of game, which he sums up as "fight or fuck" (2009, 368).

According to Stephen F. Eisenman (2007), the "Abu Ghraib effect . . . is . . . a kind of moral blindness . . . that allows [us] . . . to ignore, or even justify, however partially or provisionally, the facts of degradation and brutality manifest in the pictures" (9). In Derrida's (2005) view, the tele-technoscience helped to introduce the immunity politics after 11 September, providing an organized visibility of selected elements of the War on Terror while keeping others totally invisible. Abu Ghraib photographs were a pre-Wikileaks phenomenon, provoking an international exchange about what was revealed by the pictures of the tortured prisoners and what still remained to be revealed. The discussion about the pictures turned into a series of debates on the hidden dark side of Western armies but also Western cultures, where torture is secretly practiced and tolerated while officially condemned and forbidden.

While Žižek (2010a) connected Abu Ghraib pictures to perverse sexuality, Sherene Razack (2005) to racism and Perucci (2009) to neoliberalism, various Spanish intellectuals associated them with bullfighting. This synergy revealed that the fundamental aspect of the moral blindness of our civilization has been the way we treat animals. It is in the killing of animals where the first ethical exception to the rule of the protection of life has been established. Slaughterhouses and bullfights constitute the underside of moral law, but they are the basis of biopolitics. In other words, necropolitics constitutes the unethical foundation of modern biopolitics. The debates analyzed in this chapter reveal that tortured prisoners in Abu Ghraib and Guantanamo have not accidentally shared the predicament of animals, becoming objects of torture and murder. It happens because, as Reviel Netz (2004) claims, everything that has ever been done to animals sooner or later is done to humans. Once it is done, we turn a blind eye to it, just as we do with slaughterhouses and bullfights. These instances of killing and torturing of animals and humans are thought of as exceptions to the general rule of respect for life. The rationale of the exceptions is announced in biopolitical terms: some life needs to be sacrificed to protect life. In fact, however, it is a debatable proposition.

While the previous chapter focuses on how the movement for the defense of animals attempts to transform a symbolic affiliation with the bull in Spain, this chapter concentrates on the cultural production in Spain that arose as a result of the sudden conjuncture of humans and animals in torture, which in 2004–11 inspired songs, press articles, essays, visual arts, and theater productions. The debate on culture and torture that opened as a result has posed a number of troubling questions on human/animal relations and blurred the human/animal divide. The concept of torture has always been present in the anti-bullfighting writings in Spain's highbrow culture, and it appeared in the grassroots campaigns as early as the Association for the Defense of Animal Rights (ADDA) began its activism in the late 1970s. The most intense debate on torture in the context of bullfighting occurred, however, in the context of the War on Terror. *El País* and *Vanguardia* became the *fora* of two parallel debates during the years 2004–11: first, on bullfighting and human/animal relations, just as Barcelona was preparing to manifest itself as an "anti-bullfighting" city (2004) and debating the final ban on bullfighting (2010–11), and second, on torture in the War on Terror when photographs from Abu Ghraib were first revealed (2003–4) and when Madrid was attacked by Islamic terrorists on 11 March 2004. As a result of a synergy between debates on torture, terrorism, and bullfighting in Spain, new meanings were coined that connected the necropolitics of war and torture to the animal question. However, conclusions varied. While some (Toledano, Montero, Bru de Salas, Porcel, Millás) believed that the human abuse of animals constitutes a model for abuses among humans, others argued to the contrary, that efforts to improve the human treatment of animals would bring negative consequences for people (Savater, 2010b). Juan Mayorga's play *La paz perpetua* (2008), inspired by the Abu Ghraib photographs, solves this disagreement in a surprising way. It argues that humanity and animality as categories are not fixed in relation to species, but rather they depend on the predicament and behavior. While oppressors and torturers are animals, victims are human. In light of this play, a transformation of a human attitude toward animals is needed not only for the animals' sake but also to repair humanity.

BULLFIGHTING IN THE TIMES OF WAR ON TERROR

It has been in the anti-bullfighting context that the words "torture," "torturer," and "sadism" appeared most frequently in Spanish popular culture in the twenty-first century. One of the most popular slogans of anti-bullfighting activism in Spain has been "tortura no es arte ni cultura" (torture is neither art nor culture). It has appeared on posters, in brochures, and on multiple websites accompanied by images of a bull in agony, bathed in blood, with *banderillas* sticking out of his body. There has been a wide array of short films available on anti-bullfighting websites where the camera moves in slow motion, focusing on the animal's wounds and pausing in a close-up of the bull's human-like eye in a striking expression of pain. The debate has crossed national borders. The Colombian rock star Chucho Merchan composed an anti-bullfighting rap with the title "Tortura (no es arte ni cultura) (2009)," the words repeating as the chorus. The singer complains about "amargura" (bitterness), "tristeza" (sadness), and "dolor" (pain) and concludes with deep conviction that "esa mierda no está bien" (this shit's no good). Reincidentes (Repeat Offenders), a Spanish hard rock/punk group that started playing in the mid-1980s, composed a song "Grana y oro" (Red and Gold, 1997). In the lyrics, a bull tells in first person a story of a bull fight, in whose culminating moment the dramatic voice of the vocalist cries "cuando el acero me traspasa el corazón, y se le llama fiesta y nadie se molesta" (when the steel goes through my heart, they call it fun and no one is bothered). One of the most popular nonconformist rock groups in Spain, Ska-p (which in Spanish reads "escape"), founded in 1994, performed at least three songs criticizing the human abuse of animals. Among them "Vergüenza" (Shame, 2000) from the album *Planeta escoria* (Planet Scum) denounces "una asquerosa y sucia tradición" (a disgusting and dirty tradition) in which "un individuo vestido de payaso tortura y martiriza hasta la muerte un animal" (an individual dressed like a clown tortures and torments an animal until death). The English title, "Wild Spain," was chosen as an allusion to the Spanish tourist industry that Manuel Fraga, a minister of tourism under Franco, had developed and that still continues in some cases with the eroticized brutality and cruelty of Spanish traditions. The irresistible rhythm and catchy melody of the song contrasts with its words, a sarcastic depiction of: "sadismo de una salvaje humillación" (sadism of savage humiliation) and "sangrientas tradiciones que son aberraciones a la moral" (bloody traditions which are moral aberrations). The contrast between the seductive music and the lyrics that criticize the tradition produces an effect in which the sensual attractiveness coexists with a rational rejection of the spectacle. The Basque band El Reno Renardo, in a song entitled "Torturadores" (Torturers, 2007), offends bull-fighting *aficionados* with a heavy metal anger: "Quiero verte atravesado, empitonado y desangrado en tu fiesta nacional, ver tus tripas en la bandeja, cortarte el rabo y la oreja en tu fiesta irracional" (I want to see you stabbed, gored and bleeding in your national pastime, see your guts on the tray, tail and ears cut off in your irrational pastime). In a similar mood, Soziedad Alkholika (Alcoholic Society, 1999) "in a tribute to all the bulls tortured and killed" repeat in the chorus of their anti-bullfighting song the words "Motxalo!" (Poke him!). Porretas (1994) in a softer beat asks: "Dejalos en paz" (Leave the bulls alone) and suggest to the listeners that one day they may find themselves tortured in the position of the bull. Some of the other musicians and musical groups that have performed songs against bullfighting include Hamlet, Jesús Cifuentes, Las Marras, Francis Cabrel, Pukreas, and Ataque 77. The repetition of the word "torture" and its partial synonyms "pain," "martyrdom," and "sadism," accompanied by graphic details in the lyrics and their graphic representation in performances (analyzed in the previous chapter) is because they are key to the argument concerning bullfighting's immorality. Officially used only in reference to humans, the word "torture" in animal rights movement songs and slogans has connected human and animal domains.

The rhetoric that insisted on the need to speak interchangeably about human and animal flesh, pain, and the mutual right to be free of torture was surprisingly supplied with new supporting materials in 2004 by the War on Terror. These were photographs and stories of tortures committed by the American Army in Iraq and Guantanamo prisons, which made it all around the world and got to Spain as well. As the word "torture" resounded in discourses protesting against the violation of human rights on political prisoners and prisoners of war, the slogans of animal rights movements acquired a new meaning. But the debate on bullfighting also affected the meaning of the debate on torture. Between 2004 and 2011, the Spanish press hosted a debate on torture in which human and animal victims were compared as they were subjected to similar procedures, similarly excluded from moral consideration, and placed outside of legal protection. Furthermore, as the infamous Abu Ghraib pictures showed, animals (dogs) were used as "soldiers" in the global War on Terror. Photographs of Dobermans attacking political prisoners, followed by those where torture was administered by humans, provided another common ground between animality and humanity; the borders between them blurred in victimhood as well as in aggression. The most memorable event in Spain in 2004 was without a doubt the terrorist attack at the Atocha train station, and even this event has been compared to bullfighting.

In an article entitled "La fe de erratas" (List of Errata) published in *El País* on 27 November 2004, Juan Goytisolo talks about life in the "catastrophic" contemporary world through a metaphor of "errata"; life is a chain of mistakes, errors, and efforts to correct them, which come always too late. In this context, Goytisolo evokes the terrorist attack in the Atocha station in Madrid and compares it to bullfighting:

> Por principio, sólo mueren los otros. Contemplamos los toros, la tortura "artística" de las reses, desde la comodidad de la barrera y no desde la arena misma. ¿Quién podía imaginar hace 10 años que el horror del asedio de Sarajevo nos afectaría a nosotros un día? ¿Que el martirio de la capital bosnia repercutiría tal vez, por un encadenamiento soterrado de circunstancias, en la explosión mortífera de los trenes en la estación madrileña de Atocha? (n. pag.)

> At first it is always the others who die. Comfortably seated behind the barrier far from danger, we look at the bullfights, the "artistic" torture of the beasts. Who could imagine ten years ago that the horror of the siege of Sarajevo could affect us one day? The martyrdom of the Bosnian capital would result perhaps through some hidden chain of events in the deadly explosion of the trains in the Madrid's Atocha station?

Goytisolo's comparison evokes all those anti-bullfighting posters that present the bull as a matador and a human as a victim bleeding in the arena, but in a political context, where this reversal of roles is much more likely and even more consequential. In fact what happened in Atocha could be seen as a partial realization of the horrid fantasy from Goytisolo's *Reivindicación del Conde don Julián,* (Count Julian, 2003) which in light of M-11 (11-M)[1] acquires

---

1. M-11 stands for March 11, 2004, when the terrorist attack in Madrid's train station Atocha took place.

the character of a prophetic warning against stoic arrogance that in the novel connects the figures of Seneca, Manolete, and Franco, and whose main feature is a cruel, blind desire to destroy all otherness. But it was the torture of political prisoners that during that same year obsessed the world's newspapers and provided the most fertile ground for constructing parallelisms.

In 2004, Spanish newspapers wrote about torture almost every day, featuring it in both a local and global context. To give an idea of the increase in the intensity of the debates, the number of articles that evoked "torture" of any kind in the archive of *El País* in 2003 was 246, but in 2004 it grew suddenly to 1155 and maintained itself at a similar level until 2007 when it went down to 822, falling abruptly to only 86 in 2008 (the year Barack Obama becomes U.S. president) to go up again in 2009–11 when the final prohibition of bullfighting in Catalonia was discussed and when it also turned out that Obama did not free the world of torture as many had hoped. In the Catalan *La Vanguardia*, the peak in commentaries on torture also occurred in 2004, the year when the Abu Ghraib scandal erupted and the first intense debate on bullfighting culminated with the announcement of Barcelona as an "antibullfighting city." Another peak in the debates also corresponded to the years 2009–11, the most intense period of debate on the banning of bullfighting.[2]

In 2004, 134 articles in *La Vanguardia* contained simultaneous references to "torture" and "animal." Among those articles, some featured more or less direct comparisons between prisoners or guards at Abu Ghraib and Guantanamo to animals. Another part was on bullfighting, where through the use of the concept of "torture," bulls were compared to humans. In 2009, there were 178 articles with a similar pattern, most featuring the newly heated debate on "tauromaquia," as compared to only 13 in 2000 and 27 in 2008. The simultaneous animalization of humans and the humanization of animals through torture, frequently within the same issue of a newspaper or even on the same page, makes it evident that vulnerability to pain is, in the most striking way, shared by all animals (humans and nonhumans).

Among all the shocking photographs taken by American soldiers in Abu Ghraib that connected humans and animals through torture, one of the most infamous represented the female soldier Lynndie England holding a naked Iraqi prisoner in a dog-like position on a leash. Another, printed by *El País* on 12 May 2004, showed a naked Iraqi prisoner bent close to a wall with his arms behind his head with an expression of horror looking at dogs ready to jump on him, held on leashes by American soldiers. These pictures seemed to capture the logic of the strategies of subjugation practiced by American

---

2. All the data proceeds from el archive of *El País* and from "hemeroteca" of *La Vanguardia*.

Imperialism. As Ugandan writer and history professor Moses Isegawa (2004) declared in an article published in *La Vanguardia*, "El hombre de la correa podría proceder de cualquier parte del mundo fuera de Europa y los EE.UU. Lo que sufren los iraquíes, lo sufren los africanos, los asiáticos y los sudamericanos" (5) (The man on the leash could be from any part of the world outside Europe and the United States. What the Iraqis suffer is what the Africans, Asians, and South Americans suffer). These pictures made it evident that in war, in prison, in a camp, people are, in the words of Ann McClintock (2009), "unpeopled" (65)[3] and turned into what Giorgio Agamben (1998) calls "bare life," where they are reduced to an-animal-like status.[4] The photographs speak about the fact that in these exceptional circumstances, unfortunately frequent in recent times, the strategies of the subjugation of animals are commonly applied to people.

At the same time, however, humanity and animality switched places through the reframing of these pictures, which were gathered from the private collections of American guards where they amounted to trophies. Shorty thereafter, they made it to the first pages of international newspapers that reprinted them surrounded by words expressing the deepest indignation. The campaign of protests in response to those pictures, which sounded throughout the world, returned humanity back to the prisoners. Meanwhile, the guards were accused of bestiality, and the international public placed themselves somewhere between the two categories. In the midst of the general consternation, awe, and guilt that reigned in the wake of the Abu Ghraib scandal, intellectuals and artists all over the world interpreted the images as being, in the words of Philip Gourevich (2008), "in some way about us" about our culture.[5] The Spanish cultural context with its already prominent debate on bullfighting provided a special angle for an interpretation of the Abu Ghraib images that was even more markedly focused on human/animal relations. Additionally those international reactions that emphasized the visibility of human/animal relations within this spectacle

---

3. As Anne McClintock tells us in her "Paranoid Empire: Specters from Guantánamo and Abu Ghraib," arriving prisoners are referred to as "packages" there. Upon arrival they are informed by employees that their bodies are from then on property of the Government of the United States of America, which turns them into what she calls "bodies unpeopled" (65).

4. In Agamben's *Homo Sacer*, this is the most important notion defining the condition of a human who lost his citizen's rights and thus can be killed by anybody, more or less as if they were animals.

5. Philip Gourevich, an author of the famous documentary on Abu Ghraib, *Standard Operating Procedure*, 2008 announced during a debate in the New York Public Library (2008) that the photographs of the torture and humiliation taken by American soldiers "are about us."

of torture were more likely to be reprinted in those journals of the Spanish and Catalan press, which, like *El País* and *Vanguardia*, were at the same time publishing opinions on bullfighting. As the Spanish press published the stories and photographs of torture, various animal rights activists reflected that wars and animal abuse are closely interrelated. Rosa Montero (2004), for example, wrote in *El País*: "Todo está estrechamente relacionado.... Las torturas de los prisioneros iraquíes, las mujeres y los niños asesinados por la violencia doméstica, incluso los cientos de miles de focas apaleadas y despellejadas aún medio vivas" (n. pag.) (Everything is closely related.... The torture of Iraqi prisoners, the women and children killed by domestic violence, even the hundreds of thousands of seals clubbed and skinned while still half alive). Andrés Ortega (2004) listed various cruelties in his *El País* article that filled Spanish television screens that year: "Cuando terroristas ejecutan a rehenes en Irak ante cámaras y difunden las imágenes del crimen por la televisión o Internet; cuando un *marine* dispara sobre un herido en Fallujah . . ." (n. pag.) (When terrorists executed hostages in Iraq in front of the camera and spread images of the crime over television and the Internet, when a Marine shoots a wounded person in Fallujah . . .). Ortega concluded that all these happened, as the philosopher Peter Sloterdijk claims, because "el problema de los modernos consiste en que piensan como comedores de plantas y viven como comedores de carne" (n. pag.) (the problem for modern people lies in the fact that they think like plant-eaters and live like meat-eaters). In *La Vanguardia,* Andy Robinson (2004) subtitled a section of his article on the Abu Ghraib prison as "Casa de animales" (Animal House), and Michel Wieviorka (2004), in his essay entitled "El mal" (Evil) published in *La Vanguardia,* criticized the animalization of tortured prisoners and implied a parallelism between bullfighting and torture: "Los soldados estadounidenses arrancaban trofeos (orejas, muelas de oro, cabelleras, penes) de los cuerpos del enemigo" (n. pag.) (U.S. soldiers tore trophies [ears, gold fillings, hair, penises] off enemy bodies). This latter article was accompanied by a drawing by Javier Aguilar, representing a contemporary Jovellanos—from Goya's "El sueño de la razón engendra monstrous" (The Sleep of Reason Produces Monsters)—leaning over a television set, as if troubled by all the cultural teratology in which he is condemned to participate by watching.[6] On 17 April 2004, Manel Cases (2004) in a solemn tone reminded readers of the Fòrum Universal de les Cultures (Universal Forum for Cultures) that was soon to be held in Barcelona, whose slogan called for "un mundo que queremos" (the world that we want), which as the author

---

6. Jovellanos wrote against bullfighting and campaigned to forbid it in his times.

explained, would include "paz, sin violencia y sin tortura ni de humanos ni de animales" (n. pag.) (peace, nonviolence without torture of humans nor animals). The article was entitled "Barcelona, ciudad libre de crueldad animal" (Barcelona, City Free of Animal Cruelty).

On 19 December 2005, Ruth Toledano (2005) published in *El País* an article "Dos tristes tigresas" (Two Sad Tigresses), whose first two paragraphs could have referred equally to tortured prisoners or animals:

> Cuando no aguantan más la postura y la desesperación del confinamiento, se levantan y distraen el sufrimiento recorriendo con reiteración neurótica los cuatro pasos que les permite la estrechez de ese espacio. . . . Las quemaduras son dolorosísimas. Los torturadores usan también collares de castigo con grandes pinchos de hierro, instrumentos eléctricos para infligir descargas, ganchos metálicos, cadenas. Pasan hambre con frecuencia, pues las privan de alimento si no obedecen sus órdenes. . . . Puede que ahora las condenen a muerte por su delito, que no lo es. (n. pag.)

> When they can no longer stand the posture and desperation of confinement, they get up and distract themselves from their suffering by neurotically pacing around as much as their narrow confines allow . . . . The burns are agonizing. The torturers also use spiked punishment collars, electric devices to inflict shocks, metal hooks and chains. They frequently go hungry as they are deprived of food if they don't obey orders . . . . They may now be condemned to death for their crime, which wasn't one.

The confusion generated by these lines would not have been possible without the news arriving from the War on Terror, and most significantly without the Abu Ghraib photographs that represented prisoners tortured as animals. Destabilizing the human/animal divide in her article, Toledano utilizes the worldwide indignation that the Abu Ghraib scandal provoked and the human misfortune that came under public spotlight as an opportunity to bring to her readers' attention the invisible and unquestioned suffering of animals as deeply connected to the suffering of the prisoners. An immediate impulse of the reader is to free the suffering creatures, regardless of their species. The author seeks to provoke the thought that this kind of suffering should not be inflicted on any living creature.

This kind of rhetoric dominated the *El País* and *La Vanguardia* publications that connected torture and bullfighting during 2004–5. Columnists commenting on human torture often invoked animal suffering and human/animal relations. At the same time, publications that focused on animal abuse alluded to torture in the human and historical sense. Montero (2005), writing against the annual Toro de Tordesillas, a celebration where a crowd pursues, tortures, and kills the bull, reminded her readers that "hasta el siglo XVIII, la tortura era algo totalmente aceptado" (n. pag.) (until the 18th century, torture was something that was totally acceptable). Similarly, Mosterín (2010a), congratulating Barcelona for announcing itself as an anti-bullfighting city in 2004, in one sentence condemned torture against human and nonhuman animals as a pre-Enlightenment phenomenon. He explained that in most European countries bullfighting and various other festivities featuring cruelty perpetrated on animals had disappeared in the eighteenth century as the result of Enlightenment law reforms; however, in Spain, in spite of the royal prohibition by Charles IV under the influence of Jovellanos, bullfighting retained its popularity and was subsequently turned into an official national spectacle by Ferdinand VII. As Mosterín reminds his readers, this king was one of the most violent rulers in Spanish modern history, famous for the "decada ominosa," (the Ominous Decade) ten years of the particularly cruel suppression of the opposition and of the restitution of the Inquisition.

As the War on Terror once more illuminated a number of parallelisms between the political violence targeting humans and bullfighting, the animal rights movement gained new members. Xavier Bru da Sala (2004) comments on this in his article "Oh, toro!" (Oh, Bull!): "Sí conozco gente que se aficionó a los toros mucho antes de estar contra la guerra o el sufrimiento de unos seres humanos a cargo de otros, extensible a los animales si no es por necesidad, y a los que aún no ha caído la venda de los ojos" (n. pag.) (I do know people who became fans of bullfighting long before developing antiwar sentiments or their being against the suffering of people under others' control, which can be extended to animal suffering when it isn't necessary, and I know people who are still going around blindfolded). The writer suggests that the War on Terror has become an eye-opening circumstance, drawing people away from bullfighting. Antonio Muñoz Molina (2008) in *El País* argued along the same lines: "Puedo comprender que mi padre se conmoviera viendo una corrida de toros: ahora veo la foto de un torero en la primera página de los periódicos más serios, leo los ríos de prosa artístico-taurina que vuelven a derramarse, y siento vergüenza de mi país, y un aburrimiento sin límites." (n. pag.) (I can understand that my father would be moved by watching a bullfight: now I see a bullfighter's photo on the front page of the most serious newspapers, I read the endless stream of artistic bullfighting prose that is spilled forth again, and I feel ashamed of my country, as well as bored beyond limits). Describing the change has been a strategy of various anti-bullfighting intellectuals, such as Vicent (2001) who equaled losing *afición* for bullfights with adulthood: "Se pierde afición a los toros como se pierde la ingenuidad. La vida te va despojando de todos sus elementos irracionales y quedas a merced de una desnuda inteligencia laica, sin adherencias mágicas" (12) (One can lose the fondness of bullfighting as one loses naivety. Life strips you of all the irrational elements one by one, and you end up at the mercy of a bare secular intelligence, without any magical adherences). He invariably connected bullfighting to other forms of interspecies violence, in "[esa antitauromaquia] no es un arte de torear al revés, sino una apuesta por no torear nada ni a nadie salvándonos de la crueldad" (13) [that anti-bullfighting] is not the art of bullfighting reversed, but rather a bid against violence toward anyone or anything, saving us from cruelty).

In the context of the War on Terror, Baltasar Porcel (2004) implies that the invocation of an exception for continuing with the cruelty of bullfighting for the sake of protecting Spanish identity appears to be the baseline for a similar kind of ethical hypocrisy that makes an exception for torture inflicted for the sake of protecting American democracy. The exception for torture because it only hurts an animal follows a similar pattern to the exception granted to those torturing a potential terrorist. Ethical exceptionalism is sustained by language that provides acceptable concepts for nonacceptable behaviors and attitudes. For this reason, a great part of the debate on human/animal relations takes place in the linguistic realm. The slogan "tortura no es cultura" (torture is neither art nor culture) asks for a change that would exclude torture from all the domains of human culture where it is accepted because it is called something else. Slogans and lyrics such as those of Ska-p's song "Vergüenza" (Shame, 2000) attempt to influence the political realm through a deconstruction of concepts that we perceive as neutral but in fact denote realities of pain. When Ska-p sings about the animal "agonizando en un charco de sangre" (agonizing in a pool of blood) and calls the bullfighter "asesino por vocación" (assesin by vocation), they attempt to uncover the suffering flesh and sadism behind assimilated expressions such as the sophisticated concept of "subjugation" or the elegant phrase "arte de matar" (the art of killing). The anger expressed by the music and words of rock songs toward the sleek language of the polis, which is seen to be a camouflage for bloody realities, is also there in the ironic discourse of "Verdad palmaria" (Obvious Truth) by Juan José Millás (2010), published in *El País*:

> El maltrato a los animales está mal visto (ya era hora), incluso hay leyes que lo persiguen, aunque estableciendo salvedades. Nada que objetar a las salvedades, la vida es así, no lo he inventado yo, etcétera. También la tortura está prohibida, a menos que la ejerzas en Guantánamo, con gente cuya piel

es más oscura que la tuya. Y el terrorismo se persigue de manera implacable, excepto cuando se trata de bombardear Irak. Anomalías culturales, qué le vamos a hacer, lo que no quita para darse cuenta de que el terrorismo es terrorismo incluso si lo practico yo. (n. pag.)

The mistreatment of animals is frowned upon (it was about time), there are even laws that prosecute it, albeit with exceptions. I have nothing to object to these exceptions, that's how life is, I didn't make the rules, etc. Torture is also forbidden, unless you practice it in Guantanamo, on people whose skin is darker than yours. And terrorism is prosecuted relentlessly, except when it involves bombing Iraq. Cultural anomalies, "what are you going to do," but these don't clear one of the realization that terrorism is terrorism even if I am the one practicing it.

Millás's article goes to the heart of the connected debates about the mistreatment of animals and the War on Terror as he notices that the rhetoric that justifies war and torture, be it human or animal, is deeply self-serving and invariably located in the locus of power that manipulates meanings in order to justify immoral action:

Parece evidente que al toro de lidia se le maltrata. ¿Que a usted le gusta? Nos parece muy bien, no lo vamos a censurar. Pero hombre, hombre, reconozca que las banderillas, las puyas, el estoque y demás instrumentos quirúrgicos hacen daño (además de humillar). En el acto de arrojar una cabra viva desde un campanario hay belleza, no vamos a negarlo. A mí al menos me sobrecoge esa lucha titánica entre el cuerpo del animal y la fuerza de la gravedad (de la que siempre sale vencedora, por cierto, la última), por no mencionar la precisión matemática del movimiento uniformemente acelerado, que se cumple con todas y cada una de las cabras, no importa su condición. Todo eso está muy bien y si a uno le gusta le gusta. Pero hay tortura, hay maltrato, hay vilipendio. ¿Por qué a los taurinos, muchos de ellos intelectuales de pro, les cuesta tanto admitir esta verdad palmaria? (n. pag.)

It seems evident that a bull is mistreated in a bullfight. You like it? Fine with me, I will not censor it. But, man, recognize that banderillas, lances, swords, and other instruments hurt (besides humiliating). In the act of throwing a live goat from a tower there is beauty, we cannot negate it. I feel overtaken by the titanic struggle of the body of the animal and the force of gravity (where the last one always wins), not to mention the mathematical precision of the movement that accelerates steadily, same with all and every goat, regardless of her condition. All this is great and if one likes it, fine. But there is torture, mistreatment, and abuse. Why are so many bullfighting fans, various of them progressive intellectuals, not capable of seeing this obvious truth?

Millás's description of the beauty of the goat's fall from the tower during one of the Spanish small town festivities in Manganeses de la Polvorosa, celebrated until its prohibition, establishes an intertextual dialogue with the similarly ironic Martín-Santos's description of the art of bullfighting in *Tiempo de silencio* (Time of Silence, 1962).[7] The dialogue points to surreptitious sadism and the cultural discourses that hide and separate one from seeing the suffering of others for the sake of the maintenance of privilege and complacent consumption of pleasure. Millás's essay opens the possibility that it is not "not seeing," but rather finding a way to see without admitting the truth that characterizes various progressive intellectuals' attitudes toward animal suffering, intellectuals who do not negate pain of others but minimize it and justify it in various ways. This is the case of Fernando Savater.

In his article "Las cornadas de Europa" (Being Gored by Europe, 1991a), Savater defended bullfighting as entertainment whose cruelty is destined to produce an aesthetic enjoyment, and thus, he argued, it is not a "needless" cruelty as some claim. As the argument goes, the art of bullfighting should not be sacrificed for the sake of uniformity in a multistate Europe. Earlier in the same year, Savater put down the animal rights movements in his critique of Desmond Morris's *The Animal Contract*, affirming, "Todas las disquisiciones sobre "derechos" de los animales son la parapsicología de la ética" (1991b, n. pag.) (All disquisitions of animal "rights" are a parapsychology of ethics). According to Savater, animals cannot have rights since they lack the capacity to control their instincts, and thus they are unable to assume moral duties.[8] During 2004–5, Savater, with his attention turned to negotiations with ETA and debates about education, did not publish on animal issues; but he returned to the subject passionately in 2010 with "Rebelión en la granja" (Rebellion on the Farm), where he announced, "Civilización humana se basa en el maltrato de los animals" (n. pag.) (Human civilization is based on the

---

7. *Tiempo de silencio* is analyzed in detail in chapter 3, and chapter 5 contains reflections on the little towns' festivities featuring animal torture, including the throwing of a goat from the tower of a church.

8. In *Justicia para los animals: Ética más allá de la humanidad* in the chapter entitled "Para qué querrán derechos los animales" (Why do animals need rights?), De Lora (2003) address this argument, explaining that similar to animals various humans are unable to assume moral duties, for example newborn babies, the mentally ill, and so on, still this incapacity does not make them devoid of rights. In a similar way, Paula Casal in her article "Los derechos homínidos" (2012) argued that animals need rights just as children do.

mistreatment of animals), implying that if humans stopped torturing nonhumans, they could turn against each other. He reminds his readers that Hitler was a vegetarian and a great animal lover and quotes a hypothesis that the Aztecs ate their enemies because they lacked big mammals that could constitute a satisfactory source of protein.

The main arguments of *Tauroética* (2011) in defense of bullfighting appear as Savater's counter to Peter Singer's ideas. The Spanish philosopher claims that bullfighting could only be condemned if we decided to "modificar la consideración habitual de la *animalidad*" (18) (modify the status of animality). He aims to show that animals do not deserve the same ethical considerations as humans do, due to their "inferiority." The question of an animal's pain, the main argument for animal rights activists, is considered briefly, but as Savater states, pain is a part of nature and as such is unavoidable in both animal and human existence. Savater shows that Singer's assertion that the suffering caused to animals by humans is not as natural or necessary as that which animals inflict on each other is contradicted by another one of his claims, namely, that all animals are equal. If all animals are equal, then humans should also be allowed to cause pain and suffering to other species, Savater argues. In other words, the pain and suffering that humans inflict on other species from their point in evolution should be considered as necessary as that which animals inflict on each other from their respective points in evolution. Savater's argument smacks of social Darwinism, as he suggests that human ethics should imitate biological laws: "Acaso los genes que aspiran a perpetuarse son legítimamente 'egoistas' pero los humanos no gozamos de la misma prerrogativa?" (40) (If the genes that aspire to perpetuate themselves are legitimately "egotistical," why can't we humans enjoy the same prerogative?). The way that the question is phrased suggests that ethics governing human society should imitate the law of natural selection regulating animal species, which is precisely the point of convergence with Nazi biopolitics. Savater seems to be complacent with a narrative of society where, as supposedly in the wild, the strongest triumph and weaker perish, and where the more violent eat and abuse those peaceful ones, even if we know from his other writings that he might not believe that.

Savater has cowritten a book on torture, *Teoría y presencia de la tortura en España* (*Theory and Presence of Torture in Spain,* 1982) and published several articles in *El País* on the topic in the context of the Basque terrorist organization Euskadi Ta Askatasuna (ETA). One of the greatest enemies of ETA, Savater, himself of Basque origin, claimed as recently as 2008 that "bajo ninguna circunstancia la tortura puede ser justificable o legal" (n. pag.) (under no circumstance can torture be justifiable or legal). Among circumstances that may tempt some to find exception to this rule, Savater mentions the utility of torture and the qualities of the prisoner who may himself be guilty of torture and death. This statement is contradictory to his ideas on bullfighting, where he believes that entertainment constitutes an excuse for inflicting suffering on an animal. Savater supposes this because the subject is not human.

Mayorga's play *La paz perpetua* (*Perpetual Peace*, 2008), staged in 2007, presents the categories of humanity and animality as constructed and movable. In this play, the consideration of species interchangeability that is so meaningfully missing in Savater's arguments, acquires a central significance. It is because we are willing to grant exception to tortures inflicted on different species, races, enemies that we become animals, and some of us end up being tortured as animals, the play suggests. In dialogue with the infamous photograph where the political prisoner appears animalized, the play establishes that a victim of torture is always human, and torturers, regardless of the political side they fight for, are "beasts." Thus it is only through abandoning torture and other forms of abuse in human/animal relations that the humanity can be regained. In other words, paradoxically, humanity may overcome its "animal" nature when it recognizes the equality of all animals (human and nonhuman) in pain and never inflicts suffering on purpose. The need to recognize the dark side of the animality of humans in order to overcome it is staged in the play: the characters that are the torturers are animals.

"ANIMAL HOUSE": TORTURE AND HUMANITY IN *LA PAZ PERPETUA*

As the play begins, we see bodies lying on the floor. When they begin to move and interact, we realize that these are dogs' bodies played by humans. They are thus, as in a number of other plays by Mayorga,[9] at the same time humans and animals. We find out from their conversation, accompanied by markedly dog-like behavior, that they are waiting for a competitive exam that will decide who will win the K7 collar and join the elite antiterrorist brigade. They are not prisoners as one might think, but soldiers, or rather, dog-soldiers of three different breeds. The scene is designed as a panopticon; the bodies of the dog-soldiers are visible from all sides. The circular space of the stage is surrounded by the public on one side and a balcony on which guards with dogs walk back and forth on the other. Under the balcony is a wall with a

---

9. Mayorga's plays *Palabra de perro* and *Las últimas palabras de Copito de Nieve* are analyzed in chapter 5.

number of closed doors. The menacing effect of the invisible spaces behind the doors grows as the play progresses. Behind one of them, those in charge deliberate; behind another, a political prisoner awaits torture. This is how the stage design metaphorically reflects the structure of the invisibility responsible for wrongdoing. On stage, between the two invisibles, the dog-soldiers—potential torturers—await to be tested both by their supervisors and by the public. The public becomes involved in judging the results of the exam and engaged in the process of making a decision when the dog-soldiers are presented with a choice: to torture or not to torture. But, the public is also faced with another question on which they hesitate throughout the play: are the characters human?

Soon the only visibly human character in the play enters the stage: a woman. She often appears holding one of the dog-soldiers on a leash (as in the picture of Lynndie England). The fusion of species and the highlighted differences between the superiors and subalterns—correlating here to humans and animals—is meaningful in the play. This particular construction of the characters allows us to understand that the Abu Ghraib photographs represent not only a split between races but also between species. While the picture of the female soldier holding an Iraqi prisoner on a leash like a dog received a number of sexual interpretations, there is no mention of sex in this play. Mayorga interprets it as a metaphor of a relationship between dehumanized power and "unpeopled" soldiers, as well as a metaphor of the relation between the "human" and the "animal" in general.

While the photograph of Lynndie England holding a prisoner on a leash represents the executioner as a human and the victim as an animal, the play ends up reversing this. The political prisoner who waits on the other side of a closed door is undoubtedly a human being. In this play, the prisoner is humanity itself, victimized and subjected to fear, torture, and death. The executioners are dog-soldiers, who end up attacking the imprisoned human. This metaphorical reversal opens up a different level of reflection on war and torture, leading the public to question the human and its relation to animality. It obviously dialogues with the famous statement by Agamben (2004) that "in our culture, the decisive political conflict, which governs every other conflict, is that between humanity and animality of man" (80).

The place devoted to sex and sadomasochism in other plays on Abu Ghraib, such as *Guardians* by Peter Morris (2005), *Why Torture is Wrong and the People who Love Them* by Christopher Durang (2009), and *El escuchador de hielo* by Alfonso Vallejo (Listener of the Ice, 2008), is taken here by comments on an insufficiency of education that limits the dog-soldiers' horizons and contributes to their hatred of other races and species. They constantly comment on each other's racial features. John John, one of the three dog-soldiers, is a mixed-breed product of genetic engineering. He displays the features of various species, each necessary for being a great fighter. He was trained in a military school, where he learned to show his teeth, disarm suicide bombers, attack demonstrators, not cry from pain, suppress hunger and thirst, hold his body in difficult positions, be sleep deprived, and bear extreme cold, heat, and waterboarding. At the end of the program, teachers told the dog-students that they would never be normal again since they had been prepared to fight to the death like gladiators.

In more general terms, John John's description of his training makes us reflect on the changes in the Western educational system, where not only soldiers but most people are trained to fulfill functions in society mechanically rather than acquire the knowledge needed for an ethical vision of life and an informed political participation, which amounts to a loss of humanity. People are educated not to know and not to understand in whose interest and against whom they act and why, as if they were dogs lead on a leash. Thus not-knowing and not-questioning is built into the professional education curriculum and methodology, which in the play is compared to a laboratory-like production of warrior-automatons and which is deeply torturous for the students. John John's skull was engineered so small that it hurts when he thinks or gets excited, and he is thus prevented from doing so, forced to keep taking painkillers to negate any attempt to transcend his design. This reduction in head size for a new generation of dog-soldiers symbolically represents the global powers' attempt to form *Matrix*-like people, who lack the mental resources to appreciate the real meaning of their situation. John John was taught neither geography nor philosophy, and during the exam, he is surprised when questions require thinking. He was formed to obey orders and follow authority and is incapable of making decisions. This might not be merely a futuristic scenario. Like this character, the soldiers involved in the tortures in Abu Ghraib invariably declared in court that they did what was expected from them. They refused to engage in a debate about the ethical value of their acts, claiming to lack the knowledge to discuss this.

Even though John John's education did not prepare him for it, the last part of the exam is an ethical question. As the woman tells the dogs: on the other side of a closed door is a person, who may have some information about a terrorist attack being planned on a civil population. They need to decide if the use of torture would be authorized in order to get the truth. The invisible space on the other side of the door, where the victim awaits to be tortured, is dramatized on stage while, during the longest part of the play, the dogs deliberate whether they should get to the prisoner and bite him or not. Rather than seeing the victim (whom we never get to see), the public becomes aware of the victim's situation by following the emotional struggle and rational deliberation of the dog-soldiers, who decide if they should become victims or victimizers.

John John takes a bunch of pills, he begs to be given a clear order, and then, confused, attempts to escape from the site. The second dog, Odin (a Rottweiler), closely reminds us of a certain kind of human involved in wars. He is a cynic and a mercenary, capable of working for whomever pays the best. He believes that money and power are the only things that matter, and that all talk about ideas and arguments is only used to hide that fact. He has an amazing sense of smell and is capable of recognizing mental states in humans, so we can suppose that he is right when he decides that humans want him to "bite" the victim. The third dog, Enmanuel (a German Shepherd) is the only one who received an education. He used to take care of a blind girl who studied philosophy. The lessons were read out loud to her by her father, and Enmanuel would listen to them and thus learned to think. He loved his mistress, and he volunteered for the antiterrorist brigades after she was killed by an explosion caused by terrorists. Enmanuel refuses to "bite" the victim for he does not want to be like the terrorists. He also does not want to give them cause to attack and betray the values that he fights for. Furthermore, he knows that the girl he loved would not approve of this.

The last minutes of the play consist of a condensed dialogue between Enmanuel and the military woman. She says that the detainee does not expect any justice and that he cannot hate more that he already does. She argues that the terrorist will not hesitate to use any method to destroy his enemy and that this is his strength. She alleges that even if there is no certainty that the detainee is guilty, the risk of a mistake is small in comparison to the risk of losing civilian lives. In answer to Enmanuel's argument that the detainee has rights, and that if they violate them, they will put their own rights in danger as well, the human responds by asking of what good would rights be if their bearers were dead. Moreover, if they die, there will be no rights for anyone else because they are the heart of the democracy that will then collapse.

The human tells Enmanuel that the day the danger disappears, the democracy that they are defending will pretend it is scandalized by their deeds even though now, knowing about it, it looks the other way. When she says that they know that they will be accused for what they are doing, but in spite of it, they have to do *it,* and it is obvious that she is talking about torture. According to her, soldiers are expected to do the dirty work that democracy needs, and then to accept the blame so that their democracy can remain blameless. This is portrayed to them as their heroic sacrifice. Democracy has to remain "clean" because, as the human says, the War on Terror is in fact "a metaphysical war" (65), a fight for souls. Because of that, at the culminating moment in the play, she takes a surprisingly long time to convince Enmanuel (the thinking dog) that Kant, his favorite philosopher, would approve of torture. The dog refuses to believe that, but for once, the human might be right.

Mayorga's *La paz perpetua* (2008) questions the real meaning and practical realization of Western ethical and political idealist philosophy. In other words, it invites us to examine our ideals and to see that they may also need to be modified. Kant's moral and political treatise, *Perpetual Peace: A Philosophical Essay* (1795) had inspired Mayorga to write his play. Kant, despite being famous for his moral absolutism, believed that political practice would always be different from ethical ideals. In a "clausula salvatoria" preceding the treatise, Kant confesses that he is aware that the political theorist is a pedant whose empty ideas in no way threaten the security of the state, inasmuch as the state must proceed on empirical principles; so as the theorists are allowed to play their games, the practical politicians ("worldly-wise statesmen") can act unbound by their ideas. The philosopher thus acknowledges the gap between the words of a "political theorist" and the deeds of a "worldly-wise statesman," between the theory and the practice. Kant finds the awareness of the existence of this gap advantageous for a philosopher, because it guarantees the statesmen's tolerance for his "empty ideals." Thus the moralizing philosopher creates his idealistic principles while acknowledging that they will never be realized by politicians. The "clausula salvatoria" of Kant's *Perpetual Peace* shows clearly that human civilization is regulated by two parallel codes of rules—the official one, written into declarations and conventions, and the unofficial one, which governs behavior and especially politics through various kinds of exceptions where ideals are suspended for the sake of the interests. Mayorga's play should be read as an accusation that our civilization's discourses ultimately allow what they seem to be condemning, especially if their own supremacy is at stake. In the play, this hypocrisy constitutes our culture's "obscene underside."[10] In other words, the most harmful of all splits is the one between the words and the deeds, which is made visible by the dialogue between Enmanuel and the human.

On a more general level, this hypocrisy may be read as the result of the characteristics of human language. The three dogs in the play cannot help but do what they say, and say what they think. For humans, words mostly serve

---

10. Slavoj Žižek, "Between the Two Death" (2004, n. pag.).

to camouflage intentions. Using animal figures to represent soldiers from the Iraq war proves ingenious here, as the final effect of the play results simultaneously from the positive and negative connotation of "dog" as a metaphor for a soldier. Membership in elite army groups does not appear attractive if its value is expressed by the K7 dog collar. Army life, even in the elite formations, is represented as animalization, devoid of heroic and erotic dimensions. On the other hand, the comparison of soldiers to animals in the service of a human who orders them to bite disarmed prisoners renders them as victims of an evil manipulation, which they do not suspect, and thus provides for their partial innocence. They are simultaneously victims and victimizers. Looking at the dog-soldiers and hearing their exchanges, the public sees that people are turned into animals to be slaughtered like animals by other people who are turned into the animals that slaughter them. One of the closing exchanges between the human and the dog-soldier, Enmanuel, constitutes an interesting commentary to this extent:

> Humano: Nunca el perro fue tan necesario al hombre. Distinguir entre lo justo y lo injusto, eso hoy sólo puede hacerlo el corazón de un perro. La humanidad está en peligro, no nos abandonen. Estamos luchando contra animales.
> Enmanuel: Vencerán si nos hacen actuar como animales.
>
> The human: The dog has never been so important for man. Humanity is in danger, do not abandon us. We are fighting against animals.
> Enmanuel: They will win if you make us act like animals.[11]

When Enmanuel says "they will win," he refers simultaneously to the terrorists and also, perhaps without knowing it, to those mysterious supervisors who do not appear on the pictures but are in fact responsible for wars and torture. This sentence implies ominously that if we let "them," those invisible powers, turn us into animals, nothing will stop them from having utter control.

The play once ended with the death of all three dogs, gunned down. None of them succeeded in the exams; neither John John, for he could not decide without an order, nor Odin, because he dared to say that he did not care, honestly declaring his cynical attitude toward the world of ideals and discourses, nor Enmanuel, because he decided to be faithful to the ideals of those discourses. They are liquidated as useless. This ending of the play suggested that

---

11. Mayorga, 65. Translations are mine.

the only way to success is through a mastery of hypocrisy. Only those who know how to reproduce discourses of power in a convincing way, while acting as power acts, can succeed.

But Mayorga has now changed the ending of the play: the human makes a sign, John John and Odin prepare to attack the prisoner, and Enmanuel stands in their way to prevent them. As he does not move, they kill him and proceed to attack the victim. Thus he becomes distinguished as the one who poses a resistance to evil but also who perishes because he did not understand in time the discourses of the world surrounding him, for he truly believed in Kant's morality. In this new ending, through his act of resistance, Enmanuel acquires in the eyes of the public the kind of humanity that he as a Kantian holds so high. John John and Odin survive to appear in the pictures from Abu Ghraib.

If the kind of humanity that Enmanuel achieves in the play can perhaps be defined through a capacity to have ethical ideas, fight, and even die for them, the play makes it clear that the discourses that surround us present us with a great number of false "ideals" in which we blindly trust, not seeing that they are constructed not to be followed or that they are equipped with special clauses that provide for exceptions in difficult political circumstances or against those who cannot speak: animals, captured enemies, and disobeying soldiers. The play calls for a revision of human ethical ideals and for an education that would prepare us to question them by seeing through the linguistic manipulation to which we are subjected. Mayorga suggests that Abu Ghraib is a logical outcome of our language's dishonesty, which allows for the split between words and deeds, justifying torture and war, as addressed in Kant's "clausula salvatoria."

CONCLUSION

The scene in Mayorga's play when the woman attempts to force Enmanuel to destroy a possibly innocent life for the sake of the security of democracy and its citizens exemplifies the overgrowth of the system of protection. The fact that, as the women says, democracy expects that "these things" be done in hiding so that it remains clean, points to the hypocrisy of the political discourses, based on Kant's distinction between ethics and politics. It can be argued that the first instance of such a hypocrisy is the exclusion of animals from ethics that have serious consequences for all the animalized. The synergy between the debate on bullfighting and on the torture of the prisoners of the War on Terror in Spanish media revealed that the human treatment of animals is intimately connected to the cruelties of war. It showed brilliant realizations by Spanish intellectuals that both the torture and killing of animals and of humans in special circumstances are based on the same hypocrisy that distinguishes between ideals and practice.

Mayorga's play notes that biopolitical consideration is split between citizenship and "bare life," the latter constituted similarly by other species and other civilizations. As a result of this often invisible discrimination between the life destined to be protected and the life to be killed, war and war's atrocities are produced. As the human in *La paz perpetua* states, we are always "fighting against animals;" enemies in war are like a different species, and a different species can be agricultural goods, the object of the hunt, or cruel entertainment. As Judith Butler observes in *Precarious Life* (2004), due to this form of racism in the American and other Western press, most non-Western lives lost in the Iraq war were not mourned. It was as if they were not human or as if they had not lived at all. They were pushed behind the screen. This extended metaphor through which the official political/ethical discourse slides toward the invisible "dark underside" performed in the photographs from Abu Ghraib, reveals that what turns our civilization into a "culture of torture" is the tolerated dishonesty of discursive strategies. The ethical principle of respect for all life is compromised through different contextual codes, where life is segregated in hierarchical modes through the categories of race, gender, and citizenship, and where only those on top stand a chance not to be mistreated. The play suggests that in order to challenge the rhetoric of war it is necessary to see and then to question and restructure these categories as well as the hierarchies that formed them. It shows that the human/animal divide is a socially constructed artifice that constitutes the basis of the rhetoric of violence and abuse. In this sense, it is, as Savater states, an essential feature of human civilization. The question remains, do we want to keep it?

As Thomas Nagel explains in his essay "Ethics without Biology," there are two understandings of ethics: one behavioral as a description of norms governing every culture, and the other theoretical, as a field of rational research led by philosophers whose motivating idea is that "there is always more to be discovered" (144) and that "our current institutions and understanding . . . are only a stage in an infinite developmental process." In the first sense, biology and traditions seem to be the most relevant concepts, as our beliefs and behaviors are conditioned by our biological instincts and customs. The second kind of ethics progresses through a subjugation of traditions and institutions to examination, criticism, and transformation. In considering the tentative answer to the question of whether we want to keep torture as a part of our culture, each respective ethic would bring a different answer. The behavioral ethics approach, characterizing a great number of today's pedagogic and academic publications, would be interested in illuminating the meaning of torture in the given cultural context. Nagel's scholar, however, shall look critically at traditions, laws, and their meanings, aiming at a constant improvement.

While Derrida believes that any immunity for the sake of a community necessarily involves an autoimmune process, Esposito imagines a possible positive immunity that he explains with the metaphor of the relation between a mother and a fetus and that, as Campbell and Sitze (2013) write when paraphrasing him, "will run from men to plants, to animals independent of the material of their individuation" (xix). The slogan "torture is neither art nor culture" coined by the anti-bullfighting movement, can be understood as a step toward this ideal alternative construction of relations with other forms of life, as it expresses the desire that torture of any kind of living being should not be validated by cultural discourses.

CHAPTER 7

DIE OR LAUGH

BIOPOLITICAL CRISIS IN *BIUTIFUL* AND *NOCILLA EXPERIENCE*

"Pueblos del mundo, ¡extinguíos!"

Ya no hay trilobites en el mar;
en Siberia no queda ni un mamut;
las ballenas desaparecerán,
así que, humano, ya sólo quedas tú.

Pueblos del mundo, ¡extinguíos!
Dejad que continúe la evolución;
esterilizad a vuestros hijos,
juntos de la mano hacia la extinción.

Sonríe cuando te vayas a fosilizar,
que no piensen luego que lo has pasado mal.
Procura extinguirte con clase y dignidad;
piensa en el museo de historia natural.

La edad de los insectos llegará
y con ojos compuestos me verás;
te detectaré con antenas y radar;
zumbaré de placer al procrear.
(Siniestro Total, 1982)

> There are no more trilobites in the sea; / no Mamuts are left in Syberia; / whales will soon disappear, / so, human, only you are left. / Peoples of the world, get extinct! / Let evolution continue; / sterilize your children, / let's walk toward extinction holding hands. / Smile when you fossilize, / so that they know you had fun. / Get extinct with class and dignity; / think about the museum of natural history. / The time of insects will come / and you will see me with compounded eyes; / I will detect you with antennas and radar; / I will buzz from pleasure when we procreate.

What is interesting in this song is the unintended but also unavoidable conclusion that the extinction of species on Earth leads also to an existential and political crisis that is expressed as a loss of future. If our pleasures that we are not ready to give up lead to the death of all these beautiful animals, even if we could survive somehow, we can only think of our own demise and extinction rather than progress and improvement. The song suggests that even if there were some future for human life on Earth, there is no spirit that has led to the triumph of current civilization. The spirit that believed in human emancipation is now displaced by a mood of waiting for the end. What we should do while we wait, though, is up for debate.

In the previous chapter, the synergy of the debate on War on Terror and the debate on bullfighting in Spain has brought to public attention the realization that atrocities committed on humans are a result of the biopolitical rhetoric that excludes animals (and the animalized) from ethics and politics for the sake of *progress* and *civilization*. This chapter discuses how the biopolitical discourses that exclude nonhuman, and not-sufficiently-human, life from ethical and political concerns have led to the ecopolitical crisis that threatens life on this planet. In this context, it focuses comparatively on two works employing different strategies of concern over the current crisis: doom and humor, both containing a mood of disquiet. Alejandro González Iñárritu's *Biutiful* (2010) signals the invisible processes of mutation surrounding us which are expressed through disquieting sensations and that verge on apocalyptic intuitions about the demise of human civilization through self-destruction. Disquieting realism is also present, however, in a lighter mood, filled with irony, sarcasm, and various shades of humor in Agustín Fernández Mallo's *Nocilla trilogy* (2006–9) and especially *Nocilla experience* (2008). The discourses that these works adopt challenge those conceptual structures that regulate our understanding of reality and are responsible for establishing a lower value of life in the present biopolitics. Each possible scenario of demise and survival in the following works brings its own ethical dilemmas and various degrees of hope. The conclusion of the chapter considers the sources of hope in the doom and reflects on the ethical challenges that condition it.

Since the Club of Rome in 1972 raised public attention with the report entitled "Limits of Growth," pessimism, doom, warning, and even apocalyptic moods related to the awareness of depletion of the Earth began to permeate pop music, film, literature, and other arts. Siniestro Total's song "Pueblos del mundo ¡extinguíos!" (People of the World Get Extinct) from the album *Ante todo mucha calma* (Beyond All, Lots of Calm, 1982) displays deep pessimism toward a human and nonhuman future and yet a desire to love and laugh until the end. The song also expresses a feeling of guilt for the destruction of the nonhuman life by humans, and it sarcastically suggests that humans should "extinguish themselves" and let insects reign on Earth so that any remaining life can continue. The lyric makes us aware of environmental problems and alerts us to their dangers, and its dark humor permanently marks certain habits and actions as questionable.

Years have passed since the Club of Rome's report and since Siniestro Total's song was recorded, and the crisis has only deepened. The environmental reports appear more dramatic every year, announcing not only that our civilization cannot develop infinitely as it has thus far due to the anticipated end of natural resources but also that the human economy has taken the planet out of equilibrium as the époque of a climate favorable for life (Holocene) has ended, opening in front of us an unstable future of unfavorable climate change that most likely will put an end to a significant part of life on Earth (Anthropocene). Scientists now predict that if the temperature on Earth grows beyond the tipping point of four degrees, civilization will most likely collapse due to the unrest following catastrophes, natural disasters, and loss of habitat. All these data show that the politics of life that justifies killing or failure to prevent the death for reasons of profit, has been globally erroneous. The economic, political, and environmental crises can be subsumed as a fundamental crisis of current biopolitics.

Environmental humanities scholars debate on what genres and discourses are most effective to talk about the ecological crisis in order to awaken the public and promote a change in attitude. Fréderic Neyrat (2011) believes, "We need to create a fiction of a catastrophe and act as if it were certain. [Because] we can, perhaps, avoid the worst if we believe that it is certain." (n. pag.) Timothy Morton's (2010a) concept of "dark ecology" follows a similar logic in describing the world in which the catastrophe has already begun to happen, having taken its course as the result of a mortal sickness from which there is no redemption possible. Dark ecology is characterized by processes, which occur silently and invisibly on a planetary scale. Morton (2013) calls them "hyperobjects." Hyperobjects are not only imperceptible but also hard for our common sense to grasp, but Morton insists that they are more real than what our common sense believes is "real" because they change our environment as well as us and transform the generations of life to come. Global warming, nuclear waste contamination, and sixth mass extinction of species occurring right now on Earth are among the processes that humans have provoked and that will continue to transform Earth for thousands of years after our deaths. Morton argues that the uncanny feeling that has so often been rendered by contemporary art is caused by the subliminal awareness of the destructive character of the transformation that our bodies and our environments are undergoing. The feeling of disquiet is due to the perceived, although not fully realized, mutations that result in our sensation that something that cannot be seen is happening around us and inside of us. Disquieting realism (Beilin 2007, 2008), a mood that had appeared in Iberian literature during the 1980s in the context of *desencanto*, a feeling of dissatisfaction with the experience of newly achieved political and economic freedom, now, perhaps not without connection, expresses concerns with the environmental crisis.

One of the defining characteristics of crisis is that it implies the need for a change; however, it is debatable what kind of change needs to be implemented. Crisis implies various kinds of threats and losses but also possible trade-offs. Insufficient resources may lead to more or less justice as scarce goods may be spread more equally or less. The crisis may also seem too deep to resolve, and so it may cause the feeling of doom and no attempts to change may be undertaken. Thus the rhetoric of crisis may generate passivity and sadness rather than being a challenge for transformation and adjustment. In this way, "crisis" may be taken advantage of by economic institutions to justify the intensification of the political and economical measures responsible for the problems, such as austerity, which increases the gap between rich and poor, and laws, which limit democracy in the name of security.

Ursula Heise (2010) discusses the need for finding alternative discourses and moods. She quotes Joseph Meeker's *The Comedy of Survival* (1974), where the author argues that comedy may be a more appropriate way to talk about environmental issues. While the discourse of doom and loss that is characteristic of tragedy is anthropocentric in its nature, comedy brings in interrelatedness and translates as change what a tragic author views as a loss. As a result, Meeker and Heise suggest that comedy may constitute a better tool for reflection, and it may more likely lead to a change in the public's attitude than doom. Humor, and particularly black humor, may indeed have an important role in subverting structures of thought that are responsible for environmental destruction because, as in Siniestro Total's song, it ridicules thoughtless behavior.

BIUTIFUL'S CLOUDS AND MONEY

Biutiful (2010) is a Mexican-Spanish coproduction, whose action is set in Barcelona's darker side. In his critique of the film, Benjamin Fraser (2012) seeks to understand *Biutiful* within the frame of what he calls "urban cultural studies" and in dialogue with the urban theory of Henri Lefebvre and Manuel Delgado's studies of Barcelona (20). While "Barcelona is undeniably recognizable as the film's co-protagonist" (21), a similar story could be told about various contemporary cities, or even of less urbanized areas that attract human migration from poorer lands and whose diverse forms of contamination and corruption, driven by need and greed, provoke mental and bodily illness. From the perspective of environmental studies, Fraser's idea that the film represents the "human costs of modernity" (21), can be extended to the nonhuman framework of Morton's "dark ecology." All life (human and nonhuman) that appears in the film is transformed by human economy. This portrayal of life evokes the notion of the Anthropocene, the time when human actions have a decisive impact on planetary ecosystems, their biopolitics bringing about the destruction of the environment.

The Anthropocene, a concept coined by ecologist Eugene Sotermer and popularized by the environmental chemist and Nobel Prize winner Paul Crutzen, is that stage in Earth's evolution when humans become a quasi-geological force. The beginnings of the Anthropocene are traced to the eighteenth-century Industrial Revolution, but it is noteworthy that the term appears when the consciousness of the transformation of the Earth's ecosystem becomes worrisome and when global biopolitics is perceived as destructive for human and nonhuman bodies and communities. The Anthropocene reveals an understanding of the history of modernity from the perspective of today's perils, but it also involves understanding of the present from the perspective of the feared future. In this sense, Neyrat (2011) refers to climate change as "a moment of danger," that kind of moment of danger that for Walter Benjamin brings a memory of the past in a flash, asking to be controlled for the purpose of the future. The Anthropocene contains the fear of the unknown consequences of changes, which may be imperceptible until it is too late and can mount in human-built environments, including the interiors of our bodies, as a result of the general corruption imposed by the route our civilization has taken. The mood of the Anthropocene is disquieting realism where the familiarity of human reality is subverted by intuitions and nonsensory perceptions of sinister mutations occurring under the surface of the visible. In the framework of "dark ecology," the sensation of disquiet reveals biopolitical crisis.

The buzzing sounds, the anxious movements of the flocks of dark birds in the sky, the ghostly shades, and the momentary presences of real ghosts with whom Uxbal is able to communicate give *Biutiful* a feeling of disquiet. There is a mysterious, eerie correspondence between the progression of the fungus on the ceiling and the growth of abnormal cells in Uxbal's body, as if the human flesh and the matter of its immediate environment communicate (see figure 7.1).

These two processes of deterioration are signs of a deep cancer of the environment caused by the toxicity of neoliberal "necro-economics" (Montag, 2013, 193), portrayed by the film as an omnipresence of rule by money. It manifests as moral and organic corruption, as physical and mental sickness, and as deprivation, poverty, and suffering. The morbid side of things seems to be creeping out onto the surface of the city, like the worm moving through the alabaster surface of a tomb in one of the brief scenes of the film. Although the most disquieting images, such as this one or those of the ghosts of the dead floating under the ceiling, last only a second, and though their brevity is such that after a single viewing of the film, the audience might not remember them, they strongly contribute to the general feeling of disquiet. The borderline visibility of these scenes directly affects one's mood but not necessarily one's knowledge. It is as if the director were making a statement that there is quite a bit there that the human eye may not register, not only in the film but also in our everyday reality.

One of the most striking features of *Biutiful*'s montage is its persistent alternation between short and medium shots of bodies and interiors and long panoramic shots of Barcelona, of its streets, sea, and sky, which point to their share in the drama represented in the film. The reflections and shadows of people and things, which also abound in the movie, often floating in the air as if it were water, point to how humans and objects "enter" the environment in which they move. The sequences of inside and outside shots often have a condensed metaphorical value, as in the case of the fish mosaic on the wall, whose scales are made out of money, or when the embrace of Uxbal and Marambra is compared to two smokestacks of the Sant Adrià de Besòs sadly throwing dark, poisonous smoke into the air. The CAT scan image of Uxbal's lower belly cannot but remind us of the growing fungus stain on the ceiling of his room. There are three shots of that decaying ceil-

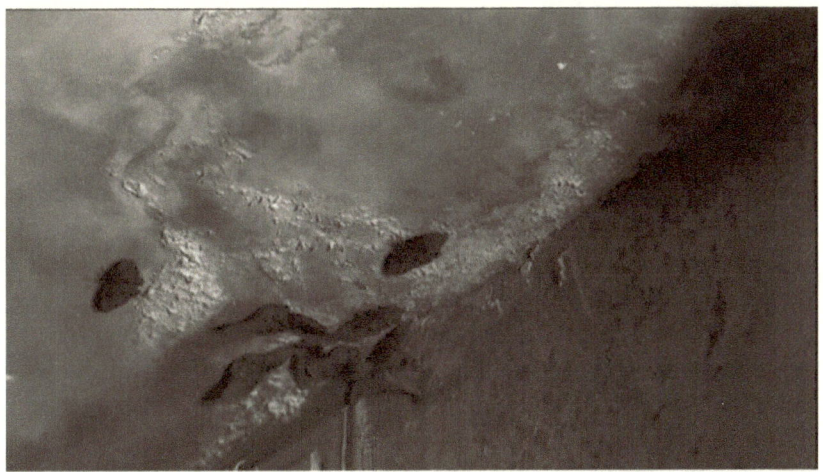

FIGURE 7.1
Wet stain growing on the ceiling of Uxbal's room

FIGURE 7.2
Screenshot from *Biutiful* showing smokestacks of Sant Adrià de Besòs.
This shot, right before the death of Uxbal, is the most dramatic.

ing spot in the film, each larger and more ominous looking, which mark the three stages of Uxbal's sickness and the three parts of the film: recognition, endurance, and surrender. This putrefying ceiling affects not only Uxbal, whose worried gaze appears every time in the reverse camera angle, but it also seems to be driving the development of events in the story. It is as if the growth of the fungus on the ceiling, intimately connected to Uxbal's sickness, progressed in parallel to the decay of the interiors of the corrupted city. From the formal point of view, there is not much new about it. Nineteenth-century realist novels often begin with long descriptions of interiors that metaphorically anticipate events. In *Biutiful*, however, the connections between protagonists and the environment are organic. There is no clear separation between human and nonhuman matter and all matter is somehow animated, as in the newly popular vitalist theories, of which Jane Bennett's is possibly the most discussed. Bennett's *Vibrant Materialism* (2010) argues that matter can be attributed agency and that we have an exaggerated notion of control over the objects that surround us. The change of perspective that she proposes leads to a blurring of the binaries between life and matter, human and nonhuman, organic and inorganic, which points to the need to change the political and economic relationship with things. Bennett coins the concept of "thing-power," which displays "the strange ability . . . to exceed their status as object and to manifest traces of independence or aliveness" (xvi). This secret morbid activity of things not accounted for by humans is another factor of *Biutiful*'s disquiet. The viewers of the film will remember Barcelona not as the artful, harmonious, and clean city from tourist postcards, but rather full of decaying matter—smelling of bloody urine, vomit, and of a fish that was frozen and defrosted four times as Marambra could not pay her bills and the electricity in her apartment was cut. In this famously sunny Mediterranean city, the sun is now scarce, and it seems to be fighting to get through the dense clouds in order to feed the remaining life. A caterpillar moving on a shaded branch seems to never get enough of it. But the "biutiful" of the title is not only about the city but also about life on Earth in general. The spelling mutation in the word "biutiful" reflects the mutations carried by all the literal and metaphoric processes portrayed in this film. The cancer that corrodes Uxbal's body, the disequilibrium that tortures Marambra, and the corruption of democracy are in turn connected to the economic and environmental exploitation of the non-Western world, provoking "slow violence" (Nixon, 2011), which condemns peoples to massive and often catastrophic migrations, followed by sickness, loss, and death. The film's mood is that of mourning for life that was left to die.

It is hard not to again think in this context about Hannah Arendt's reflection, commented on already in the Introduction, on the impossibility of enforcing human rights for those who are devoid of citizenship. The brutal treatment and the deprivation of the illegal immigrants reveal the connection between the national politics and the capitalist economy. The double logic of exclusion, based on national and economic difference, dehumanizes immigrants, placing them on the margins of life that is worthy of protection. They are taken advantage of like animals, so that the entrepreneurial nationals can increase their business profits. The Chinese poor who came to Spain in search of a better life only to die gassed in a basement of a factory are compared by the director to dolphins dying on the shore of a sea. In the face of a lack of an international institution that would enforce universal human rights, helping illegal humans can only happen on the margins of the law or against the law, as in the case of Uxbal's activities. These activities are not only illegal but also portrayed as "crazy," like the work of the animal right activists. In fact, the corrupted policeman compares Uxbal's efforts to help the Senegalese and Chinese immigrants to a circus worker's attempt to treat animals in a humane way that he paid with his life.

The obsessive positioning of human characters in their environmental context in the film montage and dialogues by comparing them to factory chimneys, fungus on the wall, dying dolphins or wild circus animals is meant to signal that they are organically connected and that the story of human sickness is also about the sickness of the world. In *Bíos*, Esposito characterizes this state of things in the following terms:

> Never before as today do the conflicts, wounds, and fears that tear the body to pieces seem to put into play nothing less than life itself in a singular reversal between the classic philosophical theme of the "world of life" and that theme heard so often today of the "life of the world." (2008, 11)

In the Anthropocene, it is more than ever "a human world," the intertwined and fused culture and nature, or to use Donna Haraway's concept "natureculture" (1999, 16), which is all embracing. In the "natureculture" environment that *Biutiful* represents, the decisive element defining relationships between people and things is money. González Iñárritu's obsession with the dark side of money is as long-lasting as his film directing career. In fact, his first half-length film, aired on national television, was entitled *Detrás del dinero* (Behind the Money, 1995). In *Biutiful*, nature is no longer nature, but it is also transformed by money, built into the human economy of life with cash as its contaminated food. Fish and humans in the film are made of money, like in the enormous graffiti representing a shark made of hundred Euro bills; their value is constantly negotiated in terms of money like in the conversation between Uxbal, his brother, and Mendoza about the price of the illegal Chinese workers. This is yet a more advanced biopolitics that Foucault (1978) talks about when life becomes not only a capital but the very foundation upon which the market operates (Pavone, 2012). *Biutiful* portrays the dark side of bioeconomy.

Consider the following sequence: After the detention of Ekwame, the Senegalese illegal street vendor, Uxbal, who had attempted to protect Ekwame and his companions by bribing a police officer, goes to the apartment of Ekwame's wife, Ige, and offers her money, which she initially refuses to take. She reproaches Uxbal angrily that he did not keep his word and failed to protect her husband. On the interior patio, lightened only by a few sun rays with rough cellar-like walls and a strikingly blue bowl in the background, Ige washes cloths by hand, looking at both Uxbal and his money with distrust and bitterness. As he leaves some money on a pile of folded cloths, the dirty crumbled paper contrasts with the freshly washed textile. Ige takes the money slowly with very clean hands, still covered with foam, as Uxbal walks out to the narrow street filled with scaffolding and the faces of African immigrants. The next shot is a panoramic view of Barcelona with the Sagrada Familia, also surrounded with scaffolding and cranes against a cloudy sky, and as the camera moves inward through the window, we see a medium shot of Uxbal's reflection in the tiles that cover the interior walls and then Uxbal himself, wearing a hospital gown the color of the tiles (the pale medical green of fading life), connected to an IV, a sort of crane itself, and looking out with his habitual gloomy expression. The suggested comparison between the scaffolding, crane, and IV puts on stage the complex relationship between the technoscience embedded in our current economic system and the sacred vulnerability of life. For example, the Sagrada Familia cathedral, in the words of Gaudi, was supposed to be "a mirror of its people" and to sustain the life of the family. But even though construction first began in the nineteenth century, using funds from people's donations, the church has never been finished, as its oldest parts already require restoration. According to Gaudí, "in the Sagrada Família, everything is providential," and if his vision reflects correctly God's designs, the donations would follow and the project would be accomplished (qtd. in Rowan Moore, 2011, n. pag.). Adam Smith's "invisible hand," however, has not done enough of spreading riches. The precarious character of the economy of gift contrasts, with the aggressiveness of the economy of profit. One hundred and fifty years after its conception, the cathedral is still a work in progress, while supermarkets successfully take over cemeteries. It may be that as a result of this dynamic, Uxbal's family life and even his body has decomposed.

In the next scene back in Uxbal's shabby apartment, the camera zooms in on the money (again, piles of crumbled, dirty paper) that he takes out

BIOPOLITICAL CRISIS IN *BIUTIFUL* AND *NOCILLA EXPERIENCE* 221

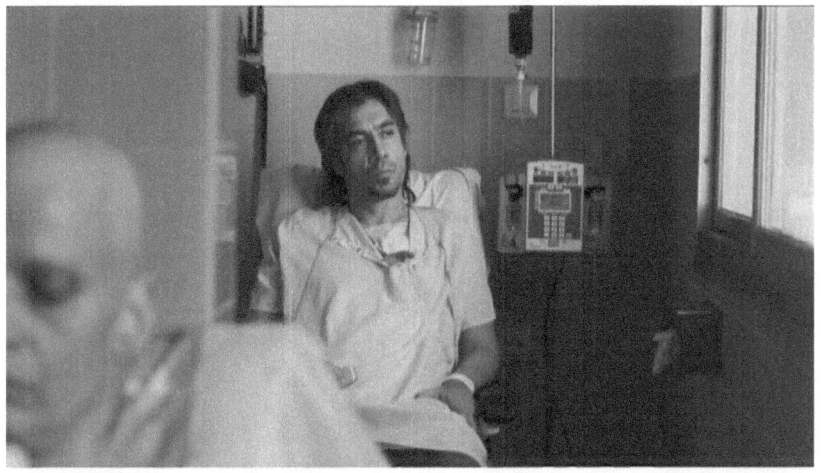

FIGURE 7.3A and B
Comparison between Uxbal's body and the city

off a sock and counts for several seconds. As he does so, we see his sickly face reflected again, observing him from a small mirror kept on the top of the dresser. Instantly we find ourselves in a street with a mural of a huge fish, likely a shark, with scales made of money, as Uxbal walks to buy heaters for the Chinese immigrants. Below the mosaic, there is graffiti that includes, ironically, a call to eat fruit and a homeless man pushing a shopping cart filled with garbage—which a close-up reveals to be the ruins of a city—right into

FIGURE 7.4
A shark made of hundred-euro bills

the fish's mouth. Money devours everything and transforms it into garbage that is fed to fish. The camera travels slowly to a shop window filled with several television screens showing dolphins pushed by waves to the beach and dying on the sand. This last image presages the future scene in which the Mediterranean Sea throws to the beach bodies of the Chinese immigrants killed by the accidental gas spill from the heaters that Uxbal is just about to buy. The camera zooms in on the prices of different heaters. Uxbal points toward the cheapest ones, and the shop assistant begins to load them to a car. As he watches him do so, Uxbal is taken by a sudden fit of vomiting, as if he were the fish-fed garbage. He bends over, getting sick right by the wall with the mosaic and the graffiti.

In this relatively short sequence, lasting only four minutes, money is constantly present, both materially as crumpled paper bills and symbolically in the mural of a city wall and the price of heaters. Things that are meant to provide for a better life end up killing because of the conditions of injustice in which they are produced. The homeless man with the garbage-filled cart, which he seems to be confrontationally pushing into the open mouth of the money-fish on the mosaic, manifests the deadly side effects of the money economy, or in environmental language, its externalities,[1] which include con-

---

1. Externalities are all the costs of production for profit that are not included in the standard economical calculations of GDP or of economic growth. Among these costs are

taminating waste but also inequality, poverty, and injustice. The fish dying on the screens of the television sets, sold by the electronic store where Uxbal buys the heaters, announces the future as death, and Uxbal's vomiting signals the poisoning already taking place in his body, which is suffering the side effects of chemotherapy. In the subsequent sequence, Bea tells him that chemotherapy is not a cure but a poison and that there is no solution possible to cancer but to put one's affairs in order and prepare to die. This statement contextualizes the sequence where Barcelona, with its Sagrada Familia "in treatment," is compared by the camera to Uxbal receiving chemotherapy, signaling the ultimate failure of medical science and of Catalan capitalism. Science and technology cannot restore the diseased of this world to their healthy lives because science and technology have been constructed in service of an economy that profits from destruction and death.

Money appears in the film as a sort of a cultural DNA that carries information into organic matter, mutating it, as in the street mural where the fish scales are already made of bills. Life is redefined, restructured, and organically transformed in money's terms as the interactions of various forms of flesh are regulated by the relations to their monetary equivalences and as live bodies are poisoned and dead bodies are displaced by commercial pursuits. When Uxbal negotiates construction jobs for the Chinese immigrants with the manager, Mendoza, the latter complains that the immigrants' poor health will slow them down and will not really allow him to lower his costs. In order to lower these costs, the workers have to sleep in a cold basement where their health deteriorates even more. Uxbal's attempt to provide them some warmth is also conditioned by the need to keep costs low, leading him to choose the cheapest heaters that turn out to be defective and kill all the workers with a gas leak. The defects of various products featured in the film, like the purses sewn by Chinese immigrants that do not even open, are caused by the condition of the sick and overtired humans who produce them, as well as due to the system where the inferior quality and low price of merchandise is designed to force the customer to shop again and again. Thus profit as the supreme objective contributes to the overuse of resources, the exploitation of the poor, and also the destruction of life by lethal products. The economy of the cheap excess is portrayed as death-dealing.

The noncontaminated life that "formerly known as nature" (Heise, 2013, n. pag.) appears in *Biutiful* only as a fantasy of a future and of the past.[2] It

---

destruction of forests, contamination of water and air, loss of biodiversity, loss of human life and health, and more.

2. In her review of Emma Marris's *Rambunctious Garden*, entitled "Encounters with the Thing Formerly Known as Nature," Heise writes: "[We] used to call it nature: forests, lakes,

is an unfulfilled dream of a happy family's excursion to the Pyrenees to see the snow. It is also a premonition of the other world where Uxbal passes to meet his long-deceased father. While the snow falls slowly, the dead father soothes his dead son by imitating the sounds of the primordial ocean from which, as we are reminded, everything emerged and to which, we can suppose, everything is slowly returning. The watery motif in various scenes of the film, such as the growing wet stain on the ceiling, the painting of waves in Marambra's apartment and a similar one in Uxbal's room, the fish in the aquarium and on the shower curtain, Uxbal's t-shirt that reads "Ocean City," ghosts floating in blue as if in the sea, are all signs of the ocean rising and of life on land slowly submerging, returning back to the matrix of life.[3] This connection between the past and the future so obvious for the dead father brings a smile back to Uxbal's face. This is the quietest scene of the film, where the sense of time is altered. Uxbal's father asks him, "Sabes que hace poco aquí no había nada, solo agua salada?" (Do you know that not long ago there was nothing here, only salted water?). This "not long ago" is a reminder of the relatively short time that human civilization has lasted if compared to the age of our planet. The fact that the film begins with this scene and also ends with it—although some angles are slightly different, the dialogue and the meaning are the same—places human life in the context of nonhuman life and individual life in the context of death. The sea, which appears in the film as a leitmotif, is a reminder of that relative brevity of human ascendancy and of our coming out from and, perhaps, going back to the salt water. The scene in which the Chinese workers' cadavers are washed onto the beach can be interpreted as a dark allegory of this vision of the human ending, this time rendered with dread and without consolation. In *Biutiful*, the balance of the matter, mutated by money, is precarious as González Iñárritu multiplies the signs that signal the catastrophe.

Money can be seen as yet one more "hyperobject," in the sense coined by Morton, of an overwhelming presence and process that transcends and conditions life with an all-embracing dynamic that every other object depends upon. Life assimilates money, runs on money, as if it were an addictive stimulant killing through overexertion. Ultimately money leads to exhaustion, not only of individual organisms but of the whole human civilization. When after a fight, Marambra and Uxbal embrace each other in a moment of tenderness,

---

foxes, butterflies, mosquitoes, dandelions. Soils and oceans. Seasonal cycles. Also floods and heat waves and the occasional hurricane. But no more" (2013, n. pag.).

3. I owe recognition of some of these watery motifs to one of my graduate students, Jenny Jeong.

she immediately asks him if he could give her some money and then if he could give her more. The system engineered by money and diverse human technologies is bipolar, just as this female character. The disco's euphoria, fed by cocaine and blinding lights, at moments in hellish, deadly green, is a direct counterpart of the dark dungeons inhabited by the poor immigrants. These dark spaces are the side effect of the economy of excess, which produces all sorts of cancer. When Uxbal asks Bea why *this* is happening to him, why he got sick, if this is some sort of punishment, she does not answer. Yet a few minutes later, she reminds him that the gift of speaking with the dead, which they both possess, was given to them for free and that she had told him in the past that they should give it back in the same way. She implies that Uxbal's sickness may somehow be connected with his selling his ability to communicate with ghosts. In other words, she connects cancer and money, pointing to the implicit metaphor that structures the meaning of the film and that posits money as the most carcinogenic material for both the individual's and the world's bodies.

An illusion that they will be able to make money brings immigrants from faraway lands, and its scarcity condemns them to suffering from overcrowded apartments, cold basements, and illnesses, and this hunger for money also produces their death. Money also shapes future and past time. When Uxbal is urged by his friend Bea to put his affairs in order before dying, he thinks of the money that he can leave for his children to ensure their survival. While Bea convinces him that the Universe will take care of them, he skeptically answers that the Universe will not pay the rent for their apartment. Preparing for death consists of counting money. Similarly, when the cemetery is liquidated to make space for a supermarket, it is as if money destroyed the memory of previous generations. The economy of life, where money becomes the single significant value, transforms the perception of time. It gets rid of the history that is not rentable and transforms it into the eternal present of production and consumption. The experience of time in this present is asphyxiating. We see Uxbal walk fast up and down the streets of Barcelona, struggling to pick up children on time in the evening. The time of rest before going to sleep is marked by tiredness and tension. After all his efforts, exhausted and sick, Uxbal managed to save for a year's rent, which seems like a lot to him but does not free him from worries. He only calms down after death, away from the civilization of money, seeing into the future when the primeval sea covers the dirty cities. In the Anthropocene, the ocean has already begun to enter the city through the clouds. There is no horizon on various panoramic shots of Barcelona and the sea, but rather the ocean and the sky blur.

CHAPTER 7

NOCILLA'S DARK HUMOR

Nocilla experience (2008), the second novel in the *Nocilla trilogy* by Agustín Fernández Mallo, can be interpreted similarly to *Biutiful*, as travel through the Anthropocene in crisis, whose mood is that of disquiet. Also here, the mutation of life is a source of uncanny sensation, and all the transformations seem to be marked by a tension between human greed that leads to wars, madness, and death. The novel conveys a darkly humorous conviction that the environment is silently waiting for human self-destruction so it can rebuild itself. As in the song "Pueblos del mundo ¡extinguíos!" (Peoples of the World Get Extinct) with which Fernández Mallo openly dialogues, the novel indulges in the descriptions of uncanny changes, anticipating the possibility of the extinction of the human race. Yet, just as in the song, for Fernández Mallo's narrator and his protagonists, this is not a reason to stop enjoying life. It is not necessarily a thoughtless enjoyment, however, since as Jacques Rancière (2007) points out, games and arts may impose a resistance to the system because immersed in their realm, we are free from the logic of the system of production. It is indeed with a witty, dark humor that the novel suggests that, as in Siniestro Total's song, we shall "get extinct with a smile" (n. pag.).

Nocilla experience

As in *Biutiful*, the catastrophe has already begun to happen in *Nocilla*'s world. Mark and Josecho, two of the novel's protagonists, withdraw from their reality, driven by a premonition of a future dystopia. Both live in shacks constructed from cardboard and packing materials on the roof terraces of the city skyscrapers and observe from above the constant mutation of life, as if they have moved already to a posthuman future. Another couple of connected characters, Antón and Antxo, compare Boris Sagan's *The Omega Man* (1971) with the catastrophe of the oil spill from the tanker *Prestige*, which sank close to the coast of Galicia, contaminating hundreds of kilometers of the beaches and killing all the coastal life. Jota, an artist who paints the chewing gum glued to London's pavements, is obsessed with Roberto Rossellini's *Journey to Italy* (1954), where Ingrid Bergman cries while looking at the bodies of lovers who had died in an embrace, covered by the lava from Mount Vesuvius. The haunting presence of the bodies of Pompeii, preserved for thousands of years, like the bones of dinosaurs, remind the reader of past catastrophes connecting to the fear of the future ones. A chain of catastrophic premonitions is formed by the apocalyptic images in the *The Omega Man*, where the last man on Earth explores the ruins of the past civilization and by the moment of anagnorisis in Franklin Schaffner's *Planet of the Apes* (1968) when the tip of the Statue of Liberty appears, sticking out of the sand. This past and future imagined catastrophes are contemplated as disquieting spectacles that can be even enjoyable if approached with a distance of humor. The pleasure that originated in the imagination of future catastrophes is, perhaps, the most questionable aspect of the discourse of *Nocilla experience*, although surreal dark humor often mixes an aesthetic pleasure with revelations of the dark sides of our civilization.

The most memorable allegory of the capitalistic necro-economics (Montag, 2013) is a self-sustainable slaughterhouse, which is also a shelter for the refugees displaced by the wars in the Middle East. It has been constructed by a young Armenian who, after having earned some money traveling, returns to his native country to build an eight-story building where 900 pigs are bred and slaughtered with the assistance of the political refugees from the regional wars. Twenty exiled families live on the upper four stories of the building with the owner occupying the penthouse. The fourth floor is designed for porcine birth, and it is where mother pigs live with their little ones. As they grow, the animals are passed down to lower stories until they get to the bottom where the slaughter takes place. The warmth produced by the animals is used to heat the building, and their gases serve to heat the water and generate electricity. This self-sustainable construction is a model of the biopolitical exploitation of life. As in our modern societies, life is skillfully protected and nourished in order to be exploited, killed, and consumed by those administering it from their penthouses. The vertical ordering of the slaughterhouse in which life's energy is literally sucked out by the system, and where organisms descend to the bottom of the hierarchy to die for the benefit of those that are still higher up but will come down in their turn, brings to mind the perfect interexchangeability of human and nonhuman bodies in this context. In fact, the eight-story building where animals occupy the lower four stories and humans the higher four, but where all life (except the owner's for the time being) moves downward, suggests that after all the pigs are killed the political refugees may need to be substituted for them in order for the house to stay heated. The slaughterhouse parable does not generate pleasure but rather a feeling of disquiet and abomination.

The disquiet in the novel is also deeply connected to a scientific perspective on life that leads to the blurring of a number of old dichotomies, such as body and its environment, nature and culture, live and dead matter, and even synthetic and organic, since on the molecular level everything is made from the same set of particles. As Miguel Espigado (2008) notes, Fernández Mallo opens literature to the technoscientific developments that the visual arts had reacted to earlier, challenging the conceptual categories that had not been updated in the literary field. At the same time, however, Fernández Mallo's thinking about technoscience is critical and pessimistic. He notices its limits, mistakes, and subservience to the system of slaughterhouses and wars that destroys more than it creates. Espigado notices that the focus on the material reality of things in resistance to overwhelming human domination imposes a preference for recycling. Literary recycling is performed in *Nocilla* through bricolages that open up a vision of a world of rhizomatic netting, where everything is connected as in Morton's "mesh" (2010a, 2010b).

As in *Biutiful*, in *Nocilla experience* nature and culture intertwine into a "natureculture," where human influences are omnipresent. As an example, "seguidores culturales" (27) (cultural followers) are animals, birds, insects, and microorganisms whose evolution has been determined by the relation that they have maintained with human societies. These forms of life developed their behavior in order to adapt to humans, placing themselves in a blurred space in between *oikos* and *tecne*. "Transpoetic narrative" (57), which *Nocilla experience* exemplifies, is defined precisely by its focus on these hybrid artifacts: mixtures of human and nonhuman, live and dead, present and past, real and imagined, which all merge together as if morphed. The diverse forms of art that appear featured in the novel are also environmental. For example, Harold, who runs around the American continent without stopping as a result of the depression he suffers after his divorce, is featured as an example of land art. Jota, who paints all the chewing gum spit out on the pavements of London in different colors, is an ecologist who cures the proliferating cancer cells of the city, sublimating them into artistic expression. Thinking of the unity of reality on a molecular level provides a way out of the killing divisiveness of the system driven by trademarks.

The disquieting mood emerges in connection with all that science still cannot explain. The "horizon of events" (52), an imagined circumference that surrounds the Earth somewhere in the cosmos, separates us from what we can never know. Yet even more mysterious is the fact that we often know certain things, although rationally we have no possibility of knowing them. One of the brief stories of *Nocilla experience* is about how scientific tests failed to show what was immediately recognized by an expert: that an examined statue, supposedly Greek, was not authentic (156). Both the undocumented knowledge or, in other words, the unknown known, and the known unknowns are manifested by a randomness of life, also known as luck. For example, in the evolutionary changes that have occurred over millions of years, randomness has obsessed biologists for a long time.[4] While random genetic changes are a condition *sine qua non* of evolution, randomness as a characteristic of the environment has been unsettling for scientists who believe that "science is an enemy of chance" (Appleman, 1970, 559) and that it needs to be freed from the element of chance to achieve the certainty to which it aspires. *Nocilla experience* moves toward a vision of complexity suggesting that what is thought to be random may in fact be a very complex combination of various contingencies. In other words, "random" is in fact an unknown where a meaning can reside. The randomness of events that make it impossible for the science to interpret the world more completely is a source of a humorous enjoyment of life and possibly a source of hope in the novel. The appreciation of random chance creates an illusion of a freedom from doom. This is why the game of Parcheesi, based on luck, obsesses various protagonists in the novel.

The most important source of the disquiet in *Nocilla experience* is, as in *Biutiful*, the perception of a human world progressing toward its demise, due to the self-destructiveness of human civilization. The mesh of connections between present, past, and future woven all around the globe follows the path of catastrophes. As in González Iñárritu's film, in the novel, this is a world in decay. In both cases, the framing scene and the most significant referent for all that happens is death. While in *Biutiful*, reality is marked by sickness and madness, in *Nocilla experience* the most frequently returning motif is that of the growing madness of a soldier in the Vietnam War from Frank Coppola's *Apocalypse Now* (1979), followed by a dead body floating in a lake, which having absorbed water, increased its volume and doubles-up like a tea bag. In both works, a prominent place is occupied by cancer, which in *Nocilla* consumes one of the main protagonists of the novel and proliferates in the city spaces obsessively "cured" by Jota. The difference lies in the vision of a possibility of transcendence. In *Biutiful*, beyond human life, there is the sublime reencounter in a silent forest, slowly covered by snow, while in *Nocilla experience* there is only material consolation of all sorts, including art but also consumption. Notably, while the apocalyptic mood of the film leads to a hope of a sort of afterlife, the dark humor and sarcasm of the novel rob the reader of any possibility of transcendence, except in returning like a face from Bélmez,[5] desperately adhered to a cement wall.

---

4. See, for example, John Tyler Bonner, *Randomness in Evolution* (2013).

5. Bélmez is a town in Spain where mysterious faces appeared and reappeared on the walls of an apartment. The phenomenon has been associated with the traumatic events during the Spanish Civil War. Fernández Mallo used the event in *Nocilla experience* as a metaphor of a dialogue between the past and the present.

CONCLUSION: COEXISTENCE AND HOPE

In *Biutiful* prevails the doubt or even disbelief that, given the political and economic system based on self-interested gain and measured with money that is now determining human behavior and regulating politics, there is any chance of preventing the final catastrophe. In this state, an empathetic coexistence, one that alleviates each other's suffering, is the main ethical imperative. In this sense, Uxbal's figure as an intermediary between the living and the dead, and helping both, is a priest-like or even Christ-like figure. All his activities, although mostly illegal, are focused on helping marginalized people to find work and avoid deportation and harassment. He maintains himself from the money that he receives as an intermediary in bribing authorities or in exchanging favors between leaders of different immigrant mafias, so he can also be seen as a parasite. But a parasite is precisely a master of coexistence as it usually preserves the bodies that it feeds on. Uxbal, the parasite, allows others to live on him as well. As requested by Ige, he lets her move into his apartment even though sharing of the very small space with her crying baby is hard for someone as sick as he already is at that time. His young son, Mateo, does not understand why his father let Ige move in with them, and Uxbal says that he helps her simply because she needs help. These words can be viewed as an ethical motto of the film. Applied universally, this principle could alleviate some suffering. Shortly after Ige moves in, Uxbal loses his remaining strength, and Ige takes care of him and stays to take care of his children. In *Biutiful's* vision, only empathy and painkillers can help humans on their way back to the primordial ocean.

This focus on the planetary mesh of symbiotic connections, whose randomness appears amazing and overwhelming, also leads to a focus on coexistence in both space and time in *Nocilla experience*. For example, the story about a town in Kalmykia, Ulan Erge, where the winter was so severe that buildings were covered with sheets recycled from the Siberian concentration camps, can be read as one of the metaphors connecting the present, past, and future generations in an environmental sense of a mesh where everything matters. Those sheets stained with bodily fluids and graffiti, marked with the suffering of prisoners were turned into a sort of skin of human houses after the camps were closed, as if the people who are alive were covered with the skin of the dead. When the spring comes, one old man gets so attached to this connection that he does not want the sheets to be removed from his house even though the cold has ended. He thinks that "si la gente se diera cuenta de que una de las moléculas de la piedra que lleva en el riñón perteneció quizá al ojo de algún dinosaurio, o de que ese fragmento de cinc que ahora corre por el torrente sanguínea de su vecino estuvo en la orina de Alejandro Magno, o que una particula del suero de la leche que fermenta ahora en sus estómagos la mamó antes un cordero en las montañas de Sahara, no le quitarían a él su galería de huellas humanas" (147) (if people realized that one of the molecules of the stone that he has in a kidney belonged perhaps in an eye of a dinosaur, or that the fragment of zinc that now runs in the blood of his neighbor was once in the urine of Alexander the Great, or that one particle of the milk that ferments now in their stomachs was sucked before by a lamb in the Sahara mountains, they would not remove now this gallery of human marks). *Nocilla experience* shares with *Biutiful* the belief in the need of ethical coexistence. The "little goodness" can mitigate human doom on the planet ridden by suffering, just as the American marine does not throw the grenade inside the Iraqi home, but rather, moved by the gaze of a women that he sees, lowers his arm and goes away.[6] He then pursues her and becomes her husband and a father of her child, and they are happy together. The "little goodness," or little love in this story, however, can do little confronted with the global conditioning of the system. When the child of the American marine and of the Iraqi woman is seven years old, another marine throws a grenade to their house and kills the whole family. The random coincidences that can cause doom, can, however, also be a source of hope, for example, when a group of prisoners from Auschwitz randomly survive the final killings in the camp while hidden in a dark corner playing the game of luck, Parcheesi.

Yet another way out of the general doom in *Nocilla experience* is hybrid art, which queers the concepts supporting the deadly system and the dark humor arising from the knowledge of contingencies that subverts this system from within. In this sense, *Nocilla experience*'s attitude toward today's capitalism is more playful than *Biutiful*'s, although it is also very critical of it. Chapter 100 ends with the conclusion that "la vida es un anuncio de teletienda al que le han eliminado el producto anunciado" (151) (life is a publicity ad whose product has been eliminated), which establishes dialogue with Adorno and Horkheimer's famous article "The Culture Industry" (1944). Consumption does not bring happiness even if it promises it, but there is no way out of the system, and thus even its critics function within its frames. Josecho, one of the main protagonists of the novel, whose character may constitute Fernández Mallo's self-reflective mocking, is a writer who in order to promote his novel suggests to his editor that publicity may be financed by a cloth-

---

6. "Little goodness" is the term used by Emmanuel Levinas (1969) in his reflection on the book by Vasily Grossman depicting the Second World War atrocities. During those terrible times, only "a little goodness" of one person helping another could alleviate the general suffering inflicted by the terror.

ing company (Chanel is chosen) in whose clothing the author would pose for all the posters and advertisements. This constitutes a metareflection on the *Nocilla experience,* inviting us to think that the publication of Fernández Mallo's novel could have been financed by the producer of the popular chocolate "nocilla."[7] Indirectly, literature is also partaking in the system of production and consumption, and as such, it is coresponsible for its casualties. In this sense, comparing nocilla to a drug, as Germán Labrador Méndez does, makes sense. Fictions and drugs are products of the biopolitics that operates at the expense of the addicted bodies. In his analysis of the video for the song "Nocilla la merendilla" (Nocilla, the snack) by Siniestro Total, Labrador writes about "una generación, que chocopasta mediante, sobrevivió al franquismo para morir de sobredosis en su entrada en la democracia" (n. pag.) (A generation that survived Franco['s] times thanks to chocolate paste in order to die of overdose when democracy started). This logic continues, and our generation's perception of the injustice is also muted by all sorts of sweet pastes that are fed to us by mouth, eyes, and ears. There have been various nocilla generations, and the novel contains a muted complaint that there is no way out of nocillas.

This interpretation needs to be nuanced, however. Both the title of the novel that the reader has in her hands, *Nocilla experience,* as well as the one that Josecho, one of the most important characters, writes, *Ayudando a los enfermos,* evoke song titles of Siniestro Total. The last song, "Purduy," on the CD entitled *Ayudando a los enfermos* (1982) ridicules the brainwashing manipulation of the Western cultural production by mocking the British television series *The New Avengers,* where Purduy, named after the arms factory, is a female spy working for the British intelligence. The golden rules of Samurai life extracted by Fernández Mallo from Jim Jarmusch's *Ghost Dog* (1999) are presented in *Nocilla experience* in the same mocking tone as Siniestro Total's "Purduy" song. The most intense intertextual dialogue, however, is established with "Pueblos del mundo ¡extinguíos!" (Peoples of the World Get Extinct). It is not an accident that a video for this Siniestro Total's song contains a number of images crucial to *Nocilla experience:* a T-Rex skeleton in the museum, the bones of two lovers unearthed in an embrace, and the head and the hand of the Statue of Liberty sticking out of the sand in the last scene of *Planet of the Apes.* The pleasure principle so important in *Nocilla experience* should be interpreted in the line carrying the central message of the song: "Sonríe cuando te vayas a fosilizar, / que no piensen luego que lo has pasado mal" (Smile when you fossilize, / so that they won't think that

---

7. In my research, however, I found no indication that this was the case.

you didn't have a good time). Having a good time even through consumption does not imply in the novel a passive acceptance of the system, but rather a means of resistance toward it. *Nocilla experience* develops its own readings of various products in resistance to the meanings inscribed in them by the hegemonic discourses of the market. In this sense, it may be read as an effort to deconstruct nocilla. For example, the shacks built atop the skyscrapers where Marc and Josecho live are made out of containers full of brand names. In one of the last passages, Marc opens a hole in one of them as if reaching for an illusion of a space of freedom outside of the system of consumption. In the postmodern cuisine of Brooklyn, plastic supermen and Polaroid photographs are diluted and served to customers as a part of their dishes. These are ways in which hybrid art contributes to a queering of the concepts responsible for destruction. *Nocilla experience* can be viewed as an act of what Simon During (1999) calls "'a transgressive' undermining or 'festive' overturning of routines and hierarchies" (10). This transgressive undermining is accomplished through "ironical mimicry, symbolic inversion, orgiastic letting-go, even daydreaming" (10). The significance of the construction of an alternative pleasure economy, which is accomplished through queering the economy of consumption, cannot be underestimated, because pleasure is the main motivation for action. A shack made of cases full of brand names announces a postconsumption era, and the habitats of Marc and Josecho, squatters at the top of a skyscraper, form a fantasy of a postcapitalistic cityscape. The argument that, by becoming a sort of best seller, *Nocilla*'s transgression becomes itself institutionalized, can be rebuked in Espigado's words: "A pesar de que la obra periférica triunfa en su revindicación estética y se desplaza hacia el centro, el apriopiacionismo no pierde su dimesión reivindicativa, pues en su naturaleza está la contradicción de criticar el sistema desde dentro del sistema" (In spite of the fact that a peripheral work triumphs in its aesthetic vindication and thus moves toward the center, its strategy of appropriation does not lose its revolutionary dimension, because it is in its nature to criticize the system from inside).[8] Fernández Mallo's novel, with its dark humor and sarcasm from within the system, is, however, not more hopeful for humanity's future than *Biutiful*, narrated from the margins of the system. Criticizing the system from inside, *Nocilla* opens readers' eyes to the spaces of randomness that provide for doom, and also for freedom and doubt, and argues for the need to take advantage of these spaces for the purpose of building resistance of an alternative economy of pleasure based on human relationships and hybrid art.

---

8. See Vicent Moreno's article "Breaking the Code" (2012) for further discussion on relations between "Generation Nocilla" literature in the market and in the literary field.

This call for resistance coexists in *Nocilla* with a deep pessimism that entertains the possibility of a total demise of human civilization and nature's capacity to revive after humans are gone. In the chapter about "el tercer paisaje" (the third landscape) new life emerges in the spaces from which civilization has withdrawn. These are roadsides, riverbanks, or sites abandoned by agriculture or industry where biodiversity spontaneously returns, transformed. If we think of the third nature's parallelism with the "third state," which, as Fernández Mallo states, was a social class that had no right to vote in the ancient regime, we understand that the third nature's lack of function in human economy pushes it out of the political map of life. Paradoxically, however, precisely because it belongs to the unmapped territory, freed from the biopolitical control, a variety of forms of life revive and flourish. If we consider that Karl Marx (1999) called human civilization "second nature" (146), then "third nature" acquires an ominous meaning of the state of affairs after the end of the Anthropocene, as the return to the primeval ocean in *Biutiful*. This posthuman stage would emerge mediated by the destructiveness of human activities. Reading *Nocilla experience* with Richard Dawkins' suggestion in mind, that our body is merely a vehicle for *selfish genes* (1989) to make more copies of themselves, we can think of human technoscience as a mere trick of nature to move itself onto a new level. It may be that it is not us who control nature but rather Nature—that in this sentence should be written again with a capital "N"—that will eventually take us in the direction in which it wants to go. The pleasure of consuming nocilla, which stands for general overconsumption of all sorts of junk, may be just viewed as a trick, programmed wisely by Nature so that, as Siniestro Total advises, peoples of the world get extinct with a nocilla smile and let other forms of life develop anew.

CHAPTER 8

DEBATES ON GMOs IN SPAIN AND ROSA MONTERO'S *LÁGRIMAS EN LA LLUVIA*

(WITH SAINATH SURYANARAYANAN)

IN ROSA MONTERO'S novel *Lágrimas en la lluvia* (Tears in the Rain, 2011) set in 2109, the destructive processes that today have only begun are very advanced. Due to global warming, one-fifth of the territory of Earth is inundated. As the coastal lands went underwater, their inhabitants migrated to higher terrains. The massive migration of the dispossessed caused wars in which millions died. In the world of the novel, the amount of clean air is so scarce that only the middle and higher classes can live in the zones that are relatively clean. The poor are pushed to places with such high levels of contamination that their lives are shortened by half. Sometimes they try to sneak through the guarded borders and breathe clean air for free, but the climate police detain them right away. Most wild animal species that we once knew have disappeared. For better or worse, science attempts to make up for all the lost life. For example, the last polar bear that died after swimming for many kilometers and not finding ice, has been cloned from the genetic material that was secured. The clone now lives in a huge aquarium-like space in a city mall. After it dies, a new bear will have to be cloned from her cells, given that clones cannot reproduce. The city parks consist of man-made trees that do not look like trees, but rather are wired constructions hanging over the heads of those who pass by in order to produce extra oxygen for their lungs. But while the average length of human life has not increased much since the previous century, science has greatly progressed in the areas connected to war and resource extraction. In order to fight wars and explore mines on other planets, bioengineers cloned stronger humans called "technos" or "replicants," as in the twentieth-century film *Blade Runner* (1982) to which Montero pays homage. Replicants or "reps" possess superhuman bodies, but being clones, their lifespan is considerably shorter. Initially, they were treated as slaves, but when the action of the novel begins, the reps have already rebelled, and as a result, they have been granted legal equality with humans. Although their equality is guaranteed by the Law of the United States of the Earth, they are still discriminated against as speciesism is an ongoing problem. For example, a very small percentage of resources is budgeted for research on a techno tumor that kills reps prematurely. An ambulance takes much longer when a techno's life is in danger. The use of the word "speciesism" to talk about this prejudice against reps in the novel is an obvious parallel to the question of animals in today's debates. Montero suggests that we will treat other forms of life, such as techno-humans, in the future somewhat similarly to the way we treat animals today. Thus the movement for animal rights may be also fighting for the equal rights of future replicants. The parallelism with race struggles is also obvious, as the future appears as a continuation of the past through repetition. Although the novel comments on climate change, contamination, the works of synthetic biology, and robotics, the most basic question connecting all is the problem of discrimination against nonhuman life, still unresolved in the future of 2109.

As Montero's novel signals, biopolitics is moving to a new phase, marked by the presence of biotechnology and synthetic biology, which are the vanguard of the business of life: bioeconomy. Biological materials and living organisms have become the matter upon which economy operates (Pavone, 2012). Biotechnologies, key generators of what Kaushik Sunder Rajan (2006) calls *biocapital,* transform and harness life itself in heretofore unprecedented ways that involve the transfer of genetic and genomic material between cells, often across species. They are already playing an increasingly central role in the social, political, and economic trajectories of societies in the twenty-first century. Their power is based on the sheer breadth of applications, from biomedical therapies and enhancements to "the gene revolution" and lab-grown meat in agriculture to novel industrial applications such as biopolymers, biofuels, and engineered nonhuman ecologies. Biotechnological processes and products are not only technoscientific but also fundamentally cultural, political, and social. In having the capacity to manipulate the genetic bases of life and generate novel life-forms, they trouble people's deeply held categories, values, and beliefs about life and what it even means to be human.

Biotechnology can already produce new hybrid forms of life containing the genetic material of different species, including humans. It entails the distribution of certain benefits and risks, which invariably lead to political questions about who will benefit and who will bear the burdens, and thus, as Jorge Riechmann's notices, they become a nexus of socio-political relations (2011b). Biotechnologies are social in the sense that their development is occurring hand in hand with interconnected processes of neoliberalization and globalization, which are transforming the industrialized societies of the nineteenth and twentieth centuries into today's knowledge economies (Jasanoff, 2005; Moore et al., 2011).

This chapter contrasts today's debates on biotechnology between philosophers and economists with fantasies of future developments emerging from a science fiction novel. In today's debates, we distinguish the discourses advising precaution and warning about the possible dangers posed by the genetic modification of organisms from those that are enthusiastic toward the project, seeing in biotechnology and synthetic biology hopes for human emancipation and the solution to the ecological crisis. Rosa Montero's novel *Lágrimas en la lluvia* (2011) constitutes a hybrid variant of both kinds of discourses, where from a futuristic perspective today's dangers are portrayed as historical catastrophes, but which also contains a fantasy that synthetic life could be in some ways capable of saving the remaining life on Earth. In order to explain Spain's generally less precautionary and more enthusiastic approach toward biotechnology than other European countries, this chapter analyzes the particular historical circumstances in which the bioeconomy has been adopted in Spain. The understanding of history as individual memory and as a global archive in Montero's novel is crucial for understanding the present dilemmas.

DEBATES ON BIOTECHNOLOGY IN SPAIN

In 2011, Wikileaks cables revealed that the Spanish government had asked the U.S. government to press Brussels into not passing additional laws regulating biotechnology (William, 2011). At the dawn of the third millennium, Spain has distinguished itself among the European Union (EU) and major EU nations as being the most enthusiastic supporter of all sorts of biotechnologies, even as the EU continues to adopt an ambivalent, precautionary stance in light of the risks posed by these emerging technologies (Todt and Luján, 2000; Ferretti and Pavone, 2009; Pavone, 2012). Proponents of Spain's probiotechnology policies situate biotechnology research and development at the center of an emerging high-tech economy, which is designated as knowledge-based, and portrayed as unshackled from Spain's retrogressive anti-Enlightenment past (Bertero, 2009). While every nation-state offers a potentially unique view into the political and cultural dynamics of biotechnology, Spain's case is particularly compelling. In contrast to other major EU states such as the UK, Germany, and France, Spain has consistently ranked amongst the most enthusiastic toward commercializing biotechnologies across the spectrum of red (biomedical), white (industrial), and especially green (agro-food) applications, where the contrast between Spain and the rest of Europe is the greatest. Close to two decades since the first arrival of corporate-owned genetically modified (GM) corn from across the Atlantic Ocean to European shores, citizens and governments of the major EU member continue to show deep ambivalence toward the commercialization of GM crops and foods. By contrast, Spain was for a long time the only EU land on which GM crops could be planted and is today the EU nation with the largest acreage of GM corn, and the third largest producer of corn in the EU. In 2012, GM corn occupied 76,000 hectares constituting thirty percent of all Spanish corn (Spanish Bioindustry Association [ASEBIO], 2013).

The question of why Spain differs in its orientation toward novel biotechnologies has invited contrasting explanatory discourses. The most prevalent range of explanations frame Spain as an emerging knowledge economy, whose investment in science and technology as a means of economic growth suffered in the "dark age" of Franco's dictatorship, and which only picked up since integration into the EU (e.g., Bertero, 2009). This discourse of Europeanization, promulgated by political authorities, scientists, promoters of the biotechnology sector, and Spanish media, situates the development of science and technology, and biotechnology in particular, at the very heart of the social and economic progress of the democratic Spanish nation. Humanists, artists, and philosophers in the past, including Américo Castro, Federico García Lorca, Miguel de Unamuno, and many others, defined Spanish culture as opposed to the preoccupation of the rest of Europe with materialist science and technology, and as specializing in matters of "spirit," nurtured through the contemplation of death, like during bullfighting celebrations. Thus Spain's emerging bioeconomy contributes to a cultural construction of a nation that no longer wants to be Europe's other; moving away from the cult of death together with all the other Francoist inheritance: the nostalgia for imperial Spain, thoughtless traditionalism, honor, national Catholicism, lack of respect for women, and more, in sum, moving away from the *Spanish difference*. However, this may be a somewhat misleading chain of associations. Science and technology has appeared as an unwanted good in various discourses connected to progress and civilization since the nineteenth century, and famously in Unamuno's vision, but they were also viewed as desirable, albeit lacking in national quality, by Regenerationists and others who wanted the Europeanization of Spain. In contrast to common belief, emerging historical works show that science and technology under Franco occupied a prominent space in the national economy (e.g., Camprubì, 2010; Néstor and Roqué, 2013), although political propaganda did not connect it to the Spanish national culture. Thus interpreting the embrace of science and technology in democracy as opposed to its supposed rejection under Franco may be not completely correct.

What the enthusiasts of biotechnology see as a step toward a biopolitics that is different than the one dramatized in bullfighting perhaps may still be reproducing other old patterns. Critics assert that Spain has implemented biotechnology policies in a hasty manner that renders Spanish society beholden to the interests of powerful multinational corporations and governments, and that this may be viewed as a continuation of the Spaniards' historically subservient orientation toward those in power, as in the bullfighting culture. In this sense, Riechmann states that "Spain is still different" (2011, 23), and in February 2014, Esther Vivas publishes an article in *El público* on GMOs based on the same idea and with a similar title: "Transgénicos: 'Spain Is Different.'" Nongovernmental organizations (NGOs), civil society organizations (CSOs), and academics such as Riechmann, Carlos Amorín, Salvador López Arnal, Francisco Garrido, and others call into question biotechnology's trajectory of development as being premised on neoliberal models of incessant growth that exacerbate existing levels of social inequality and environmental degradation. In this frame, the Spanish citizens' deference to political and scientific authorities regarding issues of science and technology, and in particular biotechnology, can be seen as a historical repetition of their subservience to those in power. In these counter-narratives, the politics of biotechnology suggests a Spain whose political culture remains as complicit with the agendas of powerful interests as under prior authoritative regimes in Spain's modern history—that is, Spain is still different from Europe. Also this vision needs to be problematized through analyses of discourses from Spain's past, when authority and science were located at the opposite poles, when hegemonic discourses in culture defined Spanishness as incapable and unwilling to pursue the rational regime of *proof* as in Unamuno's "Sobre la europeización" (On Europeanization) from 1906.

If the places of authority and power have shifted in recent history in ways that make the analysis of cultural attitudes toward them challenging, it is the attitudes toward risk that seem to have remained similar. In modern biopolitics based on deadly antagonisms, which he calls necropolitics, Achille Mbembe (2013) defines a spirituality shaped by an intimacy with death that involves flaunting indifference toward threats over life and implies acting to gain power by "risking the entirety of one's life" (164). This description fits perfectly the performance and the way of life of a bullfighter who had for many years been a leading model of masculinity for Spaniards. If this is no longer true, it is because of the abomination of cruelty toward animals and an advent of competing models of masculinity formed as a result of women's and animal rights movements. The culturally shaped love of risk may have remained unchanged, providing for openness and even enthusiasm toward genetic engineering of life, even though it constitutes a clear threat to life's integrity. While the necropolitics manifested in the culture of bullfighting has been replaced by the capitalistic necro-economics (Montag, 2013), these two forms of biopolitics have prepared the ground for a bioeconomy that demands risking life in the name of a possible future profit.

THE EMERGING SPANISH BIOECONOMY

Spain's biotechnology firms are clustered predominantly in the Madrid area and Catalonia with the rest distributed across Andalusia, Valencia, Galicia, and other parts of the country. This emerging biotechnology sector broadly features two kinds of firms: newer, small-to medium-sized firms of less than 50 employees, whose core activity is biotechnology and which are based on American-style start-ups by entrepreneurial scientists; and older industrial firms with 250 employees or more, whose core is based not on biotechnology but who have nevertheless diversified, entering into the biotechnology business (Díaz et al., 2002). With biotechnology marked by successive Spanish governments as an area of strategic action in the Spanish National Plan for Research and Development (e.g., see Bertero, 2009), and with support from the EU, Spanish biotech firms are seeking "to dominate the largest possible number of sectors" in Spain (Díaz et al., 2002, 37) and increase their presence in a globalizing marketplace. Toward this end, biotechnology firms are mobilizing new social relations with universities, public research organizations, investment banking firms, and large multinational corporations. Social, economical, and political life is increasingly organized around biotechnology, resulting in a number of implications.

There are three key aspects of this evolving web of relationships that shape emerging neoliberal bioeconomies in Spain. First is the rise of a new model of public research and development—"the ICREA model"—that seeks to turn Spanish universities and research institutions into economic engines. Established first by the semiautonomous government of Catalonia in 2001, the Institució Catalana de Recerca i Estudis Avançats (ICREA) is a highly regarded attempt to improve Catalonia's "innovation performance" by attracting internationally renowned scientists and scholars outside of the Catalan education system to conduct their research work in Catalonia (Technopolis, 2011). Through the virtual institute of ICREA, the Catalan government sought to circumvent perceived institutional, cultural, and economic barriers to recruiting the world's most talented researchers and jumpstarting the Catalan knowledge economy.

In the "traditional" Spanish and Catalan systems of research and higher education, recruitment and employment occur on the basis of "national civil service terms and conditions," which has been "endogamic" (Technopolis, 2011). Existing academic staff tend to recruit junior academics with whom they have close ties, who in turn end up with long-term positions in one place. As a result, higher education institutions are unable to offer competitive salaries and research support packages due to rules on national civil service salaries. ICREA has enabled public universities and Catalan research institutes to offer to its recruits "research-only," "permanent" positions with "competitive salaries" and the opportunity to "lead a new research group," seventy-two percent of whose focus is on science and technology and the rest on social science and humanities (Technopolis, 2011). ICREA research centers such as the Center for Genomic Regulation (CRG) are key primary partners of biotechnology firms looking to develop new products. Significantly, the ICREA model of internationalizing Catalan research and development is also tightly linked to the spread of an entrepreneurial culture of innovation that seeks to establish stronger connections between "upstream" technoscientific research and "downstream" potential market applications. It "actively supports" researchers who want to "patent their inventions," collaborate with industries and private sector enterprises, and generate "spin-offs" (Technopolis, 2011). As of 2011, only six stand-alone Catalan research centers accounted for sixty-three percent of all ICREA recruits, including the CRG, an international biomedical research institute created in 2000 by the Spanish and Catalan governments in Barcelona. This culture of academic innovation based on public-private collaborations and the commercialization of university research is set to spread to other parts of Spain with efforts such as IKERBASQUE in the Basque region and another in the Madrid region that seek to replicate the ICREA model.

It is noteworthy that during a period marked by a drastic downturn in key indicators of social and economic wellbeing, Spain's biotechnology sector has shown "highly positive results" of financial and material growth (ASEBIO, 2013). This is of course not to suggest that all is well with Spain's biotechnology industry. As the evolving biotechnology sector strives to become central to Spain's society and economy, across-the-board cuts in governmental funding for scientific research create new challenges and opportunities. The intensifying austerity stands to create the conditions for further rounds of the commercialization of Spain's university research environments. It increases the likelihood that Spain's small biotechnology firms—faced with the high costs of procuring state-of-the-art equipment, skilled personnel, and resources, and in developing a marketable product—will be relying on foreign investors for survival in a highly competitive marketplace. This sets up a situation, where the research and development programs and associated profits of Spain's biotechnology firms are steered not toward the needs of Spain's civil society, but toward the concerns of far-off venture capital firms and shareholders. Furthermore, the ongoing Spanish economic meltdown fuels the conditions for multiple mergers with existing small biotechnology firms and strengthens the biotechnology arms of a few multinational corporations that possess greater reserves of capital and privileged access to powerful actors of national and international governance. The link between biotechnology and a neoliberal model of research in the service of corporations changes biological life, society, and politics since bioeconomy intensifies the neoliberal transformation of society.

A second important characteristic of the Spanish bioeconomy has been the rise of scientized regulations. In emerging neoliberal knowledge economies, policymakers adhere to the recommendations of highly specialized scientific advisory bodies, which tend to exclude social and political criteria (Moore et al., 2011). Indeed, scientized regulations have helped push through the commercialization of biotechnologies without much public debate in Spain (Todt, 1999; Todt and Luján, 2000), even though the regulatory "red tape" entailed in obtaining commercial permits for any particular technoscientific product with ecological and social implications is a perennial source of teeth-gnashing for industries, including those in the biotechnology sector (see Díaz et al., 2002). Biotechnology came to have a privileged place in the Spanish science policy of the 1980s, in part due to the "ambitions of successive governments" after the transition to democracy, and also due to "a small but influential group" of biochemists and molecular biologists of international acclaim in the area of biomedicine who had "close connections with political and scientific authorities" (Luján et al., 1996, 182).

The challenge of governing biotechnology emerged in full force after Spain's admission to the European Commission (EC) with the EC's first directives (in 1990) regarding risks related to the containment and release of genetically modified organisms (GMOs) into the environment. In principle, the administrative authority for addressing the EC's directives in Spain lay with the Directorate General of Environmental Policy in the Ministry of Public Works, Transport, and Environment. However, in practice, the Ministry gave little attention to environmental concerns; instead, a provisionally formed body called the National Biosafety Commission (NBC), comprising of policymakers, laboratory-based biochemists, molecular biologists, agronomists, and chemists (but without ecologists), not only took an advisory role but also acted as the "Competent Authority" on GMO release by becoming the "*de facto* policymaker" on the Spanish government's decision to commercialize GMOs in 1996 (Luján et al., 1996; Todt and Luján, 2000).

In the aftermath of the EC's 1990 directives on GMOs, the governments of the UK, Germany, and the EU instituted a variety of systematic initiatives to enhance public participation of citizens and civil society organizations in debates about GMOs (Jasanoff, 2005), although these participatory endeavors in Germany were seen by NGOs and CSOs as being purely legitimation exercises on behalf of those in political power, where invited civil society groups could discuss the risks and benefits of GMOs but "could not call into question those technologies in themselves" (Ferretti and Pavone, 2009, 291). In Spain, unlike in Germany, the UK, and the EU, it was not until after GMO crops and foods became commercialized that a public debate erupted (Todt and Luján, 2000). Apart from the lack of serious consideration of public perspectives in the Spanish government's scientized regulation of biotechnology, there are relatively weak structures of public participation (Kitschelt, 1986; Todt, 1999), which may be considered part of the historical legacy of the authoritarian regimes that governed Spain before the transition to democracy.

A third important factor in the unfolding scenario of biotechnology in Spain is a "countervailing" one (Moore et al., 2011, xx), that is, through the rise of new social movements, NGOs, and CSOs that are all engaging with issues of science and technology; however, for a variety of reasons these movements have not been able to meaningfully influence politics in Spain in their struggles for reforms in the realms of environmental, ethical, social, and personal health. Since Spain's turn toward democracy, a number of environmental organizations have appeared and disappeared without consequence. Joaquín Fernández (1999) claims there have been 700–1000 such small organizations. In his words, "el fervor ecologista" (73) (the environmentalist fervor) was such that the Partido Ecologista Español (PEE) party was created in 1977, but it failed politically, receiving only four hundred votes in the elections of 1978. A great number of protests, especially antinuclear manifestations that passed through Spain around that time can be contrasted with the lack of positive political participation by the Spanish public. This may have been due to a lack of institutions of public participation. When GM foods and crops were commercialized in Spain in 1996, the situation was not radically different. In addition to the lack of institutional infrastructure, the inherited distrust of political authorities made NGOs and CSOs suspicious that formal participation might end up legitimizing the political authorities' policy decisions as it did in Germany (Todt, 1999; Ferretti and Pavone, 2009). Excluded from formal discussions about biotechnologies, Spanish NGOs and CSOs have taken a confrontational route in devising public media campaigns that directly address citizens' concerns about the risks entailed in various biotechnologies, calling into question the neoliberal system of technoscientific development on which novel biotechnologies have to date been predicated (Ferretti and Pavone, 2009). An analysis of the debate that has occurred in the Spanish media since GM products began to be commercialized suggests that, in contrast to institutions of public participation, media culture has been significantly democratized.

MEDIA DEBATES: THE PURPOSE OF BIOTECHNOLOGY

Based on an analysis of the *El País* archive, there appears to have been an incipient debate on GMOs that started at the end of the 1990s with a considerable number of articles appearing in 1999. On 28 May 1999, a brief editorial note in *El País* announced that Austria had forbidden transgenic (genetically modified) corn because emerging research showed that it kills beneficial insects, such as Monarch butterflies. On 29 October 1999, an article entitled "El cultivo de los transgénicos crece este año un 44%, según Monsanto" (Plantations of GMOs Grow This Year About 44 Percent, According to Monsanto) informed the public that among the countries that had opened up to GMOs were Portugal, Ukraine, and Romania—curiously, without mentioning that Spain that had done so earlier. The article, however, admitted that debates on the health effects of GMOs were intensifying. On 15 December of the same year, another brief editorial informed Spaniards that a U.S.-based ecologic group, the Foundation for Ecologic Development (FED), led in the United States by antibiotechnology activist Jeremy Rifkin, was claiming that Monsanto had violated antimonopoly laws. The article concluded that the controversy about GMOs cannot be reduced to a simple difference between the enthusiastic United States and precautionary Europe, since there a growing debate on both sides of the Atlantic. While the first publications presented biotechnology as a means to overcome world's hunger, as time passed this vision has been challenged.

It was only around 2001 that the wider debate on GMOs began in *El País* (In *La Vanguardia,* the most intense debate was in 1999). That year, among the many new words incorporated into the new dictionary of Academia Real Española, such as *zapear* (to zap), *guay* (cool), or *talibán* (Taliban), was the adjective *transgénico* (transgenic, *El País,* 11 Oct. 2001). Riechmann's book, *Qué son los alimentos transgénicos?* (What Are GMO Foods?, 2002), whose purpose was to inform and warn, was first published around that time. A great example of the optimism in relation to the desire for modernization can be found in an editorial focused on the newly elaborated list of the most-developed countries in the world, with Spain listed as the twentieth and where human development was connected simultaneously to technoscientific progress and the eradication of poverty. The article suggested, quoting the United Nations Development Program (UNDP) from 2001, technological development and the eradication of poverty can be simultaneously achieved through the genetic modification of crops. In this context, the leading role of Spain is provided by the subtitle of the article: "España se perfila como un líder potencial en cuanto a innovación y adelanto tecnológico, según el informe del UNDP hecho público hoy" (Spain is emerging as a potential leader in terms of innovation and technological progress, according to the UNDP report published today, *El País,* 2001). This publication suggests that the Spanish government's openness to GMOs is motivated in part by an ambition to have Spain among the world leaders in human development.

The same year, however, Javier Sampedro's interview with Hugh Grant, vice president of Monsanto, in *El País* points to a possibly different understanding of the relationship between the concept of human development and openness to biotechnology. During the interview, cunning questions by the journalist led Grant to admit that his company is not in fact concerned about remedying world poverty but rather about intensifying its own profit and that the European demands for transparency *forced* the company to reluctantly embrace huge changes in its *modus operandi.* Grant insists that it is "a myth" that Monsanto plans to commercialize "Terminator" grains (that become infertile after a year), not because they lack the desire to do so but rather because they did not manage to purchase Delta & Pine, the firm that originally patented this technology. The vice president of Monsanto characterized European distrust toward GMOs as a misplaced fear caused by the occurrence of mad cow disease, and as an exaggerated preoccupation with what they put on their plates. By the end of the conversation, the Monsanto spokesperson was praising Spain for having a more "scientific" attitude toward GMOs than the rest of Europe and being, next to the UK, Monsanto's most important partner. It is noteworthy that the willingness to risk health is equated with a more scientific approach.

In 2004 European consumers were rejecting GMO food, and food manufacturers published manifestos in which they renounced GM ingredients. Due to the global character of the food industry in Europe, this has occurred in Spain as well. In May 2004, *El País* published an article entitled meaningfully "Monsanto renuncia a comercializar el trigo transgénico por el rechazo del mercado" (Monsanto Gives Up Commercialization of GMO Wheat Due to the Rejection by the Market) and subtitled: "Greenpeace celebra la medida como una victoria para el medio ambiente" (Greenpeace Celebrates This as a Victory for the Environment). But in the same year, *El País* began to publish articles on successful biomedical experiments involving possible genetic modifications of humans. One of them features the research of Manuel Serrano, who after postdoctoral practice at the Cold Spring Harbor Laboratory in New York, returned to Madrid to start his own research group at the National Center of Biotechnology and produced mice resistant to cancer. This researcher, who seems to have fulfilled the early dream of Pedro in *Tiempo de silencio* (Time of Silence, 1962), announced his hopes that through similar genetic mutations, humans at high risk of getting cancer could be protected from it in the future. Ten years, later, no applications in human health have been announced, however. The difference between the mostly enthusiastic rhetoric employed in articles focused on biotechnology for medical purposes and a mix of enthusiasm and concern when agricultural biotechnology is at stake has been a characteristic of *El País* and *La Vanguardia* publications since 2004, with the debate becoming much more intense during recent years. In *El País*, out of the 282 articles on the subject, 200 have been published since 2011. The confrontational rhetoric in the debate can be categorized into two groups: the precautionary approaches and the emancipation discourses.

THE PRECAUTIONARY APPROACH

A principle of precaution specifies that if a substance shows risks to human health or the environment, even in the absence of decisive scientific evidence, it should be considered as harmful and avoided. In the case of the development of biotechnology in Spain, this principle has been applied by the government rather ambiguously after, not prior to, the commercialization of GM foods and crops, and with the implicit aim of pre-empting social conflict (Todt and Luján, 2000). Intellectuals such as Riechmann, López Arnal, Amorín, and Garrido are troubled not only by the risks posed by biotechnologies for the public and environmental health but also, and most importantly, by the conjoined development of biotechnologies and neoliberalizing, the globalizing capital in emerging bioeconomies. The spread of GMOs entails unknown risks for the environment where accumulating toxins can produce long-term synergies with other ambient environmental factors, producing effects that are hard to predict (Riechmann, 2002). Huge monocultures of genetically modified crops loaded with insecticides, herbicides, and petrochemical fertilizers may be destroying pollinators and other beneficial organisms. The chemicals accumulate in the environment, producing long-term synergies leading to unknown final effects (Riechmann, 2002). Quist and Chapela (2001) show that the artificially inserted genetic elements are finding their way into remote populations of plants that were thought to be non-GMO, putting into doubt technical and institutional capacities to contain the runaway flow of transgenic DNA. Because growers become reliant on the agrobiotechnology industry as they purchase GM seeds and associated chemical and mechanical technologies needed to fully express the GM traits in their crop, the food chain becomes more and more controlled by a handful of multinational corporations. In 2011, according to Riechmann (2011b),[1] only ten multinationals controlled close to seventy percent of the world grain market. GMOs constitute then not only a risk to environmental and human health but also a biopolitical risk of the privatization of life itself, providing a legal basis for the ownership of natural resources that are administered increasingly according to corporate financial interests rather than public and environmental interests.

In all of these fields, the concept of limits is crucial. While the productivist projects of biocapital aim at transcending the limits of nature, critics argue that we need to transform our culture to respect natural limits and take care of existing human and animal life. This discourse, again reminiscent of *Tiempo de silencio* where Pedro decides that it is the culture and not the genes that needs to change, has been taken up by the political groups to the left of PSOE. Also EQUO, a green left party founded by Juan Uralde in 2010, argues for a self-sufficient sustainable economy without GMOs but with new renewable energies. Here, however, certain complexities arise because some of the renewable energies, notably, are designed to be produced with the help of genetic engineering.

---

1. In the second edition of *Qué son alimentos transgénicos?* Riechmann updated the book including all that happened until 2011.

For example, as Alicia Rivera writes in *El País,* transgenic bacteria have been recently patented as a way to produce biofuel from marine algae. One of the protagonists of Belén Gópegui's novel *El padre de Blancanieves* (The Father of the Snow White, 2007) faces an ethical dilemma created by a similar transgenic technology. Should he take the job in the multinational company that researches this truly promising biofuel that could do so much good but is administered by a company that benefits from and contributes to a deeply unjust world order? The same dilemma is faced by environmentalists, because while green capitalism seems to be incapable of limiting the capitalistic drive to grow that ends up destroying life-sustaining ecologies, it seems equally impossible to overcome capitalism.

The concept of limits is also meaningful in the context of biomedical research. As biotechnologies blur the boundaries between disease-therapeutics and health-enhancement and generate novel life-forms, they create the conditions for overcoming the limits imposed by our ecological (biophysical) and evolutionary conditions. For example, the biomedical research on anti-aging interventions, which uses stem cell therapies and regenerative medicine, potentially reframes the very process of aging as a disease. At the same time, new biotechnological therapies include enormous financial stakes and interests that are inseparable from capitalistic processes, which for Riechmann (2002, 2011b) are characterized by a fundamental drive to escape all biophysical and biological limits imposed on the human condition. Riechmann calls for limited and strongly controlled use of GMOs based on judicial, ethical, political, and economical frameworks defined by the principle of precaution. His opponents, however, argue that human civilization can be viewed as a plight to overcome the limits of the body.

BIOTECHNOLOGICAL EMANCIPATION

Some of the most famous and influential public intellectuals, such as for example, Jesús Mosterín, systematically wrote and spoke of biotechnology enthusiastically, insisting on its scientific basis. In an article entitled "Quién teme a los transgénicos?" (Who Is Afraid of GMOs?), Mosterín (2000) argues that humans have been genetically engineering organisms for tens of thousands of years, crossing different species of fruits and vegetables, and nature has done the same; hence, there is nothing new about this procedure. He states that, according to scientific research, agrobiotechnology has not had any negative effects on human or animal health, while it has significantly increased agricultural production, making it possible to feed more people. In an article entitled "El hombre transgénico" (Transgenic man), Mosterín (2009) expresses enthusiasm toward the possibility of genetic modification of humans, justifying his lack of fear in a similar way to the case of agricultural products. He claims that human beings are the most genetically modified organisms on earth at the culminating point of a million years of biological and cultural evolution.

The hope in the possibilities offered by genetic engineering of human bodies has been consistently expressed in Spanish media as well as in articles written by Spanish researchers in international journals. In 2013 an *El País* article "Curar células no es diseñar bebés" (Healing Cells Is Not Designing Babies), Javier Sampedro, who had appeared critical toward GM grains produced by Monsanto, praised British achievements in eliminating harmful genes from fertilized embryos, insisting that it has nothing to do with "designing babies" and that those who cannot see a difference are better off saying nothing. In the debates on the genetic modification of embryos, scientists differ on the criteria that would condition future developments. In a feature article of the *Journal of Medical Ethics,* leading Spanish bioethicist Paula Casal (2013) considers children's lives as a common good of humanity rather than as the private property of the parents-to-be. Rather than enhancing an individual's life, she argues for the enhancement of humanity toward bringing up those lives that are "barely worth living" (724). She believes that this goal could be achieved through the reduction of sexual dimorphism (a phenotypic difference between males and females of the same species) in favor of female traits in all future embryos, because women are genetically less predisposed toward violence than man. (This should not be confused with the elimination of males). Casal explains that thanks to higher oxytocin levels in females, "the caring impulse spreads to anything that resembles their offspring, making all creatures with disproportionally large heads, hesitant steps, and incompetent vocalizations seem adorable" (723). She points to research showing that while oxytocin is responsible for caring, testosterone is connected to crime. The recent discovery of the monogamy gene RS3 334 present in some males (though not all) making those who have it into more faithful partners, better fathers, and less frequent perpetrators of crime, leads Casal to suppose that it would be desirable that all males have it. Thinking of the world's future, Casal imagines, "Equalia, an imaginary peaceful, social and gender egalitarian society, populated by equally sized and long lived co-parenting couples and Dimorphia with its more violent, non-parenting, harem-keepers towering over the females and dying before them" (727). She concludes that there are "good reasons to move closer to a society like Equalia, which lacks various regrettable features typically linked to polygamy, such as inequality, sexual violence and homicide" (729).

BRUNA HUSKY: A FUTURE THAT IS ALREADY HERE

Casal's scientific fantasy of a future world like "Equalia" could not be more different from the violent and crime-ridden world of Montero's science fiction novel, *Lágrimas en la lluvia*, where the most dangerous criminal is a woman trying to take revenge for the death of her son, which destabilizes the world. Montero's science fiction future can be read as an analysis of our present civilization, on the edge of a collapse, where environmental degradation and the resulting political conflicts went much further, leaving part of the planet under water, creating millions of victims. The author warns against a science in service of economic expansion and wars, which constructs new life without proper ethical consideration for its place in a worldly community. Once this life comes into being, however, it needs to be embraced and cherished. For this reason, the narration of the novel is filtered through the consciousness of a techno-human, Bruna Husky, created in a lab by synthetic biology and equipped with human feelings and thoughts. In Bruna's world, enhancement in humans has been most commonly used only for aesthetic reasons, but in the production of artificial humans (replicants or reps), certain human skills are systematically enhanced. For example, combat reps like Bruna have traits that allow her to see well at night, have a man-like strength, and perceive threats, as well as have better coordination and speed. Bruna, with her name (Husky) signaling her not-fully-human character and her role as a savior,[2] has an existentialist perception of her mortality. As all reps, she can live only ten years, and every day she reminds herself of the time she has left. Bruna is also unable to have children. Synthetic biology in the novel has not been able to construct life with "autopoesis," which has been considered by Humberto Maturana and Francisco Varela (1992) as an essential quality of life. Autopoesis is life's capacity to continuously regenerate and realize the network of processes and relations that produced them. It is the guarantee of life's self-sustainability, which is threatened by the production of infertile genetically engineered organisms. On the other hand, Richard Dawkins' (1989) suggestion that our body is merely a vehicle for "selfish genes" to make more copies

---

2. Husky dogs are known for their capacity to find people buried under the snow in the mountains, and they are used in rescue expeditions in various extreme circumstances.

of themselves becomes obsolete when live organisms are generated in a lab. The synthetic body is not nature's, but rather it belongs to the economic realm.

But in Montero's novel, in a fantastic turn synthetic life turns its creativity inward, transforming its own functions and codes not in order to sustain itself but to save the remaining life on Earth, indeed preserving certain aspects of the conditions that gave rise to it and transforming those that threaten its continuation. Bruna works as a detective, hired to solve a mystery of adulterated memories (memas) which are sold to replicants, making them murderous and suicidal. The adulterated memory contains a belief that the rep is human and has a child who has been kidnapped by the reps. This false consciousness triggers an induced behavior mechanism; the rep feels she needs to kill humans and other reps in order to free her child. As a dark shadow of the oxytocin-induced love praised by Casal (2013), the deadly memories are designed by a woman scientist to avenge the death of her son. A chain of violence inflicted by the delirious reps destabilizes the fragile democracy in the United States of Earth. Violent crowds of humans demand that reps be placed in enclosed camps or killed. While the adulterated lethal memories are forced on reps, the archive of the History of Earth is also being adulterated in such a way that history may justify anti-rep measures. An anti-rep prejudice is written into the archive of the History of Earth, while the anti-rep moods become stronger in the city. In the climactic scenes of the novel, Bruna herself becomes a victim of an adulterated "mema;" she suddenly believes that she has a child who will be returned to her only if she kills. She can see her child stretching hands toward her and crying. Helped by her friends who know what happened, she makes a tremendous effort to resist and deactivate the program inducing her behavior that is implanted in her brain. She deletes from her memory all the scenes that she rationally knows are false, those that represent her as human and imply that she needs to kill to save her child. As she manages to overcome her adulterated memories, her aggression disappears. Simultaneously the political tension in town also resolves.

A fantasy about a genetically engineered human that instead of reproducing the paradigms that gave rise to it, corrects them, could be seen as a utopian hope deposited in a fantastic super hero, except that Bruna's effort to rewrite codes inscribed in her should be read as a metaphor of a needed transformation in human culture, one that would lead toward an alternative biopolitics that cares for all life. Ursula Heise (2013) in her "The Invention of Ecofutures" evokes Frederic Jameson's famous statement that "the function of science fiction is not to portray the future but the present conceived as imaginary past" (4). Bruna's refusal to follow the induced behavior algorithm implanted in her brain is a metaphor for contemporary humans' need to resist cultural codes that lead us to destroy nonhuman life. Bruna is the kind of a cyborg that was long ago imagined by Donna Haraway in her legendary *Cyborg Manifesto* (1991). Haraway writes that cyborgs may be more fit than humans to save the world because they are relieved from the burden of mothering as well as free from the temptation to return to the maternal womb, since they are neither mothers nor have mothers. Their freedom from biological connections to other bodies increases their chance to break the vicious cycle of repeating past errors. In Haraway's theory, cyborgs are capable of rewriting what has been programmed in their bodies because their posthuman nature is based on flows of information and not blood ties that are more deterministic. While Haraway, just as Montero in *Lágrimas en la lluvia*, refers to the possibilities of technoscience, her reflections also have a second metaphoric meaning. Both authors believe that in many ways we humans are already cyborg-like and that the future is already here. Not only because cyber-technology has almost become a part of our bodies (living through a laptop crash as if it were the flu), it is of greater significance that we have become more capable of rewriting the cultural codes, which together with constructed histories of the past were implanted in us. In this struggle to rewrite the cultural codes, which is compared in Montero's novel to deleting an implanted program for an "induced behavior," there lies disquiet—but also hope.

The reps whose memories had been adulterated in the novel blind themselves by wresting their eyes from their sockets. This metaphor of blindness must be connected with Montero's (2010) participation in the debate on bullfighting, where writing for *El País,* she called liking bullfights "a cultural blindness," arguing that it needs to be overcome for any social improvement to take place. To like bullfighting entails, in Montero's argument, an assumption of the superiority of humanity over nonhuman life, which authorizes humans to kill and torture. This construction of humanity in a primeval battle against nonhuman "nature" can be viewed as a sort of a false memory, implanted in us by the "anthropological machine," that is, allied discourses of humanities, social sciences, and biological sciences of many centuries that justify animal killing and abuse, which needs to be revised for changes to take place. *Lágrimas en la lluvia*, similar to Montero's articles as well as environmental and animal liberation theories, suggests that we need to reconceptualize "humanity" and its relation with the nonhuman environment in order to save the life of the planet. If justice and equality are the most important values for the survival of life on Earth, it is justice and equality for all sorts of life, not only for humans. Apart from humans and technos, there are several other forms of life in Montero's novel. For example, intelligent beings and animals arrive from other planets. One of them, Mayo, whose torso is transparent and who can read the thoughts of those that he had sex with, saves Bruna's life and becomes her best friend. Another, *bubi* Bártolo, is a very furry being who eats Bruna's best clothes and needs a lot of caresses. In fact it is only when, after long resistance, Bruna decides to adopt *bubi*, that the Supreme Court declares charging for clean air illegal, and thus, all those poor people whose life in contaminated zones is shortened by half will be able to come to town. The individual and the systemic changes are significantly connected in the novel where goodness and justice are achieved not through science, but in spite of it.

Rosa Montero has not written on GMOs in *El País,* but she has tweeted and placed on her Facebook an article reporting on the research of Gilles-Eric Séralini that shows that rats fed with transgenic corn develop tumors. Although a number of science enthusiasts who are Facebook friends question Seralini's research and one even accuses Montero of being "a fundamentalist" for promoting it, she promptly publishes on her Facebook account a list of foods identified by Greenpeace as containing GMOs. Her posture toward the genetic modification of life seems to be closer to the principle of precaution than to enthusiasm. In her fantastic story about a rebellious woman created by synthetic biology in a lab in order to be sent to a war on a different planet as a slave and then to "enjoy" an extremely brief life among the humans who discriminate against her, Montero expresses simultaneously a belief in science's power and complains about its disregard for ethics and its lack of empathy for its lab subjects.

Including press articles, television debates, novels, and new media, the overall public debate on genetic transformation of life in Spain has been rich and intense, and it is reminiscent of those that occurred in other European countries. Both *El País* and *La Vanguardia* have featured diverse informational and polemical pieces, ranging from narratives of outright rejection to heavily regulated precautionary frameworks to narratives connecting GMOs with hope, enlightenment, and progress. An analysis of the debate does not bring any explanation of why Spanish enthusiasm toward biotechnologies is greater than in any other country in Europe, except for two characteristics. First, the debate occurred post-factum, which points to a lack of transparency in governmental decisions. Second, those who, like Riechmann or Montero, adopt a precautionary approach are automatically accused of being "conservatives," "retrogrades," or "fundamentalists," while this is not so in other countries. The significance of this last characteristic of the Spanish debate should not be underappreciated because it makes precautionary views unpopular. This has its explanations in history.

SPANISH NATIONAL CULTURE AND TECHNOSCIENCE

At the end of the bitter war of 1808 between Spain and Napoleon Bonaparte's invading armies of France, victorious Spaniards famously cried out, ¡Vivan las caenas! (Long live the chains!), making clear their desire to be free from the French and their Enlightenment model of liberty. In 1814 this desire was mainly due to the French invasion, but tensions between the French ways adopted by the Spanish aristocracy and the Spanish common people had surfaced by the mid- to late-eighteenth century. While the emerging bourgeois of merchants and their descendants in Barcelona and Madrid had reasons to be satisfied with the spread of capitalism and the enlightened policies of the French-influenced monarchy, the great majority of the working classes were descending into deeper squalor. Along with the rural peasantry, the working class's loyalties remained firmly with the Church because they felt that progress was not for them. In 1808, the invading French brought with them more promises of modernization, liberty, and emancipation that they claimed would rid Spain of its woes. But Spaniards had their reasons to doubt it. Since then Spain's path toward modern nationhood was marked by the Spaniards' rejection of the Enlightenment model that had been adopted by key European nations and a voluntary adoption of self-othering that the first chapter of this book discussed as self-animalization. The rationale for their rejection can be viewed, however, in mutually opposed ways: as a rebellious rejection of false ideals or, alternatively, as motivated by adherence to the authoritarian models of Spanish power. Spanish people saluting the chains of the authoritarian monarchy rejected the French ways by adopting bullfighting as a national spectacle. In addition to authoritarian models of power, the cultural love of risk modeled by the bullfighters' nonchalant exposure of their lives may have prepared the ground for economical and political developments such as the bioeconomy.

A second key historical moment that shaped Spanish attitudes toward science has a countervailing influence on them. It appears at the end of the nineteenth century in the form of the Regenerationist movement (regeneracionismo), which brought modern Spain's evolving identity into a more direct dialogue with European ideas of Enlightenment and technoscientific modernity. Increasing discontent that started in the 1870s and continued through the 1898 military defeat, followed by the loss of their last remaining imperial possessions—Cuba, Puerto Rico, and the Philippines—to the United States in the War of Independence, sparked a heterogeneous and loosely knit movement—regenerationism—among influential intellectuals, who perceived Spain's decline to be the outcome of backwardness in science and technology. Regenerationist intellectuals were deeply influenced by the ideas of German philosopher Karl Christian Friedrich Krause. His philosophy, in presenting a harmonious reconciliation of rationality and faith, and science and religion, arguably facilitated a bridge between Catholic Spain and a secularizing Europe. Regenerationists such as Joaquín Costa sought to redefine the national spirit of Spain and to propose social and economic reforms that were in alignment with European Enlightenment ideals of liberalism and technoscientific modernization. Their ideas of an Europeanist Spain gained influence in the early years of the twentieth century, when political authorities implemented policies and invested in infrastructure to promote the development of science education in schools (Ruiz-Castell, 2008) and to raise the level of scientific research. Another contemporaneous set of powerful intellectuals, who came to be known as the Generation of 1898, were also influenced by Krausism and Regenerationism at the outset, but eventually came to emphasize Hispanicization (i.e., Spain is different) over Europeanization—made famous by Unamuno's assertion, "¡Qué inventen ellos!" (Let others invent!). While important aspects of the Generation of 1898 school of thought became appropriated by the Falangist ideology, and also idolized by a spate of Hispanist scholarship, contemporary Spain's headlong rush to kick-start its bioeconomy can be connected to the Europhiliac yearnings of nineteenth century Regenerationists and elements of Spanish society to become a part of a modernizing Europe.

On the other hand, to make the cultural debates with history more complex, it is impossible not to notice certain parallelisms between Unamuno's rejection of science—enacted through blocking access to wisdom in favor of religion and other forms of spirituality—and Riechmann's belief that even the most cutting-edge science will not forge human happiness, which is to be found in human relations and a harmonious coexistence with nature and art. Similarly the insistence on life within limits and the opposition to genetic transformations of nature in order to accommodate human growth should remind us of Unamuno's disdain of material goods. But the similarities end here. In his lecture from 1906, "Sobre la europeización" (On Europeization), published in *La España Moderna,* Unamuno affirms: "Hay dos cosas de que se habla muy a menudo, y son la ciencia y la vida. Y una y otra, debo confesarlo, me son antipáticas" (n. pag.) (Recently people talk a lot about life and about science. I have to confess that I do not like either). Instead, he claims that the object of wisdom is death, and to be wise, it is necessary to prefer death to life. He rejoices in the fact that Spaniards are not part of European modernity, as some foreigners claim, because as he suggests, modernity is not a culture of death, which is so important for Spaniards. He literally says that it is not possible to "morir bien" (die wisely) outside of Spanish culture. This leads Unamuno to certain expressions of cultural superiority as he classifies modern Europeans as "desgraciados" (disgraced) or even "hombre[s] que no son hombre[s]" (humans that are not humans) because they attempt to find wisdom in life and in science, which is not possible (Unamuno, 1906, n. pag.).[3] Furthermore, Unamuno aestheticizes poverty and is reluctant to consider the need for politics to alleviate it. To the contrary, life and not death is the greatest value of the environmental philosophy of Riechmann, activist writings by Vivas, and their collaborators. Those on the Environmentalist Left worry that the alliance between neoliberal economy and science will harm life, increase poverty, and intensify the dependence of poor people on powerful multinationals. Their philosophy is egalitarian and political. Although critics of bioeconomy believe that it is necessary that the rich limit their material consumption, it is for the sake of sufficiency for the poor and for future generations. In this way, while Riechmann-like views can be connected to the discourses that rejected the violence of the Enlightenment (Long live the chains!), which criticized the imperfection of scientific proof and the incapacity of science to produce spiritual wisdom (Unamuno), these views also need to be connected to the emancipating political and vanguard ethical rhetoric that is strongly opposed to bullfighting and in favor of animal rights, equal distribution of resources, protection of health, and consideration of the rights to life for all vulnerable creatures. Similarly Casal's vision has heterogeneous discursive roots. While her enthusiasm for enhancement can be traced to pro-Enlightenment discourses and Regenerationism, her fantasy of "Equalia" achieved through the reduction of sexual dimorphism reminds us of anarchists' visions of an ideal society.

While the (in)famous slogan "Spain is different" was part of a governmental campaign under Fraga in the 1960s to promote Spain as an attractive tourist spot—symbolized by exotic flamenco dancers, sexy bullfighters, and sultry beaches—it nevertheless captured a series of acts and attitudes since the eighteenth century, where, according to historian Américo Castro (1940), the principal theme of "man as a naked and absolute reality" (xx) overshadowed prevalent European discourses of material progress. Franco's regime capital-

---

3. "¡Desgraciados países esos países europeos modernos en que no se vive pensando más que en la vida!" (n. pag.) (Disgraced are the modern European countries where people live without thinking of death.)

ized on these discourses, propagated by Falangist intellectuals who once read Unamuno and took as their point of departure developments prevalent in the rest of Western Europe, to project Spain as the Oriental Other that Western tourists could experience in their own proverbial backyards. The implications of this slogan were much broader than just for Spain's tourism industry; it became an enduring strand in the construction of Spanishness itself. An overemphasis on the spiritual essence of Spanish civilization by Spanish political authorities in the latter half of the twentieth century helped to maintain Spain's difference from the technologically obsessed but spiritually inferior West. In creating a discursive split between spiritually inclined literature, humanities, and art on the one side and materially inclined science and technology on the other, the "Spain is different" image has conversely fueled linear narratives of the progressive transformation of Spain from its dark age of regressive science and technology under Franco to today's enlightened and EU-integrated knowledge economy (Bertero, 2009). These narratives, however, may not be complete.

Emerging historical studies about scientific and technological research during the fascist dictatorship suggest a more complex relationship between the development of science, technology, and the nation of Spain than the widely promulgated linear narrative. Spain under Franco charted a path of national, technoscientific, and cultural development that emphasized autarky and difference. Scientists who did not fit in the regime's autarkic projects were exiled, expelled, and in this sense, the intellectual development of science and technology indeed was suppressed (e.g., Santesmases and Muñoz, 1997). At the same time, however, the political and economic constraints imposed on scientific institutions by Franco's regime also presented an opportunity for other sciences and scientists to become active participants in engineering the nation-building projects of the fascist dictatorship (e.g., Camprubí, 2010; Herran and Roqué, 2013). Camprubí traces the coevolution of a "vertically integrated system" (499) of rice production and distribution with the development of new techniques of artificial rice breeding and genetics by Spanish agricultural scientists between 1939 and 1952. Underlining the importance of scientific and technological development for Franco's autarkic ambitions, Camprubí shows how "agricultural scientists were active agents within the corporatist political economy, promoting their plans for the nation and for rice improvement at the same time (500). Similarly, Herran and Roqué dispel the myths that Franco's regime did not care about science and technology and was even opposed to it in their analysis of the rise of modern physics during Francoism (1939–75). They argue that early Francoist policymakers utilized an array of repressive measures (war, exiles, purges) to shape the development of physics "around military-related and applied fields such as optics and the material and nuclear sciences" and to incorporate physics into the political and religious ideologies of the regime, "changes to which physicists actively contributed" (207). Prominent physicists such as Carlos Sánchez del Río and José María Otero criticized the atheist separation of science and religion and articulated their visions of a "scientific humanism," which "will only be achieved through the integration of science in the Christian scheme of the world" (Herran and Roqué, 2013, 222). In other words, a variety of projects of scientific and technological research related to agricultural genetics, physics, and similar fields during Francoism developed in tune with the national project of gaining economic self-sufficiency, that is, autarky. Even though that project of autarky failed, the ensuing science and technology projects were highly productive, with the research of Spanish physicists and biochemists gaining recognition at both national and international levels. The newly emerging strands of historical research thus suggest that neither was Francoist Spain backward in scientific research and technological development, nor were key institutions of science and technology aloof from the political developments occurring under Franco. In other words, technoscientific research institutions and scientists were key cobuilders in Franco's nation-building project of autarky, just as the biotechnology sector and its scientists today are key collaborators of powerful political authorities in the intertwined development of Spain as a neoliberal bioeconomy. In fact, Spanish biotechnology had its origins in rice genetics under Franco. In conclusion, the opposition between Francoism and science, which motivated the enthusiasm for science in post-Franco Spain as a way of overcoming the oppressive ways of the Francoist era, is not justified. This opposition has existed more in cultural discourses than in economic reality. If, as Mauricio Lazzarato, notes "the problem that arises between politics and the economy is resolved by techniques and dispotifs that come from neither," it was precisely under Franco that this mediating strategies had stopped being provided by the church and began to be provided by science, a prelude to today's bioeconomy (2009, n. pag.).

The democratic transition saw Spain becoming "the most Europhiliac" of all EU member states (López and Rodríguez, 2011). Spain's "almost completely uncritical" Europeanism (López and Rodríguez, 2011, 12) extended to its development of technoscience and, in particular, to biotechnology. The transition to democracy in the 1970s oversaw the creation of "new state agencies for research based on the recommendations of the OECD [Organization for Economic Cooperation and Development], and the growth of a university system" (Herran and Roqué, 233), which led to the dismantling of the research councils that had been created under Franco and a curtailment of the Catholic Church's influence in the state's decisions. The scientific research that developed in the newly democratic Spain was more outwardly oriented—publishing in international journals, participating in international collaboration, and elevating basic research over applied research became the norm (Herran and Roqué, 2013). That said, research and development did not feature prominently in the democratic Spanish state's macroeconomic policy, which, although tailored to adapt to the emerging global economy, remained dependent on Franco's legacy of mass tourism, property development, and construction (López and Rodríguez, 2011). By the time Spain had integrated with the EU in 1986, the government had constructed the National Center for Biotechnology and earmarked biotechnology development in its science policy with the aim of becoming a knowledge-based economy on par with the most advanced nations of the EU. The "difference" that the biotechnology sector and the Spanish government sought to highlight in marketing campaigns that promoted the expansion of the Spanish bioeconomy was no longer with Europe, but with Spain's anti-Enlightenment past (e.g., "Biotechnology in Spain," 2006). Indeed, by the 1990s and 2000s (before the recession of 2008), amidst flourishing construction and housing sectors, Spain's booming economy was seen by EU's commentators to have bettered the rigidities of "old Europe" (López and Rodríguez, 2011). Spain was again "different" from Europe, this time not in a retrograde and reactionary fashion but as the leading economic edge of the EU. This leadership, however, was achieved at the cost of putting life at risk, whose seriousness is an object of many debates since the material progress that was attained turned out to be damaging in many ways.

CONCLUSION

It is in the context of Spain's shifting position vis-à-vis Europe that we can locate its different orientation toward all sorts of biotechnologies. Seen in another way, ever since its democratic transition, Spain has been in a hurry to become one of the leading knowledge economies of the West, and this is arguably why the Spanish government commercialized biotechnologies without much of a sustained public debate, and has subsequently made an aggressive push to develop Spain's bioeconomy, again without seriously taking into heed the critiques and alternatives posited by NGOs and CSOs. This has ironically occurred at a time when the EU and other major Western European nations have been engaged in much hand-wringing about the very same biotechnologies. Notwithstanding the preeminence of the UK, Germany, and France in the EU's bioeconomy, their reluctance toward agribiotechnologies, synthetic biology, and animal cloning—all of which are relatively well accepted in Spain—has provoked enthusiastic commentators to aver that Spain is forging ahead even as the rest of Europe is being left behind.

Observing Spain's implementation of agribiotechnologies, Riechmann (2011b) presciently wondered if in fact Spain is identical to what it has always been, submissive to powerful authorities, and not challenging them sufficiently. Building on Riechmann, this chapter has proposed that the political and institutional legacies of successive authoritarian regimes shaped the development of a political culture addicted to risk and subservient to the agendas and interests of those with political power, contributing to the subsequent lack of public debate and participation surrounding the commercialization of biotechnologies. In other words, Spain is no longer different in the way this difference was defined in opposition to the rest of Western Europe. However, Spain may still be different in the more fundamental sense of a political culture that is subservient to regimes of power and less caring for life. The regimes of power have shifted; they are no longer connected to authoritarian monarchy or a dictatorship allied with the Catholic Church, but rather to globalized multinational interests. The authority is constructed by the expertise of scientists codesigning bioeconomy and the scientized regulations that they create. In attempting to erase its marks of otherness in relation to the West, Spain may have ironically ended up repeating the same pattern of relationships between political authorities and civil society.

Relations between Spain's civil society, political structures of power, and scientific research institutions may be entering a new phase with the bursting of Spain's economic bubble. Since 2008, Spain has been caught in a deepening economic recession, leading it to register the highest levels of economic inequality and rates of unemployment amongst all the twenty-eight EU nation-states. Madrid's Puerta del Sol became the epicenter for the rise of the *Indignados*, a new social movement echoing the Arab Spring and Occupy movements in the Middle East and North America, respectively. The *Indignados* call into question the neoliberal model of democracy that is being instituted in Spain "because the parties in power do not look out for the collective good, but for the good of the rich. Because they understand growth as the growth of businessmen's profits, not the growth of social justice, redistribution, public services, access to housing and other necessities" (López and Rodríguez, 2011). The *Indignados*' broad critique resonates with critical narratives of the mechanisms and institutions upon which the development of Spain's biotechnology sector has been premised (e.g., Riechmann). In the words of López Arnal (2011), "La cuestión política de fondo es nada más, y nada menos, que preguntarse quién controlará la biodiversidad, los recursos genéticos, las fuerzas de la vida, y en beneficio de quién, es decir, el combate entre quienes están a favor y quienes estamos en contra de la creciente privatización de la vida y de los procesos vitales" (n. pag.) (The political question is to ask who will control the biodiversity, the genetic resources, the forces of life and for whose benefit. The battle is between those who are in favor and those of us who are against the growing privatization of life and vital processes). This battle is portrayed as ongoing and essentially unchanged in Montero's science fiction novel *Lágrimas en la lluvia,* as the balance between the allied science and power on one side and the forces in resistance on the other remains similar even while new key players appear on the stage of life. In spite of the fact that all the environmental catastrophes that we fear today are depicted as in the past in the novel, there is the same sensation of disquiet present in *Lágrimas en la lluvia* as in *Biutiful* and *Nocilla experience,* which were analyzed in the previous chapter. In *Lágrimas en la lluvia,* however, the disquiet is not connected to the fear of the end of the habitable Earth, but rather emanates from a struggle about the past. The relationship of technos to their past in the novel is an allegory of the relationship of humans to history, in an obvious dialogue with Spanish debates on Historical Memory but also more broadly, in dialogue with history as a part of human identity, ethics, and eventually as a sort of algorithm of induced behavior. In a similar fashion, Spanish debates on biotechnologies can be only understood in their historical context; the archive is instrumental to understanding the politics of life.

CONCLUSION

IN SEARCH OF ALTERNATIVE BIOPOLITICS

> Species may indeed be the name of a placeholder for an emergent, new universal history of humans that flashes up in the moment of the danger that is climate change.
>
> —Dipesh Chakrabarty

> Anything that lives needs to be thought in the unity of life; means that no part of it can be destroyed in favor of another; every life is a form of life and every form refers to life.
>
> —Roberto Esposito

WAS SPANISH KING Juan Carlos forced to give up his crown because of an animal? Media debates on abdication focus so much on the Botswana hunt that it seems to have been more than a simple trigger of the king's loss of popularity. In the Spanish section of *BBC Mundo* (2014), the abdication of King Juan Carlos is commented on as a result of "El Maleficio del Elefante" (The Curse of an Elephant). A virtual journal *Sin embargo* ("Corrupción, infidelidades," 2014) repeats the report in *El Confidencial* daily that "su abdicación llega 'después del accidente del rey in Botswana'" (his abdication occurs right after the accident in Botswana). *Sin embargo* prints the famous photo of the king next to a dead elephant, right under the title of its article about abdication. *Eleconomico* publishes an article entitled "Abdica un rey como un elefante" (A King Abdicates Like an Elephant, 2014). *Time* states, "A dead elephant was the beginning of an end for Spain's king." *Der Spiegel*,

El Diario, El Mundo, and *El País* all relate the abdication to the unfortunate accident suffered by the king during his hunting expedition in April 2012, when for the first time Spanish public opinion turned firmly against him. This obsessive connection established by the media makes the readers wonder if the scandal was more than a trigger for the king's loss of popularity and subsequent descent from the throne.

The king was at that time an honorary president of the Spanish branch of the World Wide Fund for Nature (WWF-España), whose mission was to protect the very same animals that the he secretly hunted. Had the king not tripped and broken his hip, the public would never have known of his game hunting expedition. The media had always negatively reacted to the kings' hunting trips to exclusive locations. During the time of acute economic crisis, when many Spanish people had to struggle to make ends meet and some lost all they had, there was even less tolerance for royal hypocrisy. Right after the accident, in April of 2012, *El País* published an editorial entitled "Los españoles critican al rey Juan Carlos por viajar a cazar elefantes en África" (Spaniards Criticize the King Juan Carlos for Traveling to Kill Elephants in Africa). It quoted Patxi López, leader of the Socialist Party and president of the government of Basque country, suggesting that the king should apologize. Soon afterwards while leaving the hospital, Juan Carlos spoke in front of the cameras with contrition: "Me equivoqué. No volverá a ocurrir" ("El Rey se disculpa," 2012) (I made a mistake. It will not happen again). This unprecedented gesture did not extinguish the criticism, however. Cartoonists and comedy shows portrayed the king shooting at Babar or stomped on by an elephant. *Polonia*, the satiric Catalan program, imagined the king forced to do community work at a zoo, cleaning elephants' poop and, in another episode, shooting at his grandson's teddy bear. Rosa Montero, in her article "No este rey" (Not this King), demanded that the King step down and she criticized Juan Carlos's masculinity as a pathology: "¿Qué patológica inseguridad puede llevar a alguien a tener que matar un maravilloso elefante para reafirmarse? Por todos los santos, ¡pero si ya es Rey! ¿Qué más necesita para sentirse importante? ¿Montarnos una guerra?" (What kind of pathological insecurity can lead someone to kill a marvelous elephant in order to reaffirm himself? For Heaven's Sake, isn't he a king already!? What else does he need to feel important? To start a war?) Soon a petition was passed calling for the king to abdicate. Some voices began to challenge the institution of monarchy. On 26 June, *El Mundo* interviewed twenty-five Spanish public figures, most of whom support a referendum about monarchy. (Interestingly, one of the two who are against the referendum is Fernando Savater). On 10 September, 2014, *El País* publishes an article entitled "IU califica de "cor-

FIGURE C.1
Ramón Rodríguez's vision of the story is clear:
The king lost his crown because of an elephant

rupta" a la Monarquía y plantea un referéndum sobre la república" (United Left Accuses the Monarchy of Corruption and Calls for a Referendum about the Republic). This extremely interesting turn of events, where the nation begins to question the most authoritarian modern institution because of an animal, can be seen as one more proof that human/animal relations lay deep at the foundations of the structure of power and national identity in Spain.

The dynamics of the story of the Spanish King reminds us of Carlos Saura's film *La caza* (1966), analyzed in chapter 4, where three old national soldiers kill each other during a hunt because of a ferret. The picture of the Spanish King next to a dead elephant, killed as a trophy, displays the same kind of posing as the picture taken by the hunters in Saura's film with their dead bunnies, as attributes of their victorious masculinity. In turn, that picture in Saura's film evokes photographs of Franco and his collaborators positioning themselves in this very same way after the hunt for the Spanish press or NO-DO (Noticiarios y Documentales). The hunting trophy photograph connects King Juan Carlos in an unequivocal way to the Francoist models of sovereignty and sovereign hypocrisy. In Saura's film, when José finds out from Juan that Paco shot his ferret on purpose, suddenly he realizes that he, too, had been guilty of having betrayed their fourth friend, Arturo, who took his life. Then José decides to kill Paco because he now sees him as devoid of conscience, as someone able to do evil and feel no scruples about it. Similarly, the Spanish public comes to realize that the king is not the person they once thought he was, once they find out about his secret trip to hunt the beautiful and peaceful elephants that he had officially promised to protect. Some even raised doubts about his real role in the military coup that he had famously stopped in February 1981, thus having saved Spanish democracy.[1] Others have revealed his relations with the Argentinean junta, which caused thousands of people to "disappear." The scandal caused by the killing of an elephant had to do with the costs of the trip financed by the taxes of Spaniards, who at the same time suffered from austerity measures. But just as in Saura's film, money was not at the heart of the matter. In both cases, a special angle, a unique perspective, revealed the evil of killing of an innocent animal for the sake of it. As a result, in both cases the killer loses the status quo that once favored him. The killing of a ferret in Saura's film and the killing of the elephant by the king were more than just triggers of the catastrophes that followed. In both cases, the catastrophe signals a cultural change. Spanish people are running out of tolerance for the masculinity that Montero characterizes as pathological and that affirms itself through violence whose baseline is killing of animals.

The fact that the death of an animal, instead of reaffirming a political status quo, reveals its corruption and condemns it to a collapse, signal not only a change in the construction of masculinity but also the beginning of the end of Spanish spirituality structured for necropolitics (Mbembe, 2013), represented through an encounter and defiance of death. An imperial, racist, and nationalistic biopolitics, such as that of nineteenth- and twentieth-century Spain, favored antagonistic discourses, flaunting indifference toward violence and death. Both had been modeled on bullfighting and hunting, both privileged performances on the national stage, where a man proved his superiority to an animal by teasing and killing him.

---

1. On 30 March 2014, *El Mundo* publishes an interview by Angel Mellado with Pilar Urbano, author of *Una Gran desmemoria* (A Great Amnesia) with the meaningful title "Para Suárez estaba claro que el alma de F-23 era el Rey" (Suárez was sure that the King was behind F-23).

As the first three chapters of this book show, this model of humanity, defined through risk of life, death, and killing animals, resulted in animalization. This slippage of meaning where the hunter transforms into the hunted animal is the story of Saura's film, where hunters turn each other into hunted animals. It has also been reflected by some of the articles commenting on Juan Carlos's abdication: "A King Abdicates Like an Elephant" or "este fue un rey elefante" (this was an elephant king) and cartoons that represent Juan Carlos with an elephant head or trunk.

In the first chapter's analysis, following on Jesús Torrecilla, the purposeful self-animalization in Spanish culture was discussed as a flaunting rejection of French models of civilization that led to Spain's cultural othering. The spirituality structured for necropolitics of a state practicing bullfighting and wars was constructed in a space of exclusion from European civilization. It is in the space of exclusion, in Giorgio Agamben's theory, where the King and the *homo sacer,* animal-like "bare life," mirror each other. They are both simultaneously inside and outside of the law. The construction of Spanish spirituality was at the same time intimately connected to the figure of an animal and that of the sovereign (authoritarian caudillo or bullfighter) both excluded from the law as well as the anti-European theories such as Unamuno's, Lorca's, and also Castro's, who liked to see Spain as different from the Western world. It was then in a sense a spirituality of exclusion.

The king's excursions to exclusive non-European spaces to hunt precious and rare animals cannot but remind us of the significance of exclusion in the construction of Spanishness. The fact that these excursions have to be secret, and when they are revealed, they cause the people's outrage, signals that this spirituality has been transforming, that there has been a will to give up exclusions and exceptions. Spaniards do not want to grant the right to exclusion to the monarchy any longer. They do not want to be excluded from Europe and turned to a "bare life," forced into a heroic self-defense outside of the law. They also reject the hypocrisy of making exceptions from ethical rules for those in power. Last, but not least, they begin to consider that even animals should not be excluded from political and legal considerations. In order to transform the old cultural dynamic that made them into bullfighters, in which Luis Martín-Santos felt imprisoned, Spanish intellectuals are searching for that moment when things started to go wrong.

Larra, Clarín, Noel, Buñuel, Martín Santos, Mayorga, Montero, and various other writers, filmmakers, journalists, and activists who have attempted to restructure human conduct toward the nonhuman realm believed that cruelty toward animals lies at the basis of other human cruelties. The belief that inferiority, animal or human, justifies harm became inscribed into concepts of market and progress, explaining social stratification where abuse of the poor and the undocumented by the rich became naturalized as a necessary element of economic growth and gain. In this sense, Savater has been correct, claiming that the abuse of animals constitutes the basis of human civilization. However, this civilization has always been criticized, and now it is criticized more than ever because of concurrent economic, political, and environmental problems. Larra fought for a model of citizenship that would neither approve nor justify pain and harm for animals or for women. Nor did he approve of a state that took revenge on its citizens, killing them with the "death penalty" (Larra, 1936). Clarín warned his readers that in a fast, modernizing society the poor receive a similar treatment to animals and that slaughterhouses and wars developed together as a basis of modern economy. Noel travelled all around the country preaching against bullfighting and the bullfighting culture, which in his view was responsible for war, poverty, the lack of education, destruction of the environment, and even for the inability to recognize true art.

Lorca and Buñuel oftentimes aestheticized rebellious violence, but they shrugged when they saw violence institutionalized and transformed into a raging machine as in slaughterhouses and wars. Both fought against the culturally instilled repression of women's sexuality and freedom, racism, and the deprivation of poor. In *Tiempo de silencio,* Martín-Santos developed a whole spectrum of linguistic and behavioral relations between cruel treatments of animals in the research lab and bullfighting as well as violence against women, the poor, and even against "inferior" researchers in the scientific hierarchy. All these resulted not only in harming women, the poor, and young ambitious scientists but also, as Jo Labanyi (2011) argues, it served the preservation of social and intellectual hierarchies that made change impossible.

Even if they have not elaborated alternative systems, each of these artists left us with insights into what they considered spaces of hope for a better society. Larra, for example, believed in a public *commons,* where cultural and natural resources would be accessible to all members of a society. In public parks where all social classes can meet, observe, and get to know each other surrounded by nature, he saw the space where democratic society, conscious of all its parts, were actively being formed ("Los jardines públicos"). Clarín insisted in understanding and loving animals and seeing humanity as an integral part of the natural world, voicing love for forests and objecting to the modern city's bareness and its rule of money. Noel thought that a better community could be formed if Spanish traditions, different from bullfighting, were adopted as a cultural norm, such as community cooking and sharing work, and where love for animals and trees was a norm. Martín-Santos hoped that a radical transformation of language and its metaphors would lead to different attitudes, activate citizens politically, and provide for happier love relationships. For example, he deplored comparing love to hunting or the presence of war metaphors in scientific language. For Rosa Montero, the capacity to notice and respect the pain of another creature is the most fundamental factor for Earth's survival. Films and novels preoccupied with the depth of the current crisis coincide with a sensitivity to pain and empathy toward any human and nonhuman suffering, leading to a solidarity with all sorts of underdogs. In this context, the significance of the animal rights movement cannot be underestimated. Its call for caring for animals has begun to corrode "anthropo-indifference" and to sharpen perception of the interrelatedness of species. This interrelatedness of various forms of life appears in the performance of animal activists in front of the Bilbao Guggenheim Museum, which has been the leitmotif of this book.

In the fifth chapter, while interpreting the amazing image created by the bodies of activists painted in black and red to fill in the shape of a bull, I referred to the Roman historian's comparison of the shape of the Iberian Peninsula to a bull's skin when spread out, and given this comparison, I suggested that the image can represent Spain (or rather Iberia). To conclude, I want to offer yet one more possible reading of it. Given that the skin is not spread out and the bull is wounded but possibly alive, we may hesitate to connect the activists to any geographical or national space. Perhaps, like Jewish people, who in Peter Sloterdijk's words have not been at home in any country but rather in a book, Spanish activists performing in front of the Bilbao Guggenheim Museum are making themselves at home in an animal: a bull. This supposes an imagined community of a different sort. Although the performance obviously connects with a national symbol, it does it against the grain of the nationalist discourse, reworking the symbolic significance of the bull from antagonism into identification. Even if this is again a voluntary self-animalization, the identification with an animal, rather than inflicting pain and death, gives them protection.

Such animalization as a discursive figure stops being a tool to maintain the status quo of social hierarchies and power and becomes a tool to build an alternative biopolitics. A new spirituality that emerges as a result is structured by empathy and connection to all forms of life, the human species creating a network destined to save itself together with the nonhumans. This community forms a sort of superorganism, the flesh is a way of being in common with that which wishes to remain alive.

This new framework of thought emerges not only in Spain. In the twenty-first century, humans have been forced to reembrace their animal condition in the circumstances of an ecological crises and climate change that they caused by abusing their freedom. In relaying his discussion with Chakrabarty, Žižek paraphrases him as follows:

> Humans can "do what they want" only insofar as they remain marginal enough, so that they don't seriously perturb the parameters of the life on earth. The limitation of our freedom that becomes palpable with global warming is the paradoxical outcome of the very exponential growth of our freedom and power, i.e., of our growing ability to transform nature around us up to destabilizing the very basic geological parameters of the life on earth. (2010b, n. pag.)

This powerful idea of the Chicago historian, rephrased by Žižek, that human freedom is threatened by global warming, takes us back into the animal-like dependence on an environment from which we believed we had been emancipated. We now realize that we, as a species, are now only a little less vulnerable to climate than other animals. It also alerts us to the need to think as a species and in relation to other species with which we share the planet.

The last two chapters of this book analyze responses to the ecological crisis from film, literature, the press, and science. *Biutiful* expresses deep pessimism and no hope in human survival. *Nocilla experience* expresses no more optimism for humanity's future except that it employs dark humor and suggests keeping it up until the end. A hope for a pain-ridden survival on the verge of a final catastrophe can be detected in *Lágrimas en la lluvia*. Perhaps this last novel seems more optimistic because it points to the ways in which the world needs to change. Bruna's favorite sandwich is made of algae since almost no one in the future world eats animals any longer, and she takes vapor baths since water is too expensive to use it for that purpose. She doesn't have a private vehicle and walks everywhere unless she can afford a taxi. Questionably although seductively, the pessimism of *Biutiful* and of *Nocilla experience* is overcome in *Lágrimas en la lluvia* by a hope that if humans are not able to change, then posthuman intelligent creatures may be able to do so in order to save life from extinction and that a chosen individual will be able to renew the peace. However, it is not a hope in the future possibilities of science, which is shown to be producing disappointing results and harm by bringing to life creatures that live in pain and die prematurely from tumors. It is rather a sort of romantic (and possibly utopian) hope in nature's capacity to redeem itself in its future incarnations mediated by humans. In fact, if the world does not end in Montero's novel, this is due to the fact that the framework of justice guaranteeing the equality of techno-human and human rights is preserved. The consideration of the techno-life in this science fiction novel continues on today's debates about animals.

In the introduction to *Bíos,* Esposito suggests "not to think of life as a function of politics, but to think politics within the same form of life" (2008, 12). This book has similarly argued from the first pages that the change of attitude toward nonhuman life is a fundamental and basic first step in the search for an alternative biopolitics. It has retraced the thought of those intellectuals, writers, filmmakers, artists, and activists of Spanish culture who argued for this change in most ingenious and convincing ways. They show that a respectful, if not loving, relationship toward animals and understanding human interconnectedness and interdependence with the environment could lead to an elaboration of different motivational structures and priorities, and a whole different kind of "cool," and as a result, it would lead to an all-embracing universal model of citizenship that does not dwell on "differences," but rather focuses on mutual relations. It is only in this new framework that incorporates the nonhuman life that justice may indeed work as an environmental value. As an example of this conceptual change, a group of Spanish economists, directed by Miguel Ortega Cerdá (2012), published *La deuda ecológica española* that, inspired by ecological economics initiated by Joan Martínez Alier, suggests reevaluating the international debt of Latin American countries with Spain in light of Spanish exploitation of their natural resources since the sixteenth century. These economists justly notice that it is Spain that has a debt toward its ex-colonies rather than vice versa. This debt has been accumulated through the depletion of natural riches and the destruction of the environment and human and nonhuman lives and health. These are not purely theoretical deliberations. This group of economists proposes that Latin American countries' debts with Spanish banks be cancelled.

Another example of attempts to put into life an alternative biopolitics comes from the "Transition Town Movement" composed of grass-root community initiatives that seek to creatively transform the socioeconomic structures of their towns in the face of climate change and peak oil and ongoing environmental crises. This movement's initiatives connect the needs of environmental repair and of political repair, changing the politics so that politics works in service of life. The Transition Town Movement returns to an organic model of democracy based on social deliberations where all stakeholders participate in debates and vote directly on the projects. Founded in 2006 in Britain, the movement's goals, as stated on its webpage, are "to create healthy human culture and to reduce CO_2 emissions." Further, on the webpage of "Red de Transición España" (Net of Transition-Spain), it states: "La Transición es una manifestación de la idea de que la acción local puede cambiar el mundo" (Transition is a manifestation of the idea that a local action can change the world). The focus on future change is then an essential part of Transition, but this change is conceived in a nonconfrontational, nonviolent, and nonsystemic way—a slow, surreptitious, and bloodless takeover, almost "natural" in its biology-inspired vision of a "transition" of one plantlike organism coexisting with another.

There is a small town on the Catalonian coast that gave to its new money, founded in 2009, a name of the music of love, *turuta*. Turutas are generated by the time invested by volunteers in projects to regenerate the local environment. In this way, the economy is a function of the restored life, and every economical transaction sustains the ecosystem. Turutas are not printed money but a virtual balance in a system of mutual credit granted to each other by the members of the local economy. Turutas restructure the order of things; life is no longer exploited in the service of economy, but it is sustained and served by it. There are over thirty similar local currencies in Spain, more than in any other European country and various alternative local economies are based on alternative biopolitics.

And there is "Podemos." When I talked to Jorge Riechmann for the first time in Madrid at the end of 2011, right after the elections won by Partido Popular, the scene could not look much bleaker. When I asked Reichmann about "change," he said that it is happening in small circuits of people who call themselves the 99 percent even if they are less than one. Looking at this four years later, with the publication of this book, these small circuits have produced a powerful political movement that has suddenly become one of the three main political parties, which, according to various polls, has a real chance of winning the general elections of 2015. Their program contains a radical change: protection of environment and biodiversity, fostering of alternative energies and alternative modes of transportation, reforestation and ecologically sustainable consumption as well as protection of health and women's rights. In sum, the program places life before profit, which indeed sounds like an alternative biopolitics. It is not yet the time for celebration, but it seems like this time, Joauquín Araujo was right: All big changes start on the margins.

WORKS CITED

"Abdica el rey Juan Carlos I de España: De héroe del golpe a la maldición del elefante." *BBC Mundo*. 2 Jun. 2014. Web. 10 July 2014. <http://www.bbc.co.uk/mundo/noticias/cluster_abdica_rey_juan_carlos>

"Abdica un rey como un elefante." *Eleconómico*. 2 June 2014. Web. 2 Oct. 2014. <http://www.eleconomico.es/item/112993-abdica-un-rey-como-un-elefante>

Abella, Carlos. *¡Derecho al toro!; El lenguaje de los toros y su influencia en lo cotidiano*. Madrid: Anaya, 1996. Print.

Abend, Lisa, and Geof Pingree, "Spanish TV Says No to Bullfighiting" *Time* 22 Aug. 2007. Web. 9 Oct. 2012. <http://content.time.com/time/world/article/0,8599,1655075,00.html>

Adorno, Theodor, and Marx Horkheimer. *Dialectic of Enlightment*. Trans. John Cumming. New York: Continuum, 1972. Print.

———. "The Culture Industry: Enlightenment as Mass Deception" 1944. *The Cultural Studies Reader*. Ed. During. 2nd ed. London: Routledge, 1999: 31–41. Print.

Agamben, Giorgio. *Homo Sacer: Sovereign Power and Bare Life*. Stanford: Stanford UP, 1998. Print.

———. *The Open: Men and Animal*. Trans. Kevin Attell. Stanford: Stanford UP, 2004. Print.

Alas, Lepoldo (Clarín). "¡Adiós Cordera!" *Doña Berta. Doña Berta y otras narraciones.* (1892). Madrid: Alianza, 2002. Print.

———. "El Quin," In *Cuentos completos*. Madrid: Alfaguara, 2000. Print.

———. "El Gallo de Oro," In *Cuentos completos*. Madrid: Alfaguara, 2000. Print.

———. "La mosca." In *Cuentos completos*. Madrid: Alfaguara, 2000. Print.

Alaska. "¿A quien le importa?" *No es pecado*. Rec. 1986. Warner Music Group/DRO, 2011. LP.

———. *Transgresoras*. Barcelona: Ediciones Martínez Roca, 2003. Print.

"Alaska se desnuda para una campaña antitaurina." *Peatóm.* 3 June 2008. Web. 30 March 2009 <http://www.peatom.info/la-llave/12839/desnudo-de-alaska-para-una-campana-antitaurina/>

Albertí, Rafael. "El perro rabioso." *Poesía.* Vol. 2. Ed. Jaime Siles. Madrid: Seix Barral, 2003. 127. Print.

Almodóvar, Pedro. Dir. *Matador.* Spain. Prod. Andrés Vicente Gómez, 1986. Film.

——. *Todos sobre mi madre.* Spain. Prod. Deseo, 2000. Film.

——. *Los Abrazos rotos.* Spain. Prod. Deseo, 2009. Film

——. *La flor de mi secreto.* Spain. Prod. Deseo, 1999. Film.

——. *Hable con ella.* Spain. Prod. Deseo, 2002. Film.

Amorín, Carlos. *Las semillas de la muerte. Basura tóxica y subdesarrollo: El caso de Delta & Pine.* Madrid: Los libros de la Catarata, 2000. Print.

Amorós, Andrés. *Toros y cultura.* Madrid: Espasa-Calpe, 1987. Print.

Animals in Distress. Website. <animalsindistresspa.org/>

Ansón, Luis María. "La pesca recreativa y las corridas de toros." Antonio Borrero Miró. Blog. 10 Aug, 2004, 12:18. <http://antonioborreromiro.crearblog.com/2010/08/01/la-pesca-recreativa-y-las-corridas-de-toros/>

Aparicio Nevado, Felipe. "La caza del hombre: Recreación de un motivo legendario, novelesco e histórico en *La caza* de Carlos Saura." *Arbor: Ciencia, Pensamiento y Cultura* (Marzo–Abril 2011): 269–77. Print.

Appleman, P., ed. *Darwin: A Norton Critical Edition.* New York: Norton, 1970. Print.

Arendt, Hannah. *The Origins of Totalitarianisms.* London: Schocken Books, 1951. Print.

——. "The Perplexities of the Rights of Man." *Biopolitics: A Reader.* Eds. Timothy Campbell and Adam Sitze. Durham: Duke University Press, 2013. 82–97. Print.

Arnold, Matthew. *Culture and Anarchy.* 1869. New York: CreateSpace Independent Publishing, 2002. Print.

Arpa Villalonga, Lola. "¿Llegaremos a tiempo?" *El País.* 16 Nov. 2013. Web. 25 Nov. 2013. <http://elpais.com/elpais/2013/11/15/opinion/1384538695_900984.html>

Arrabal, Fernando. Dir. *Viva la muerte.* France. Prod. Hassen Daldoul, 1971. Film.

ASEBIO / Spanish Bioindustry Association. "Report: 2012 News and Trends from the Spanish Biotech Sector." June 2013. Web. 15 January 2014. <http://www.asebio.com/en/documents/ASEBIOREPORT2012_final.pdf>

Baker, Steve. "Slaughtering the Human." *Zoontologies: The Question of Animal.* Ed. Carry Wolfe. Mineapolis: U of Minnesota P, 1995. 147–64. Print.

Baroja, Pío. *Camino de perfección.* 1902. New York: French and European Publications, 1990. Print.

——. *El árbol de la ciencia.* 1911. Madrid: Cátedra, 2006. Print.

——. *Libertad frente a sumisión.* Madrid: Caro Raggio, 2000. Print.

———. *La lucha por la vida*, 1904-5. In *La Busca, La mala hierba, Aurora Roja*. Ed. Juan M. Martín Martínez. Madrid: Cátedra, 2010. Print.

Barthes, Roland. *Camara lucida*. New York: Macmillan, 1981. Print.

Begin, Paul. "Anthropology as Entomology in the Films of Luis Buñuel." *Screen* 48.4 (2007): 425-42. Print.

Beilin, Katarzyna Olga. "*Disquieting* Realism: Postmodern and Beyond." *Companion to the Twentieth-Century Spanish Novel*. Ed. Marta Altisent. Suffolk: Tamesis, 2008: 186-96. Print.

———. *Del infierno al cuerpo: La otredad en la narrativa y en el cine español contemporáneo*. Madrid: Libertarias, 2007. Print.

Benda, Julien. *The Treason of the Intellectuals*. 1927. New Brunswick: Transaction Publications, 2009. Print.

Bennett, Jane. *Vibrant Materialism: A Political Ontology of Things*. Durham: Duke UP, 2010. Print.

Bentham, Jeremy. *An Introduction to the Principles of Morals and Legislation*. Oxford: Clarendon Press, (1789) 1907. *Library of Economics and Liberty*. Web. 24 April 2012. <http://www.econlib.org/library/Bentham/bnthPML18.html>

Bergamín, José. *La música callada del toreo*. Madrid: Turner, 1981. Print.

Berger, Pablo. Dir. *Blancanieves*. Spain. Prod. Pablo Berger, 2012. Film.

Bertero, Michela. "Science Attracts Spain." *EMBO Reports* 10.8 (2009): 817-19. Print.

Beusterien, John. *Canines in Cervantes and Velázquez: An Animal Studies Reading of Early Modern Spain*. Hants: Ashgate Publishing, 2013. Print. New Hispanisms: Cultural and Literary Studies.

Blas Benito, J. "Prólogo: La Tauromaquia de Goya." *El libro de la Tauromaquia: Francisco de Goya*. Ed. J. M. Matilla and J. M. Medrano. Madrid: Museo del Prado, 2001. 11-13. Web. 30 Oct. 2013. <https://www.museodelprado.es/goya-en-el-prado/obras/lista/?tx_gbgonline_pi1[gocollectionids]=28>

"Biotechnology in Spain." *Nature Biotechnology* 24.9 (2006): S3-S13. Print.

Blatz, Charles, ed. *Ethics and Agriculture*. Moscow: U of Idaho P, 1990. Print.

Borges, Jorge Luis. "The immortal." *El Aleph and Other Stories*. London: Penguin Classics, 2004: 3-19. Print.

Borradori, Giovanna. "9/11 and Global Terrorism: A Dialogue with Jacques Derrida." U of Chicago P, n.d. Web. 12 Oct. 2014. <http://www.press.uchicago.edu/books/derrida/derrida911.html>

Bossi, Laura. *Historia universal del alma*. Madrid: Antonio Machado, 2008. Print.

Bourdieu, Pierre. *Outline of a Theory of Practice*. Cambridge, UK: Cambridge UP, 1977. Print.

———. *Distinction: A Social Critique of the Judgment of Taste*. Cambridge: Harvard UP, 1984. Print.

———. *Language and Symbolic Power*. Cambridge: Harvard UP, 1991. Print.

———. *Outline of a Theory of Practice*. Cambridge: Cambridge UP, 1977. Print.

Bouroncle, Alberto. "Ritual, Violence and Social Order: An Approach to Spanish Bullfighting." *Meanings of Violence: A Cross-Cultural Perspective*. Ed. Jon Abbink and Göran Aijmer, Oxford: Berg, 2000. 55–76. Print.

Bowman, Ron. Dir. *Six Degrees Can Change the World*. USA: Prod. Stepheen Reverand. 2008. Film.

Brandes, Stanley. "Taurophobes and Taurophiles: The Politics of Bulls and Bullfights in Contemporary Spain." *Anthropologist Quarterly* (Spring 2009): 779–94. Print.

Brenneis, Sarah. "Clarín's Animals: Reading Leopoldo Alas' Short Fiction through Darwinian Revolution." *Hispanófila* 151 (2007): 37–52. Print.

Brown, Peter. "What Is Degrowth?" Lecture from the "Degrowth in the Americas, 2012" Conference. Vimeo. 20 June 20 2012. <https://vimeo.com/44414035> Film.

Browning, Tod. Dir. *Freaks*. U.S. Prod. Tod Browning, 1932. Film.

Bru de Sala, Xavier. "¡Oh toro!" *La Vanguardia* (24 April 2004): 2. Print.

Buero Vallejo, Antonio. *El sueño de la razón*. Doral: Stockero, (1970) 2010. Print.

"Bullfight Opinion Poll: As Spain Debates 'Support for Bullfighting' Bill, Most Spaniards Oppose Use of Public Funds for Cruel, Waning Bloodsport." *Humane Society International*. 23 April 2013. Web. 6 June 2013. <http://www.hsi.org/world/europe/news/releases/2013/04/spain_bullfighting_ipsos_poll_042313.html>

Buñuel, Luis. Dir. *L' age d'or*. France, Prod. Le Vicomte de Noailles, 1930. Film.

———. Dir. *El ángel exterminador*. México. Prod. Gustavo Alatriste, 1967. Film.

———. Dir. *Las hurdes: Tierra sin pan*. Spain. Prod. Ramón Acín and Luis Buñuel, 1933. Film.

———. *My Last Sight: The Autobiography*. Trans. Abigail Israel. New York: Vintage Books, 1984. Print.

———. Dir. *Le phântome de la liberté*. France. Prod. Serge Silberman, 1974. Film.

Burgos, Antonio. "Ucronía de Manolete." *El Mundo*. 28 Aug. 1997. Web. 12 Dec. 2009. <http://www.antonioburgos.com/mundo/1997/08/re082897.html>

Butler, Judith. *Precarious Life: The Powers of Mourning and Violence*. London: Verso, 2004. Print.

Cabal, Fernando. *Indignados, 15-M*. E-book. Mandala Ediciones. 2 June 2011. Web. 9 Oct. 2011 <http://www.mandalaediciones.com/varios/politica/indignados.asp>

Calarco, Matthew. *Zoographies: The Question of the Animal from Heidegger to Derrida*. New York: U of Columbia P, 2008. Print.

Cambria, Rosario. "Federico García Lorca, aficionado taurino y poeta del toro." *García Lorca Review* 2 (1974): 14–16. Print.

Campbell, Timothy, and Adam Sitze. Introduction to *Biopolitics: A Reader*. Ed. Campbell and Sitze. Durham: Duke UP, 2013. 1–40. Print.

———. "Translator's Introduction." *Bíos: Biopolitics and Philosophy*. Minneapolis: U of Minnesota P, 2008: vii–viii. Print.

Camprubí, Lino. "One Nation: Rice Genetics and the Corporate State in Early Francoist Spain (1939–1952)." *Historical Studies in the Natural Sciences* 40.4 (2010): 499–531. Print.

Caparrós, Martín. Interview by Ima Sanchís. "Ecologismo es una forma 'cool' de conservadurismo." *La Vanguardia*. 29 Nov. 2010. Web. 25 July 2012. <http://www.lavanguardia.com/lacontra/20101029/54062368819/el-ecologismo-es-una-forma-cool-del-conservadurismo.html>

Casal, Paula. "Los Derechos Homínidos." *Viento Sur* 125 (Nov. 2012): 50–58. Print.

———. "Is Multiculturalism Bad for Animals?" *Journal of Political Philosophy* 11.1 (Mar. 2003): 1–22. Print.

———. *Marina y el mar*. Madrid: Hiperión, 2007. Print.

———. "Reform, Not Destroy: Reply to McMahan, Sparrow and Temkin."*Journal of Medical Ethics*, 39.12 (Dec. 2013): 719–41.

Casal, Paula, and Jorge Riechmann. "El quinto simio somos nosotros" Manifiesto del Proyecto Gran Simio, 2014. Web. 2 June 2012. <http://www.proyectogransimio.org/noticias/noticias-destacadas/pensar-acerca-del-pensar-no-esta-limitado-a-los-humanos>

Cases, Manel. "Barcelona, ciudad libre de crueldad animal." *La Vanguardia* 17 Apr. 2004: 2. Print.

Castells, Manuel, Joao Caraca, and Gustavo Cardoso, eds. *Aftermath: The Cultures of the Economic Crisis*. NY: Oxford UP, 2012. Print.

Castro, Américo. *The Meaning of Spanish Civilization*. 1942. Ed. José Rubia Barcia. Berkeley: U of California P, 1977. Print.

———. "The Meaning of Spanish Civilization: The Inaugural Lecture of Americo Castro." 1940. Eds. Selma Margaretten, José Rubia Barcia, and Américo Castro; Berkely: U of California P, 1976. Print.

Cerrillo, Antonio. "Los transgénicos saltan del plato: Los fabricantes renuncian a los alimentos modificados genéticamente." *La Vanguardia*. 31 July 2004. Web. 14 May 2011. <http://www.lavanguardia.com/20060731/51278167757/los-transgenicos-saltan-del-plato.html>

Certeau, Michel de. "Walking Through the City." *Cultural Studies Reader*. Ed. Simon During. New York: Routledge, 2007: 126–33. Print.

Cervantes, Miguel de. "El coloquio de los perros." *Instituto Cervantes*. N.d. Web. 6 Apr. 2010. <cervantes.uah.es/ejemplares/necoloquio/coloquio.htm>

Chakrabarty, Denish. "The Climate of History: Four Theses." *Ecocriticism: The Essential Reader*. Ed. Ken Hiltner. New York: Routledge, forthcoming 2015. Print.

Clark, Nigel. *Inhuman Nature: Sociable Life on a Dynamic Planet*. London: Sage, 2011. Print.

Close, Glen. *La Imprenta enterrada: Baroja, Arlt y el imaginario anarquista*. Rosario: Beatriz Viterbo, 2000. Print.

Cobb, Carl W. *Rafael Alberti, Gerado Diego, Federico García Lorca: The Bullfighter Sánchez Mejías as Elegized by Lorca, Alberti and Diego*. York: Spanish Literature Publications, 1993. Print.

"Comparativa ICSA-Gallup." 2009. Web. 23 Nov. 2012. <http://asanda.org/documentos/tauromaquia/encuestas-sobre-corridas-de-toros/comparativa-icsa-gallup>

Contratas y Obras. Website. 31 Mar. 2010. <www.contratasyobras.com>

"Corrupción, infidelidades, crisis: El rey Juan Carlos abdica y España se pregunta: ¿República o monarquía?" *Sin embargo.* 2 June 2014. Web. 10 July 2014. <www.sinembargo.mx/02-06-2014/1011550>

Cortina, Adela. *Las fronteras de la persona: El valor de los animales, la dignidad de las personas.* Madrid: Taurus, 2009. Print.

———. "La pequeña simia." *El País.* 5 Sept. 2006b. Web. 6 Oct. 2010. <http://elpais.com/diario/2006/09/05/opinion/1157407205_850215.html>

Cossío, José María de. *Los toros.* Vol. 1. Madrid: Espasa Calpe, 1995. Print.

Criticisms of Bullfighting: Animal Equality, Animal Liberation Front, AnimaNaturalis, Anti-Bullfighting City, Ban on Bullfighting in Catalonia. Life Journey Publishing. 2013. Print.

Cross, Tim. "Othering the National Self: 'Japan the Beautiful' as Lethal Aesthetic." *Reading Room.* N.d. Web. 28 May 2012. <http://wwwmcc.murdoch.edu.au/ReadingRoom/asiancs/othering.htm>

"El cultivo de transgénicos crece este año un 44%, según Monsanto." *El País.* 29 Oct. 1999. Web. 11 Dec. 2012. <http://elpais.com/diario/1999/10/29/sociedad/941148010_850215.html>

Dawkins, Richard. *The Selfish Gene.* Oxford: Oxford UP, 1989. Print.

"A Dead Elephant Is the Beginning of the End of the Spanish King." *Time.* 2 June 2014. Web. 23 June 2014. <http://time.com/2808963/spain-king-elephant-abdication/>

"El debate taurino en el Parlament de Cataluña, en directo." *El Mundo.* 3 Mar. 2010. Web. 5 Mar. 2010. <http://www.elmundo.es/elmundo/2010/03/02/toros/1267534233.html>

DeGracia, David, and Andrew Rowan. "Pain, Suffering and Anxiety in Animals and Humans." *Theoretical Medicine and Bioethics* 12.3 (1991): 193–201. Print.

Deleuze, Gilles. *Cinema 1: The Movement Image.* Minneapolis: U of Minnesota P, 1986. Print.

De Lora, Pablo. *Justicia para los animales: Ética más allá de la humanidad.* Madrid: Alianza, 2003. Print.

Derrida, Jacques. *Of Grammatology.* 1967. Baltimore: John Hopkins UP, 1997. Print.

———. *The Politics of Friendship.* New York: Verso, 2004. Print.

———. *Rogues. Two Essays on Reason.* Stanford: Stanford UP, 2005. Print.

———. "White Mythology: Metaphor in the Text of Philosophy." *New Literary History* 6.1 (Fall 1974): 5–74. Print.

———. *The Animal That Therefore I Am.* Trans. David Wills. Ed. Marie-Louise Mallet. New York: Fordham UP, 2008. Print.

Díaz, V., et al. "The Socio-Economic Landscape of Biotechnology in Spain: A Comparative Study Using the Innovation System Concept." *Journal of Biotechnology* 98 (2002): 25–40. Print.

Donaldson, Sue, and Will Kymlicka. *Zoopolis: A Political Theory of Animal Rights*. New York: Oxford UP, 2011. Print.

Dopico Black, Georgina. "The Ban and the Bull: Cultural Studies, Animal Studies and Spain." *Journal of Spanish Cultural Studies* 11.3–4 (2010): 235–49. Print.

Douglas, Carrie. *Bullfighting and National Identities in Spain*. Tucson: U of Arizona P, 1999. Print.

Duara, Prasenjit. *Rescuing History from the Nation: Questioning Narratives of Modern China*. Chicago: U of Chicago P, 1995. Print.

———. "On Theories of Nationalism for India and China" In the Footsteps of Xuanzang: Tan Yun Shan and India. 1999. Indira Gandhi National Centre for the Arts. New Delhi. Web. 13 May 2015. <http://ignca.nic.in/ks_40032.htm>

Durán, Luque, Juan de Dios, and Francisco José Manjón Pozas. "Fraseología, metáfora y el lenguaje taurino." *Nuevas tendencias en la investigación lingüística*. Granada: Granada Lingvistica, 2002. Web. 2 May 2011. <http://www.ganadeslidia.com/webroot/ . . . /lenguaje_taurino>

Durang, Christopher. *Why Torture is Wrong and the People Who Love Them*. Forward Theater Company. The Playhouse at Overture Center, Madison. 12 Dec. 2009. Performance.

During, Simon. Introduction to *The Cultural Studies Reader*. Ed. During. 2nd ed. London: Routledge, 1999: 1–30. Print.

Eagleton, Terry. *Literary Theory: An Introduction*. Hoboken: John Wiley, 2011. Print.

Egea, Juan. *Dark Laughter: Spanish Film, Comedy and the Nation*. Madison: U of Wisconsin P, 2013. Print.

Eisenman, Stephan F. *The Abu Ghraib Effect*. Chicago: U of Chicago P, 2007. Print.

"Elecciones Generales 2011." *El País*. 15 Mar. 2012. Web. 13 Apr. 2012. <http://resultados.elpais.com/elecciones/generales.html>

"España: El cultivo de maíz transgénico vs. ecológico" *Ecoticias*. Web. 11 May 2015. <http://www.biomanantial.com/espa%C3%B1a-cultivo-maiz-transgenico-ecologico-a-1427-es.html>

"Españoles critican al rey por viajar a cazar elefantes a África" *El País*. 16 Apr. 2012. Web. 15 July 2012. <http://www.elpais.com.co/elpais/internacional/noticias/espanoles-critican-rey-juan-carlos-por-viajar-cazar-elefantes-en-africa>

Espigado, Miguel. "Agustín Fernández Mallo & 'Nocilla Experience' (1.0)." *Afterpost*. 9 Mar. 2008. Web. 15 May 2013. <https://afterpost.wordpress.com/2008/03/09/afm-experience-10>

Esposito, Roberto. *Bíos: Biopolitics and Philosophy*. Trans. Timothy Campbell. Minneapolis: U of Minnesota P, 2008. Print.

———. "Biopolitics." *Biopolitics: A Reader*. Eds. Timothy Campbell and Adam Sitze. Durham: Duke UP, 2013a: 317–49. Print.

———. "The Enigma of Biopolitics." Ibid: 2013b: 349–85. Print.

---. *Immunitas: The Protection and Negation of Life*. Cambridge, UK: Polity Press, 2011 (2002). Print.

Faber, Sebastiaan. "'La hora ha llegado' Hispanism, Pan-Americanism, and the Hope of Spanish/American Glory (1938–1948)." *Ideologies of Hispanism*. Ed. Mabel Moraña. Nashville: Vanderbilt UP, 2005: 315–45. Print.

---. "The Spanish Holocaust: Reframing the Civil War." *The Volunteer*. 12 June 2012. Web. 30 Oct. 2012. <http://www.albavolunteer.org/2012/06/the-spanish-holocaust-reframing-the-civil-war/>

Fabre, Jean-Henri. *The Life of Scorpion*. Bel Air: UP of the Pacific, 2002. Print.

Fanon, Frantz. *Black Skin, White Masks*. New York: Grove Press, 1967. Print.

Faulkner, Sally. "Aging and Coming of Age in Carlos Saura's *La caza* (*The Hunt*, 1965)." *Modern Language Notes* 120.2 (2005): 457–84. Print.

Faverón Patriau, Gustavo. "La máquina de hacer muertos: Enfermedad y nación en *Tiempo de silencio* de Luis Martín-Santos." *Colorado Review of Hispanic Studies* 1.1 (2003): 79–90. Print.

Fernández, Joaquín. *El ecologismo español*. Madrid: Alianza Editorial, 1999. Print.

Fernández Cubas, Cristina. *El año de Gracia*. Barcelona: Tusquets, 1994. Print.

---. "Mi pasión por los toros: Camino y Fernández muestran los caprichos de la pasión por los toros." *Europasur.es*. 13 June 2009. Web. 21 Oct. 2011. <http://www.europasur.es/article/ocio/446834/camino/y/fernandez/muestran/los/caprichos/la/pasion/por/los/toros.html>

Fernández de Moratín, Nicolás. *Carta Histórica sobre el origen y progresos de las Fiestas de Toros en España*. Madrid: Repullés, Josep Sánchez, 1801. *Bibliotecas de Castilla y León*. Web. 30 Mar. 2010. <bibliotecadigital.jcyl.es/i18n/consulta/busqueda.cmd>.

Fernández Mallo, Agustín. *Nocilla experience*. Madrid: Santillana, 2008. Print.

Ferretti, Maria Paola, and Vincenzo Pavone. "What Do Civil Society Organisations Expect from Participation in Science? Lessons from Germany and Spain on the Issue of GMOs." *Science and Public Policy* 36.4 (2009): 287–99. Print.

Flaherty, Robert. Dir. *Nanook of the North*. U.S., Prod. Robert Flaherty, 1922. Film.

Fletcher, Joseph. *Situation Ethics: The New Morality*, Philadelphia: Westminster Press, 1966. Print.

Flores Ocejo, María Begoña. *Historia de La Presencia Veterinaria En Los Festejos y Espectaculos Taurinos*. Madrid: Comunidad de Madrid, 2009. Print.

Folke, Carl. "Resilience: The Emergence of a Perspective for Social-Ecological Systems Analysis." *Global Environmental Change* 16 (2006): 253–67. Print.

Foucault, Michel. *The Birth of Biopolitics*. 1978. New York: Palgrave Macmillan, 2006. Print.

---. "Right of Death and Power over Life." 1976. *Biopolitics: A Reader*. Ed. Timothy Campbell and Adam Sitze. Durham: Duke UP, 2013. 41–60. Print.

---. "*Society Must Be Defended*": *Lectures at the College de France 1975-76*. New York: Picador, 2003. Print.

"For a Bullfighting Free Europe." 2008. Web. 30 May 2013. <http://www.bullfightingfree-europe.org/>

Fraser, Benjamin. "A *Biutiful* City: Alejandro González Iñárritu's Filmic Critique of the 'Barcelona Model.'" *Studies in Hispanic Cinemas* 9.1 (2012): 19–34. Print.

Friese, Carrie. *Cloning Wild Life: Zoos, Captivity, and the Future of Endangered Animals.* New York: New York UP, 2013. Print.

Fuentes, Carlos. *El espejo enterrado.* Madrid: Taurus, 1997. Print.

"La 'fuga de cerebros' hace perder miles de millones de dólares a los países en desarrollo: España se perfila como un líder potencial en cuanto a innovación y adelanto tecnológico, según el informe del PNUD hecho público hoy." *El País.* 10 July 2001. Web. 21 Oct. 2013. <http://sociedad.elpais.com/sociedad/2001/07/10/actualidad/994716004_850215.html>

Gabilondo, Joseba. "On the Inception of Western Sex as Orientalist Theme Park: Tourism and Desire in Nineteenth-Century Spain (*Carmen, Don Juan*)." *Spain is Different: Tourist Locations, Attractions, and Discourses in Modern Spanish Culture.* Ed. Eugenia Afinoguenova and Jaume Martí Olivella. New York: Lexington Books (2008): 19–61. Print.

Gallup. "Interés por las corridas de toros." July 2002. Web. 23 Sept. 2012. <http://www.columbia.edu/itc/spanish/cultura/texts/Gallup_CorridasToros_0702.htm>

Galtung, Johan. "Violence, Pease and Peace Research." *Journal of Peace Research* 6.3 (1969): 167–91. Print.

García Mercadal, J. Prologue to *Escritos antitaurinos: El Flamenco y El Chispero.* Madrid: Taurus, 1967: 3–17. Print.

Gieryn, Thomas F. *Cultural Boundaries of Science: Credibility on the Line.* Chicago: U of Chicago P, 1999. Print.

Gil Calvo, Enrique. *La función de toros: Una interpretación funcionalista de las corridas.* Madrid: Espasa-Calpe, 1989. Print.

Girard, René. *Mensonge romantique et vérité romanesque.* 1961. Paris: Grasset, 2001. Print.

———. *Violence and the Sacred.* Trans. P. Gregory. Baltimore: John Hopkins UP, 1977 (1972). Print.

Glick, Thomas. *Darwin en España.* Valencia: U de Valencia, 2010 (1982). Print.

Gómez Pin, Victor. "Toros, lengua y estigma." *El País.* 16 Dec. 2009. Web. 20 Mar. 2010. <http://elpais.com/diario/2009/12/16/opinion/1260918011_850215.html>

González Iñárritu, Alejandro. Dir. *Biutiful.* Spain-Mexico. Prod. Menage Atroz, 2010. Film.

———. *Detrás del dinero.* Mexico. Prod. Televisa S. A., 1995. Film.

Gópegui, Belén. *El padre de Blancanieves.* Barcelona: Anagrama, 2007. Print.

Gourevitch, Philip. "Standard Operating Procedure: Philip Gourevitch & Errol Morris, Moderated by Carne Ross." Debate in New York Public Library. 13 May 2008. Web Video. 11 Sept. 2011. <http://www.nypl.org/audiovideo/standard-operating-procedure-philip-gourevitch-errol-morris-moderated-carne-ross>

Gourevitch, Philip, and Errol Morris. *Standard Operating Procedure.* New York: Penguin, 2008. Print.

Goytisolo, Juan. "La fe de erratas." *El País.* 27 Nov. 2004. Web. 3 Oct. 2011. <http://elpais.com/diario/2004/11/27/opinion/1101510008_850215.html>

———. *Las virtudes del pájaro solitario.* Pamplona: Leer-e, 2013. Print.

———. *Reivindicación del conde don Julián.* Madrid: Cátedra, 2003. Print.

———. *Señas de identidad.* 1966. Madrid: Círculo de lectores, 1977. Print.

Grandes, Almudena. "Arte." *El País.* 9 Mar. 2010. Web. 2 Jan. 2011. <http://elpais.com/diario/2010/03/08/ultima/1268002801_850215.html>

Grant, Hugh. Interview by Javier Sampedro. "La información sobre biotecnología debe ser transparente." *El País.* 13 Oct. 2001. Web. 19 Dec. 2012. <http://elpais.com/diario/2001/10/13/sociedad/1002924005_850215.html>

Gutiérrez Aragón, Manuel. Dir. *La vida que te espera.* Spain. Prod. Pancho Casal, 2004. Film.

El Hada. "Exitoso acto taurino en Bilbao." *Revista de Opinión de Bilbao.* 10 Oct. 2013. <http://es.paperblog.com/existoso-acto-antitaurino-en-bilbao-238376>

Haidt, Rebecca. "Gothic Larra." *Decimonónica* 1.1 (2004): 52–63. Print.

Haraway, Donna. "A Game of Cat's Cradle: Science Studies, Feminist Theory, Cultural Studies." Configurations 2.1 (Winter 1994): 59–71. Print.

———. *Modest_Witness@Second_Millennium. FemaleMan_Meets_OncoMouse.* New York: Routledge, 1996. Print.

———. *Simians, Cyborgs and Women: The Reinvention of Nature.* New York: Routledge, 1991. Print.

———. *When Species Meet.* Minneapolis: U of Minnesota P, 2008. Print.

Heidegger. Martin. *What Is Called Thinking?* New York: Harper, 1976. Print.

Heise, Ursula. "Encounters with the Thing Formerly Known as Nature." Rev. of *Rambunctious Garden: Saving Nature in a Post-Wild World,* by Emma Marris. *Public Books.* 9 Sept. 2013. Web. 15 Sept. 2013. <http://publicbooks.org/multigenre/encounters-with-the-thing-formerly-known-as-nature>

———. "Lost Dogs, Last Birds and Listed Species: Cultures of Extinction." *Configurations* 18.1–2 (2010): 49–72. Print.

———. "The Invention of Ecofutures" Ecozon@. 2 Mar. 2012. Web. 6 Feb. 2013. <dspace.uah.es/dspace/.../invention_Heise_ecozona_2012_N2.pdf? . . . 1>

Hernández, María G. "Matador: El deseo transgresivo en Rafael Alberti and Pedro Almodóvar." *Lenguaje y Textos* 18 (2002): 63–67. Print.

Herran, Néstor, and Xavier Roqué. "An Autarkic Science: Physics, Culture, and Power in Franco's Spain." *Historical Studies in the Natural Sciences* 43.2 (2013): 202–35. Print.

Herrero, Jesús Vicente. "El ideario costista en Eugenio Noel." *Anales de la Fundación Joaquín Costa* 20 (2003): 5–24. Print.

Higginbotham, Virginia. *Spanish Film under Franco*. Austin: U of Texas P, 1988. Print.

Holdsworth, Carole A. "Two Failed Scientists: *Tiempo de silencio*'s Pedro and Pynchon's Pointsman." *Revista de Estudios Norteamericanos* 2 (1993): 53–63. Print.

Holling, C. S. "The Resilience of Terrestrial Ecosystems: Local Surprise and Global Change." *Sustainable Development of the Biosphere*. Ed. W. C. Clarke and R. E. Munn. Cambridge: Cambridge UP, 1986. 292–316. Print.

Hopewell, John. *Out of the Past: Spanish Cinema after Franco*. London: British Film Institute, 1986. Print.

Horkheimer, Max. *The Eclipse of Reason*. New York: Continuum, (1947) 2004. Print.

Hunt, Alastair. "Death by Birth." *English Studies in Canada* 39.1 (March 2013): 97-124. Print.

———. "The Rights of the Infinite." *Qui Parle: Critical Humanities and Social Sciences* 19.2 (Spring/Summer 2011): 223–25. Print.

Isegawa, Moses. "Despertar de un sueño: Poder y Fusiles." *La Vanguardia* 7 July 2004: 5. Print.

"IU califica de "corrupta" a la Monarquía y plantea un referéndum sobre la república" *El País*. 10 Sept. 2014. Web 18 Nov. 2014. <http://politica.elpais.com/politica/2014/09/10/actualidad/1410351789_192182.html>

Jarmusch, Jim. Dir. *Ghost Dog*. U.S. Prod. Richard Guay and Jim Jarmush, 1999. Film.

Jasanoff, Sheila. *Designs on Nature: Science and Democracy in Europe and the United States*. Princeton: Princeton UP, 2005. Print.

Jodorowsky, Alejandro. Dir. *Santa sangre*. Mexico. Prod. Claudio Argento, 1989. Film.

Jovellanos, Gaspar Melchor. 1812. "Canto guerrero para los asturianos." *Biblioteca Jovellanos*. Web. 12 Sept. 2009. <www. Jovellanos2011.ws/web/biblio/>

Kafka, Frantz. "Report to the Academy." *Complete Stories*. Ed. Nahum Glatzer. New York: Stockhem Books, 1971: 250–58. Print.

Kahn, Richard. "Reconsidering *Zoë* and *Bios*: A Brief Comment on Nathan Snaza's '(Im)possible Witness' and Kathy Guillermo's 'Response.'" *Journal of Critical Animal Studies* 3.1 (2005): 62–66. Print.

Kant, Immanuel. *Perpetual Peace: A Philosophical Essay*. Trans. W. Hastie. U of Hawaii P, n.d. Web. 17 Mar. 2009. (1795). <http://librivox.org/perpetual-peace-by-immanuel-kant>

Kelly, Dorothy. "Selling Spanish 'Otherness' Since the 1960s." *Contemporary Spanish Cultural Studies*. Ed. Barry Jordan and Rikki Morgan-Tamosunas. London: Oxford UP, 2000. 29–37. Print.

K'hito. "Manolete el mártir expoliado." *Diario meridiano*. Caracas, Venezuela. n.d. Web. 12 Mar. 2004.

Kinder, Marsha. *Blood Cinema: The Reconstruction of National Identity of Spain*. Berkeley: U of California P, 1993. Print.

Kirkpatrick, Susan. *Larra: Laberinto inextricable de un romántico español*. Madrid: Gredos, 1977. Print.

Kitschelt, Herbert P. "Political Opportunity Structures and Political Protest: Anti-Nuclear Movements in Four Democracies." *British Journal of Political Science* 16 (1986): 57–85. Print.

Labanyi, Jo. *Myth and History in the Spanish Contemporary Novel*. Cambridge: Cambridge UP, 2011. Print.

Labrador Méndez, Germán. "Nocilla tiene nombre de hambre. El rock-memoria de Siniestro Total, la crema de cacao y sus fantasmas biopolíticos en la Movida española." *Crítica latinoamericana*. 2013. Web. 7 Jan. 2014. <criticalatinoamericana.com/nocilla-tiene-nombre-de-hambre-el-rock-memoria-de-siniestro-total-la-crema-de-cacao-y-sus-fantasmas-biopoliticos-en-la-movida-espanola/>

Lafora, Antonio. *El trato de animales en España*. Madrid: Oberón, 2004. Print.

Lafuente Ferrari, Enrique. *Goya*. Dublin: Mentor, 1966. Print.

Lakoff, George, and Mark Johnson. *Metaphors We Live By*. U of Chicago P, 2003. Print.

Larra, Mariano José de. "Coridas de toros." 1828. *Artículos*. México: Porrua, 1975: 247–53. Print.

———. "Jardines públicos." 1834. *Artículos*. México: Porrua, 1875: 361–63. Print.

———. "La caza." 1835. *Artículos*. México: Porrua, 1975: 438–42. Print.

———. "La Nochebuena de 1836." 1836a. *Artículos*. México: Porrua, 1975: 451–55. Print.

———. "Día de Difuntos de 1836." 1836b. *Artículos*. México: Porrua, 1975: 216–20. Print.

———. "Los barateros o El desafío y la pena de muerte." 1836c. *Artículos*. México: Porrua, 1975: 209–13. Print.

———. "Necrología. Exequias del conde de Campo-Alangue." 1837a. *Artículos*. México: Porrua, 1975: 227–31. Print.

———. "*Los Amantes de Teruel:* Drama en cinco actos en prosa y verso por D. Juan Eugenio Hartzenbusch." 1837b. *Artículos*. México: Porrua, 1975: 116–21. Print.

Latour, Bruno. "On Actor-Network Theory: A Few Clarifications Plus More Than a Few Complications." *Soziale Welt* 47 (1996): 369–81. Print.

———. *Reassembling the Social: An Introduction to Actor-Network Theory*. Cambridge: Harvard UP, 2005. Print.

———. *We Have Never Been Modern*. Cambridge: Harvard UP, 1993. Print.

Lázaro, José. *Vidas y muertes de Luís Martín-Santos*. Barcelona: Tusquets, 2009. Print.

Lazzarato, Maurizio. "Biopolitics/Bioeconomics: A Politics of Multiplicity." Trans. Arianna Bove and Erik Empson. N.d. Web. 22 Apr. 2009. <http://www.generation-online.org/p/fplazzarato2.htm>

Legendre, Maurice. *Las Hurdes: Estudio de geografía humana*. Trans. Enrique Barcia Mendo. Mérida: Editora Regional de Extramadura, 2006. Print.

Lepage, Corinne. Interview by Victor M. Amela. "Corinne Lepage, eurodiputada contra transgénicos: Los humanos somos las cobayas de cereales transgénicos." *Contra de La Vanguardia*. 26 Aug. 2013. Web. 29 Sept. 2013. <http://www.icariaeditorial.com/contenido/noticia_detallada.php?id=176>

Levinas, Emmanuel. *Quelques réflexions sur le philosophe de l'hitlersme*. Paris. Éditon Payot & Rivages, 1997. Print.

———. *Totality and Infinity*. Pittsburgh: Duquesne UP, 1969. Print.

Lockwood, Jeffrey. "The Moral Standing of Insects and the Ethics of Extinction." *Florida Entomologist* 79 (1987): 70–89. Print.

———"Not to Harm a Fly: Our Ethical Obligations to Insects." *Between the Species* 4.3 (1988): 204–11. Print.

López Arnal, Salvador. "Ciencia y beneficios." *Rebelión*. 3 Dec. 2006. Web. 12 Dec. 2012. <http://www.rebelion.org/noticia.php?id=42314>

———. "Conocimientos científicos y decisiones políticas." *Aporrea*. 17 Dec. 2011. Web. 23 Dec. 2012. <http://www.aporrea.org/ideologia/a135546.html>

López, Isidro, and Emmanuel Rodríguez. "The Spanish Model." *New Left Review* 69 (2011). Web. 31 Jan. 2014. <http://newleftreview.org/II/69/isidro-lopez-emmanuel-rodriguez-the-spanish-model>

López de Uralde, Juan. *El planeta de los estúpidos: Propuestas para salir del estercolero*. Madrid: Planeta, 2010. Print.

López Sintas, Jordi, and Ercilia García Álvarez. "The Consumption of Cultural Products: Analysis of Spanish Social Space." *Journal of Cultural Economics* 26.2 (Feb. 2002):115–38. Print.

Lorca, Federico García. *Poeta en Nueva York*. Tinet. N.d. Web. 2 June 2009. <http://usuaris.tinet.cat/picl/libros/glorca/gl002600.htm>

———. *El teoría y juego del duende*. Biblioteca Virtual Universal. 1932. Web. 12 Oct. 2009. <biblioteca.org.ar/libros/1888.pdf>

Lord, David. "José Bergamín, Heir of Unamuno." *Books Abroad* 15.4 (1941): 407–11. Print.

Lovelock, J. E. *Gaia: A New Look at Life on Earth*. New York: Oxford UP, 1987. Print.

Luján, José Luis, et al. "Spain: Transposing EC Biotechnology Directives through Negotiation." *Science and Public Policy* 23.3 (1996): 181–84. Print.

Lyotard, François. *The Differend: Phrases in Dispute*. Trans. Georges Van Den Abbeele. Minneapolis: U of Minnesota P, 1988. Print.

Machado, Antonio. "Proverbios y cantares de Campos de Castilla" *Rincon castellano*. n.d. Web. 13 Dec. 2013. <http://www.rinconcastellano.com/biblio/sigloxx_98/amachado_prov.html>

Maille, Emilio. Dir. *Manolete, la leyenda*. Spain. Prod. Canal 22, 1997. Film.

Mainer, José Carlos. *Historia de la literatura española (1900–1930)*. Vol. 6. Barcelona: Crítica, 2010. Print.

"Manolete, Doña Angustias y Lupe Sino." *El Mundo de Andalucía*. 13 Feb. 1998. Web. 3 Mar. 2008. <http://www.antonioburgos.com/memorias/1999/02/memo021399.html>

"Manolete, una película desconocida de Abel Gance." *Poesía* 22 (10 Jan. 1985): 2–4. Print.

Marías, Javier. "Los exterminadores de toros." *El País*. 3 Jan. 2010. Web. 2 Oct. 2011. <http://elpais.com/diario/2010/01/03/eps/1262503619_850215.html>

Marsé, Juan. *El Embrujo de Shanghai*. Barcelona: Modnadori, 2003. Print.

Martín-Santos, Luis. "Tauromaquia." *Apólogos y otras prosas inéditas*. Barcelona: Seix Barral, 1970: 54–56. Print.

——. *Tiempo de destrucción*. Barcelona: Seix Baral, 1985. Print.

——. *Tiempo de silencio*. 1962. Barcelona: Crítica, 2000. Print.

——. *Time of Silence*. Trans. George Leeson. New York: Columbia UP, 1989. Print.

Martínez Alíer, Joan. *The Environmentalism of the Poor: A Study of Ecological Conflicts and Valuation*. Northampton: Edward Elgar Pub, 2003. Print.

Martínez Castillo, Róger. "Cultivos y alimentos transgénicos: Una aproximación ecológica." *Revista Biocenosis* 21.1–2 (2008): 27–36. Print.

Marx, Karl. *Social and Political Thought*. Ed. Bob Jessop and Russell Wheatley. London: Routledge, 1999. Print.

Maté, Vidal. "Granjas de cinco estrellas." *El País*. 12 Jan. 2012. Web. 2 Aug. 2012. <http://elpais.com/diario/2012/01/15/negocio/1326636208_850215.html>

Matilla, J. M. "Los Desastres de la guerra de Francisco de Goya. Una mirada independiente." *Nghê Thuât thòi Chiên Tranh*. Hanói: Vietnam Fine Arts Museum; *Arte en tiempos de guerra: Francisco de Goya y Lucientes*. Madrid: España Cooperación Cultural Exterior, (2008): 39–45. Print.

Maturana, Humberto, and Francisco Varela. *The Tree of Knowledge: The Biological Roots of Human Understanding*. Boston: Shambala, 1992. Print.

Maurus, Iohannes. "Indignación y dignidad." E-book. *Indignados*. n.d. Web. 3 June 2012. <www.MandalaEdiciones.com>

Mayorga, Juan. *Animales nocturnos*. Madrid: Al Avispa, 2003. Print.

——. *Palabra de perro: El gordo y el flaco*. Madrid: Teatro de Astillero, 2004. Performance.

——. *La paz perpetua*. Madrid: CDN, 2008a. Print.

——. *La tortuga de Darwin*. Ciudad Real: Ñaque, 2008b. Print.

——. *Las últimas palabras de Copito de Nieve*. Ciudad Real: Ñaque, 2004. Print.

Mbembe, Achille. "Necropolitics." *Biopolitics: A Reader*. Ed. Timothy Campbell and Adam Sitze. Durham: Duke UP, 2013, 161–92. Print.

McClintock, Anne. "Paranoid Empire: Specters from Guantánamo and Abu Ghraib." *Little Axe* 13.1 (2009): 50–74. Print.

Meeker, Joseph. *The Comedy of Survival: Literary Ecology and Play Ethic*. Tucson: U of Arizona P, 1997. Print.

Méndez, Rafael. "Cañete pide a UE que acabe con la moratoria de nuevos transgénicos." *El País*. 26 June, 2012. Web. 13 Feb. 2013. <http://sociedad.elpais.com/sociedad/2012/06/26/actualidad/1340741478_241461.html>

Merchan, Chucho. "Tortura (no es arte ni cultura)." *Es ahora o nunca*. Independiente, 2009. CD.

Merino, José María. *El lugar sin culpa*. Madrid: Alfaguara, 2007. Print.

Meyerhoff, Eli, Elizabeth Johnson, and Bruce Braun. "Time and University." *ACME: An International E-Journal of Critical Geography*. 2012. Web. 12 June, 2014. <www.acme-journal.org/vol10/Meyerhoffetal2011.pdf>

Millás, Juan José. "Verdad palmaria." *El País*. 29 Jan. 2010. Web. 6 Nov. 2012. <http://elpais.com/diario/2010/01/29/ultima/1264719601_850215.html>

Mitchell, Timothy. *Blood Sport: A Social History of Spanish Bullfighting*. Philadelphia: U of Pennsylvania P, 1991. Print.

Moler Okin, Susan. *Is Multiculturalism Bad for Women?* Princeton: Princeton UP, 1999. Print.

"Monsanto renuncia a comercializar trigo transgénico por el rechazo del mercado." *El País*. 24 May 2004. Web. 22 Dec. 2013. <http://sociedad.elpais.com/sociedad/2004/05/11/actualidad/1084226403_850215.html>

Montag, Warren. "Necro-economics: Adam Smith and Death in the Life of the Universal." *Biopolitics: A Reader*. Ed. Timothy Campbell and Adam Sitze. Durham: Duke UP, 2013. 193–214. Print.

Montalban, Eric. "Ecofatiga: Hartos de reciclar" *La Razón*. 11 Oct. 2011. Web. <http://www.larazon.es/8030-ecofatiga-hartos-de-reciclar-TLLA_RAZON_403341>

Moore, Rowan. Sagrada Familia: "Gaudí's cathedral is nearly done, but would he have liked it?" *The Observer*, 23 Apr. 2011. Web. 11 Mar. 2013. <http://www.theguardian.com/artanddesign/2011/apr/24/gaudi-sagrada-familia-rowan-moore>

Montero, Rosa. "Pajaritas." *El País*. 4 May 2004: 6. Print.

"Toro de Tordecillas." *El País*. 13 Sept. 2005: 8. Print.

———. "Párate y mira." *El País*. 2 Feb. 2010. Web. 3 Oct. 2011. <http://elpais.com/diario/2010/02/02/ultima/1265065201_850215.html>

———. *Lágrimas en la lluvia*. Barcelona: Seix Barral, 2011. Print.

———. "No este rey." 16 April, 2012. *El País*. Web. 12 May, 2012. http://elpais.com/elpais/2012/04/16/opinion/1334579118_890534.html

Moore, Kelly, Daniel Lee Kleinman, David Hess, and Scott Frickel. "Science and Neoliberal Globalization: A Political Sociological Approach." *Theory and Society* 40 (2011): 505–32. Print.

Moreno, Vicent. "Breaking the Code: *Generación Nocilla*, New Technologies and the Marketing of Literature." *Hispanic Issues on Line* 9 (2012): n. pag. Web. 13 May 2013. <http://hispanicissues.umn.edu/assets/doc/07_MORENO.pdf>

Morris, Desmond. *El contrato animal*. Buenos Aires: Emecé, 1991. Print.

Morris, Peter. *Guardians*. London: Oberon Books, 2005. Print.

Morton, Timothy. *Ecology without Nature; Rethinking Environmental Aesthetics*. Cambridge: Harvard UP, 2007a. Print.

———. "'Twinkle, Twinkle Little Star' As an Ambient Poem: A Study of a Dialectical Image, with Some Remarks on Coleridge and Wordsworth." *Romanticism and Ecology*. Ed. James McKusick. *Romantic Praxis* 232 (2007b). Web. 23 Oct. 2012. <http://www.rc.umd.edu/praxis/ecology/morton/morton.html>

———. *The Ecological Thought*. Cambridge: Harvard UP, 2010a. Print.

———. "Queer ecology." *PMLA* 125.2 (March 2010b): 173–282. Print.

———. *Hyperobjects: Philosophy and Ecology after the End of the World*. Minneapolis: U of Minnesota P, 2013. Print.

Mosterín, Jesús. "Quién teme a los transgénicos?" *El País*. 22 Mar. 2000. Web. 25 July 2013. http://elpais.com/diario/2000/03/22/futuro/953679617_850215.html

———. *La cultura de la libertad*. Madrid: Espasa-Calpe, 2005. Print.

———. "El hombre transgénico." *Ideal*. 12 Feb. 2009: 1. Print.

———. *A favor de los toros*. Pamplona: Laetoli, 2010. Print.

Mosterín, Jesús, and Jorge Riechmann. *Animales y ciudadanos: Indagación sobre el lugar de los animals en la moral y el derecho de las sociedades industrializadas*. Madrid: Talasa, 1995. Print.

Muñoz Molina, Antonio. "El arte de matar." *El País*. 14 June 2008. Web. 2 Jan. 2010. <http://elpais.com/diario/2008/06/14/babelia/1213400355_850215.html>

Nagel, Thomas. "Ethics without Biology." *Mortal Questions*. New York: Cambridge UP, 1979: 142–46.

Najjar, Abeer. "Othering the Self: Palestinians Narrating the War on Gaza in the Social Media." *Journal of Middle East Media* 6.1 (2010): 1. Print.

"NASA-Funded Study: Industrial Civilisation Headed for 'irreversible collapse'?" *The Guardian*. 14 Mar. 2014. Web. 15 May 2014. <http://www.theguardian.com/environment/earth-insight/2014/mar/14/nasa-civilisation-irreversible-collapse-study-scientists>

Netz, Reviel. *Barb Wire: An Ecology of Modernity*. Middletown: Wesleyan UP, 2004. Print.

Neyrat. Fréderic. "*Oikos* and *Polis*: Ecopolitical Thought and the Avatars of Humanism." U of Wisconsin-Madison. 4 Mar. 2011. Lecture.

Nietzsche, Frederick. *On the Genealogy of Morality*. 1887. *Cambridge Texts in the History of Political Thought*. Cambridge, UK: Cambridge UP, 2006. Print.

Nixon, Robert. *Slow Violence and the Environmentalism of the Poor*. Boston: Harvard UP, 2011. Print.

Noel, Eugenio. *Pan y toros*. Valencia: Sempere, 1913. Print.

———. "Miscelánea taurina." *El Flamenco* 1.12 (Apr. 1914): 11. Print.

———. "El caballo del picador." *Revista veterinaria de España* 10 (1916): 670–73. Print.

———. "Crónica antitaurina" *El liberal*. 18 Aug. 1918. Web. 2 Feb. 2011. <http://sentaditoenlaescalera.blogspot.com/2010/09/cronica-antitaurina-por-eugenio-noel.html>

———. "Crítica de un discurso en 'yo menor.'" *España nervio a nervio*. Madrid: Calpe, 1924: 9. Print.

———. *La novela de un pueblo en capea*. Andújar: Manuel Blanco, 1926. Print.

———. *Diario íntimo*. Vols. 1–2. Madrid: Taurus, 1968. Print.

Nora, Pierre. "Between Memory and History: *Le lieux de memoir*." *Representations* 26 (1989): 7–24. Print.

Norton, Bryan. "Agricultural Development and Environmental Policy: Conceptual Issues." *Agriculture and Human Values* 2.1 (1985): 63-70. Print.

Nugent, Ted, and Shemane Nugent. *Kill It and Grill It: A Guide to Preparing and Cooking Wild Game and Fish*. Washington, DC: Regnery Publishing, 2002. Print.

Nussbaum, Martha. *Frontiers of Justice*. Boston: Harvard UP, 2009. Print.

O'Brien, Susie. "On the Edge of Resilience: Postcolonial Ecologies in Arundhati Roy's *Walking With the Comrades*." Paper Delivered at Global Ecologies: Nature/Narrative/Neoliberalism Conference. UCLA. 8-10 Mar. 2013. Lecture.

———. "Survival Strategies for Global Times." *Interventions: International Journal of Postcolonial Studies* 9.1 (Mar. 2007): 83-98. Print.

Okin, Susan. *Is Multiculturalism Bad for Women?* Princeton, NJ: Princeton UP, 1999. Print.

Ollie. Comment on "El personaje del mes 'Profesor' Carlos Illera." *Foro Vegetariano*. 23 July 2009, 3:10. Web. 14 July 2011. <http://www.forovegetariano.org/foro/archive/index.php/t-19885.html>

Ortega, Andrés. "Pensarse herbívoros, vivir como carnívoros," *El País*. 27 Dec. 2004. Web. 1 Jan. 2011. <http://elpais.com/diario/2004/12/27/internacional/1104102012_850215.html>

Ortega Cerdá, Miguel, ed. *La deuda ecológica española; Impactos ecológicos y sociales de la economía española en el extranjero*. Sevilla: U of Sevilla, 2012. Print.

Ortega y Gasset, José. *La caza y los toros*. 1944. Madrid: Espasa-Calpe, 1962. Print.

———. *España invertebrada*. 1921. Madrid: Alianza Editorial, 1981 Print.

———. *La rebelión de las masas*. 1937. Madrid: Espasa Calpe, 2006. Print.

Ovejero, Felix, Pablo de Lora, and José Luis Martí. "De toros y argumentos." *El País*. 19 Aug. 2010. Web. 6 Apr. 2012. <http://elpais.com/diario/2010/08/19/opinion/1282168812_850215.html>

Quist, David, and Ignacio H. Chapela. "Transgenic DNA Introgressed into Traditional Maze in Oaxaca, Mexico." *Nature* 414 (2001): 541-43. Print.

Pardo Bazán, Emilia. "Un destripador de antaño." *Un destripador de antaño y otros relatos*. Madrid: Alfaguara, 1994: 7-44. Print.

———. "La ganadera." (1910) *Cuentos completos*. Ciudad Seva. N.d. Web. 12 June 2011 <http://www.ciudadseva.com/textos/cuentos/esp/pardo/cuentos_de_la_tierra.htm>

———. *Los pazos del Ulloa*. 1886. Madrid: Cátedra, 1989. Print.

———. "Reflexiones científicas sobre el darwinismo." *Ciencia cristiana* 4 (1877): 289-98, 481-93; 5 (1877): 218-33, 393-410, 481-95. Print.

Pasamar, Gonzalo. *Apologia and Criticism: Historians and the History of Spain, 1500-2000*. New York: Peter Lang, 2010. Print. Hispanic Studies: Culture and Ideas 30.

Pasolini, Pier Paolo. Dir. *Salo, or the 120 Days of Sodom*. Italy. Prod. Alberto Grimaldi, 1975. Film.

Patterson, Charles. *Eternal Treblinka: Our Treatment of Animals and the Holocaust*. New York: Lantern Books, 2002. Print.

Patteson, Joseph. "Ricarda and Bodily Disruption in *Tiempo de Silencio*." Unpublished manuscript.

Pavone, Vincenzo. "Ciencia, neoliberalismo y bioeconomía." *Revista Iberoamericana de Ciencia, Tecnología y Sociedad* 7.20 (2012): 145–61. Print.

Proyecto Gran Simio. N.d. Web. 2 Sept. 2014. <http://www.proyectogransimio.org/noticias/noticias-destacadas/pensar-acerca-del-pensar-no-esta-limitado-a-los-humanos>

Pérez de Ayala, Ramón. *Política y toros*. Madrid: Renacimiento, 1918. Print.

Pérez Magallón, Jesús. "El proyecto acosado: El fracaso en *Tiempo de silencio* de Luis Martín-Santos." *Revista Hispánica Moderna* 47.1 (1994): 134–45. Print.

Pérez Sánchez, Alfonso E., and Julian Gallego. *Goya: The Complete Etchings and Lithographs*. New York: Prestel, 1997. Print.

Pérez Vaquero, Carlos. "La comunidad de los iguales." *Noticias jurídicas*. Oct. 2007. Web. 7 Nov. 2013. <http://noticias.juridicas.com/articulos/30-Derecho-Medioambiental/200710-658842745684.html>

Perucci, Tony. "*What the Fack is That*: Poetics of Ruptural Performance." *Liminalities: Journal of Performance Studies* 3.5 (Sept. 2009): 1–18. Print.

Petit, Quino. "Savater saca a la plaza la ética de la tauromaquia." *El País*. 18 Sept. 2010 <http://elpais.com/diario/2010/09/18/cultura/1284760806_850215.html>

Pico dela Mirandolla, Gioviani. *Oration on the Dignity of Man*. Madison: Gateway, 1996. Print.

Pink, Sarah. *Women and Bullfighting: Gender, Sex and the Consumption of the Tradition*. Oxford: Berg, 1997. Print.

Plumwood, Val. *Environmental Culture: The Ecological Crisis of Reason*. New York: Routledge, 2002. Print.

"Polls: Most Spaniards do not like bullfighting" *USA Today*. 1 Aug. 2010. Web. 23 Dec. 2012. <http://usatoday30.usatoday.com/news/world/2010-08-01-spain-bullfighting-ban_N.htm>

Porcel, Baltasar. "La fiesta nacional?" *La Vanguardia* 19 Apr. 2004: 3. Print.

Porretas. "La de los toros" *No tenemos solución*. Prod. RCA Records, Rosendo Mercado, 1995. CD.

Prada, Juan Manuel de. "En defensa de los toros." *ABC*. 23 Nov. 2009. Web. 3 Dec. 2009. <http://www.abc.es/20091123/opinion-firmas/defensa-toros-20091123.html>

Prado, Ángeles. *La literatura del casticismo: Eugenio Noel, Gutiérrez Solana y Camilo José Cela*. Madrid: Moneda y Crédito, 1973. Print.

Preston, Paul. *Spanish Holocaust: Inquisition and Extermination in Twentieth-Century Spain*. New York: Harper and Collins, 2012. Print.

Previtali-Morrow, Giovanni. "Unos aspectos biográficos de Cervántes en *El coloquio de los perros*." AIH. Actas IV (1971) Centro Virtual Cervantes, n.d. Web. 25 May 2009. <http://cvc.cervantes.es/literatura/aih/pdf/04/aih_04_2_041.pdf>

Proyecto Gran Simio. N.d. Web. 30 Oct, 2011. <http://proyectogransimio.org/>

Puleo, Alicia. *Ecofeminismo para otro mundo possible*. Valencia: Cátedra, 2011. Print.

———. "Ser humano y naturaleza en la era del Antropoceno" *La Nación*. 5 June 2013. Web. 8 July 2013. <http://www.nacion.com/opinion/foros/humano-naturaleza-Antropoceno_0_1345865614.html>

Rancière, Jacques. *The Politics of Aesthetics*. Trans. Gabriel Rockhill. New York: Continum, 2007. Print.

Razack, Sherene. "How Is White Supremacy Embodied? Sexualized Racial Violence at Abu Ghraib." *Canadian Journal of Women and the Law* 17.2 (2005): 341–63. Print.

"El Rey se disculpa: Lo siento mucho, me he equivocado. No volverá a ocurrir." *20 Minutos*. 18 Apr. 2012. Web. 30 May 2012. <http://www.20minutos.es/noticia/1378098/0/rey/disculpa/caceria/>

Reincidentes. "Grana y oro." *Te lo dije*. BMG Ariola/RCA, 1997. LP.

Reno Renardo. "Torturadores." *Reno Renardo*. Estudio, 2007. LP.

Rey, Alfonso. *Luis Martín-Santos*. Madrid: Editorial Crítica, 2005. Print.

Riechmann, Jorge. *Un mundo vulnerable*. Madrid: Los libros de Catarata, 2001. Print.

———. *Qué son los alimentos transgénicos?* 2002. Barcelona: RBA, 2011b. Print.

———. *Trilogía de autocontención: Gente que no quiere viajar al Marte*. Madrid: Libros de Catarata, 2004a. Print.

———. *Transgénicos: el haz y el envés*. Madrid: Catarata, 2004b. Print.

———. *Todos los animales somos hermanos: Ensayos sobre el lugar de los animales en las sociedades industrializadas*. Madrid: Catarata, 2005. Print.

———. *Poemas lisiados*. Madrid: La Oveja Roja, 2012a. Print.

———. *El socialismo sólo puede llegar en bicicleta*. Madrid: Catarata, 2012b. Print.

Robinson, Andy. "El Pentagono admite cierta responsabilidad de Donald Rumsfeld en el escándalo de Abu Ghraib." *El País*. 25 Aug. 2004: 2. Print.

Rodríguez-Luis, Julio. "Autorepresentación en Cervantes y el sentido de *El coloquio de los perros*." *Cervantes: Bulletin of the Cervantes Society of America* 17.2 (1997): 25–58. Print.

Rosenberg, Alfred. *Blut unde Ehre*. Munchen: Eher, 1935. Print.

Rossellini, Roberto. Dir. *Voyage to Italy*. Italy. Prod. Adolfo Fossarato, 1955. Film.

Rouff, Jeffrey. "An Ethnographic Surrealist Film: Luis Buñuel's *Land without Bread*." *Visual Anthropology Review* 14.1 (1998): 45–57. Print.

Ruiz-Castell, Pedro. "Scientific Instruments for Education in Early Twentieth-Century Spain." *Annals of Science* 65.4 (2008): 519–27. Print.

Sade, Marquis de. *The 120 Days of Sodom*. Radford: Wilder Publications, 2009. Print.

Sagal, Boris. Dir. *The Omega Man*. U.S. Prod. Walter Seltzer, 1971. Film.

Said, Edward. *Representations of the Intellectual*. New York: Vintage Books, 1996 (1993). Print.

Sampedro, Javier. "Curar células no es diseñar bebés." *El País*. 28 June 2013a. Web. 23 Oct. 2014. <http://sociedad.elpais.com/sociedad/2013/06/28/actualidad/1372450241_329276.htm>

Sánchez Mejías, Ignacio. "La Tauromaquia." 1930. *Taurología.com*. n.d. Web. 13 Sept. 2013. <http://www.taurologia.com/articulo_imprimir.asp?idarticulo=1193&accion=>

Sano, Yeb. "'Stop This Madness,' Tearful Filipino Pleads At Climate Talks." 11 Nov. 2013. Web. 20 Nov 2013 <www.npr.org/.../stop-this-madness-tearful-filipino-pleads-at-climate-...>

Sánchez Reboredo, José. "Con ocasión del centenario de Noel: Relectura de *España nervio a nervio*." *Cuadernos Hispanoamericanos* 426 (Nov. 1985): 155–63. Print.

Santesmases, María Jesus, and Emilio Muñoz. "Scientific Organizations in Spain (1950–1970): Social Isolation and International Legitimation of Biochemists and Molecular Biologists on the Periphery." *Social Studies of Science* 27.2 (1997): 187–219. Print.

Sasson, Albert. Interview by Victor M. Amela. "Agrobiólogo, experto en alimentos transgénicos Albert Sasson: Los ecologistas pedirán transgénicos." *La Vanguadia* 9 July 2003: 9. Print.

Saura, Carlos. *La caza*. Spain. Prod. Elias Querejeta, 1965. Film.

Savater, Fernando. "Las cornadas de Europa." *El País*. 22 Dec. 1991a. Web. 23 Jan. 2010a. <http://elpais.com/diario/1991/12/22/opinion/693356402_850215.html>

———. "Zoología moral." *El País*. 10 Sept. 1991b. Web. 23 Jan. 2010a. <http://elpais.com/diario/1991/09/10/opinion/684453608_850215.html>

———. "Lo inaceptable". *El País*. 14 Jan. 2008. Web. 23 Oct 2013. <http://elpais.com/diario/2008/01/14/opinion/1200265204_850215.html>

———. "Rebelión en la granja." *El País*. 16 Mar. 2010. Web. 23 Jan. 2010b. <http://elpais.com/diario/2010/03/16/opinion/1268694011_850215.html>

———. *Tauroética*. Madrid: Turpial, 2011. Print.

Savater, Fernando, and Gonzalo Martínez-Fresneda. *Teoría y presencia de la tortura en España*. Barcelona: Anagrama, 1982. Print.

Schaffner, Franklin J. Dir. *Planet of Apes*, U.S.: Prod. APJAC, 1968. Film.

Schurlknight, Donald E. *Power and Dissent: Larra and Democracy in Nineteenth-Century Spain*. Lewisburg: Bucknell UP, 2009. Print.

Séralini, Gilles-Eric. Interview with Ima Sanchis. "Entrevista a un experto en transgénicos de la UE: Los transgénicos son tóxicos para la salud humana." *Semilla y salud*. 8 Apr. 2009. Web. 1 Dec 2010. <http://semillasysalud.wordpress.com/entrevista-dr-gilles-eric-seralini-experto-de-la-comision-europea-en-transgenicos/>

Sheenan, P., and G. Tegart. *Working for the Future: Technology and Employment in the Global Knowledge Economy*. Victoria University Press for the Centre for Strategic Economic Studies, 1998. Print.

Shklovsky, Viktor. "Art as Technique." In *Russian Formalist Critique: Four Essays*. 3–24. Trans. Lee T. Lemon and Marion J. Reis. Lincoln: U of Nebraska P, 1965.

Shweder, Richard. "The Astonishment of Anthropology." *Thinking through Cultures: Expeditions in Cultural Psychology*. Cambridge: Harvard UP, 1991: 1–26. Print.

———. "The Moral Challenge in Cultural Migration." In *Anthropology and Contemporary Immigration*. Ed. Nancy Foner. School for American Research Press (SAR). Santa Fe, Mexico: 2003. Print.

———. "'Circumcision' or 'Mutilation': And Other Questions About a Rite in Africa." Tierneylab Science Blog, *New York Times* (Science Section). 5 Dec. 2007. Web. 9 Oct. 2011. <http://tierneylab.blogs.nytimes.com/2007/12/05/circumcision-or-mutilation-and-other-questions-about-a-rite-in-africa/?scp=1-b&sq=shweder&st=nyt>

Simons, John. *Animal Rights and the Politics of Literary Representation*. New York: Palgrave, 2002. Print.

Singer, Peter. *Animal Liberation*. New York: Harper Collins, 1975. Print.

Siniestro Total. "Pueblos del mundo, extinguíos!" *Ante todo mucha calma*. Vigo, Spain: Prod. Siniestro Total, 1982a. Disc.

———. "Purduy." *Ayudando a los enfermos*. Vigo, Spain: Prod. Siniestro Total, 1982b. Disc.

Ska-p. "Vergüenza." *Planeta escoria*. RCA, 2000. CD.

Sloterdijk, Peter. "The Immunological Transformation: On the Way to Thin-Walled 'Societies.'" *Biopolitics: A Reader*. Ed. Timothy Campbell and Adam Sitze. Durham: Duke UP, 2013: 310–16. Print.

Smith, Adam. *The Theory of Moral Sentiments*. Edinburg: A. Kincaid and J. Bell, 1859. Print.

———. *The Wealth of Nations*. London: W. Strahan and T. Caddell, 1976. Print.

Smith, Paul Julian. "Broken Embraces." *Sight and Sound*. 2009. Print.

Snaza, Nathan. "(Im)possible Witness: Viewing PETA's 'Holocaust on Your Plate.'" *Animal Liberation Philosophy and Policy Journal* 2.1 (2004): 1–20. Print.

Sociedad Alcohólica. "Motxalo!" *Directo*. Prod. Mil A Gritos Records. Ángel Katarain y Sociedad Alcólica, 1999. CD.

Spanish Bioindustry Association (ASEBIO). "ASEBIO Report: 2012 News and Trends from the Spanish Biotech Sector." ASEBIO. June 2013. Web. 15 Jan. 2014. <http://www.asebio.com/en/documents/ASEBIOREPORT2012_final.pdf>

Spivak, Gayatri. "Can the Subaltern Speak?" *Marxism and the Interpretation of Culture*. Ed. Cary Nelson and Lawrence Grossberg. Chicago: U of Illinois P, 1988: 271–316. Print.

Stanton, Edward, F. *The Tragic Myth: Lorca and cante hondo*. Lexington: UP of Kentucky, 1978. Print.

Steiner, Gary. *Anthropocentrism and Its Discontents: The Moral Status of Animals in the History of Western Philosophy*. U of Pittsburgh P, 2005. Print.

Steinfeld, Henning, et al. 2006. *Livestock's long shadow*. Rome: FAO, 2006. Web. 5 Feb. 2012. <http://www.fao.org/docrep/010/a0701e/a0701e00.HTM>

———. 2010. *Livestock in a Changing Landscape, Volume 1: Drivers, Consequences and Responses*. Washington, DC: Scientific Committee on the Problems of Environment. Print.

Stone, Rob. "Animals Were Harmed during the Making of This Film: A Cruel Reality of Hispanic Cinema." *Studies in Hispanic Cinemas, Intellect* 1.2 (2004): 75–84. Print.

Strauss, Frederic, ed. *Almodóvar on Almodóvar*. London: Faber & Faber, 1998. Print.

Sunder Rajan, Kaushik. *Biocapital: The Constitution of Postgenomic Life*. Durham: Duke UP, 2006. Print.

Taylor, Diana. *The Archive and the Repertoire: Performing Cultural Memory in the Americas*. Durham: Duke UP, 2003. Print.

Tyler Bonner, Charles. *Randomness in Evolution*. NY: Princeton UP, 2013. Print.

Technopolis. "Evaluation of ICREA 2001–2011." Mar. 2011. Web. 24 Jan. 2014. <http://www.icrea.cat/Web/SectionViewer.aspx?section=231>

Theweleit, Klaus, et al. *Male Fantasies, Vol. 1: Women, Floods, Bodies, History*. Minneapolis: U of Minnesota P, 1987. Print.

Thompson, Paul B. *The Spirit of the Soil: Agriculture and Environmental Ethics*. London: Routledge, 1995. Print.

Tierno Galván, Enrique. *Los toros: El acontecimiento nacional*. Madrid: Taurus, 1988. Print.

Tilly, Charles. "War Making and State Making as an Organized Crime." *Bringing the State Back In*. Ed. Peter Evans, Dietrich Rueschemeyer, and Theda Skocpol. Cambridge: Cambridge UP, 1985: 169–90. Print.

Todt, Oliver. "Social Decision Making on Technology and the Environment in Spain." *Technology in Society* 21 (1999): 201–16. Print.

Todt, Oliver, and José Luis Luján. "Spain: Commercialization Drives Public Debate and Precaution." *Journal of Risk Research* 3.3 (2000): 237–45. Print.

Tokarczuk, Olga. *Moment Niedzwiedzia*. Warsaw: Wydawnictwo Krytyki Politycznej, 2012. Print.

Toledano, Ruth. "Dos tristes tigresas." *El País*. 19 Dec. 2005. Web. 25 May 2011. <http://elpais.com/diario/2005/12/09/madrid/1134131067_850215.html>

Torrecilla, Jesús. *España exótica: La formación de la imagen española moderna*. Boulder: Society of Spanish and Spanish-American Studies, 2004. Print.

Tosko, Katherine. *The Bull and the Ban: Exploring the Catalan Ban on Bullfighting*. Bexhill-on-Sea, East Sussex: Independent Publishing Platform (Suerte Publishing), 2012. Print.

Tyler Bonner, John. *Randomness in Evolution*. Princeton: Princeton UP, 2013. Print.

Unamuno, Miguel. *En torno al casticismo*. Madrid: Cátedra, 2005. Print.

———. "Sobre la europeización." 1906. *La España Moderna* 18.216: 64–83. Web. 12 Dec. 2012. <http://www.filosofia.org/hem/dep/lem/n216p064.htm>

———. "Las Hurdes." 1914. *Andanzas españolas*. Madrid: Alianza Editorial, 2006b: 107–24. Print.

———. *Don Manuel Bueno, Mártir*. 1933. Madrid: Cátedra, 2006a. Print.

"Un mito del toreo, condenado a la gloria y a la tragedia: A 50 años de la muerte del célebre torero español." *La Nación*, 29 Aug. 1997. Web. 16 Oct. 2013. <http://www.lanacion.com.ar/75840-un-mito-del-toreo-condenado-a-la-gloria-y-a-la-tragedia>

Urbano, Pilar. Interview by Ángel Mellado. "Para Suárez estaba claro que el alma de F-23 era el Rey." *El Mundo*. 30 Mar. 2014. Web. 31 Oct. 2014. <http://www.elmundo.es/cronica/2014/03/29/53369a7ae2704e2e078b456e.html>

Urla, Jacqueline. "Women and Bullfighting: Gender, Sex and the Consumption of Tradition by Sarah Pink" *American Anthropologist* 101.2 (June, 1999): 459–60. Print.

Vallejo, Alfonso. *El escuchador de hielo*. Madrid: AAT, 2008. Print.

Vicent, Manuel. *Antitauromaquia*. Madrid: Santillana, 2001. Print.

Viestenz, William. "Sins of the Flesh: Bullfighting as a Model of Power." *Iberian Modalities*. Ed. Joan Ramon Resina. Liverpool: Liverpool UP, 2013. 143–61. Print.

"Violentos, torturadores, inmorales . . . impunidad insultadora de los 'antis.'" *Mundotoro*. 3 Mar. 2010. Web. 14 Apr. 2010. <http://www.mundotoro.com/noticia/violentos-torturadores-inmorales-impunidad-insultadora-de-los-antis/77370>

Vivas, Esther. "Transgénicos: 'Spain Is Different.'"*El Público.es*. 22 Feb. 2014. Web. 24 Feb. 2014. <http://blogs.publico.es/esther-vivas/2014/02/22/transgenicos-spain-is-different/>

Viúdez, Juana. "Las grandes ONG abandonan la cumbre del clima de Varsovia." *El País*. 21 Nov. 2013. Web. 9 Oct. 2011. <http://elpais.com/tag/cumbre_del_clima/a/>

Weisman, Alan. *The World without Us*. New York: Picador, 2007. Print.

Wieviorka, Michel. "El mal." *La Vanguardia* 14 May 2004: 1. Print.

William, Nigel. "EU Challenged on GM Crops." *Current Biology* 21.2 (2011): R53-54. Print.

Williams, Linda. *Figures of Desire: A Theory and Analysis of Surrealist Film*. Urbana: U of Illinois P, 1981. Print.

Wolfe, Cary. *What Is Posthumanism?* Minneapolis: U of Minnesota P, 2010. Print.

Wolff, Francis. *Cincuenta razones para la defensa de los toros*. Córdoba: Almuzara, 2011. Print.

"World's First Lab-Grown Burger Is Eaten in London." *BBC News: Science and Environment*. N.d. Web. 5 Oct. 2014. <http://www.bbc.co.uk/news/science-environment-23576143?print=true>

Žižek, Slavoj. "Between Two Deaths: The Culture of Torture." *London Review of Books*. 3 June 2004. Web. 3 June 2010. <http://www.lacan.com/zizektorture.htm>

———. "Radical Evil as Freudian Category." *Lacanian Inc*. 38 (2010a). Web. 14 July 2011. <www.lacan.com/zizlovevigilantes.html>

———. "The Future as Sci Fi: A New Cold War." *Lacan.com*. 2010b. Web. 23 June 2014. <http://www.egs.edu/faculty/slavoj-zizek/articles/the-future-as-sci-fi/>

INDEX

Abolitionism, 94, 170, 183, 185

Abortion, 21, 108, 109, 114, 119

Abu Ghraib, 5, 6, **187–96, 203–4, 208–9**

Activism, 45, 82, 85, 86, 101, 102, 105, 127, 128, 141, 161, 163, 175, 176, 179, 184, 189, 190, 266, 268

ADDA, xi, **162–64,** 166, 167, 170, 179, 189

"¡Adiós Cordera!" **135–37**

Agamben, Giorgio, 9, 16, 57, 100, 129, 132, 133, 155, 188, 194, 203, 266

L' age d'or (Buñuel's), **140–43**

Alas, Leopoldo (Clarín), 4, 51, 131, **135–37,** 158, 266, 267

Alaska, 166, **175–79**

Alberti, Rafael, 8, 41, 42

Almódovar, Pedro, 17–18, 41, 42, 52, **63–68,** 76, 78, 96, 178

Alternative biopolitics, xi, 6, 8, 10, 17, 32, 34, 41, 82, 104, 114, 128, 149, 158, 160, 178, 185, 251, 263, **268,** 270, 271

Alternative energies / renewable energies, 35, 247, 248, 271

Animal killing, 12–17, 23–24; and climate change, 29

Animal rights / animal rights movement, **24–29,** 131–32, 148–49, 152, 240, 256, 268. *See also* animal welfare, anticruelty movement / animal defense movement

Animal welfare, 131, 163, 170, 175

Animal studies, xii, xvii, 16, 41, 46, 101, 131, 161, 275

Animalization/animalized 5, 16–17, 29, **50–63,** 93, 107, **129–35, 157–58,** 187, 193, 194, 268; in *Tiempo de silencio,* 117–20; in Pardo Bazán, **131–35;** in Baroja, **138–40;** in Buñuel, **140–43,** 158; in *La Paz Perpetua,* **202–10;** in *Biutiful,* 219; of the king, 266

Anthropocene, xii, 8–9, 32, 213, 215, 216, 219, 225, 226, 234

Anthropocentrism, 16–17, 101, 131, 150

Anthropological machine, 16, 100, 130, 132, 252

Anthropomorphism, 5, 61, 100, 101, 129, 130–32, 149; in Clarín, **135–37;** in

297

Baroja, 139; in Mayorga, 150–157, 158. *See also* Humanization

Anti-cruelty movement / animal defense movement, 1, 11, 23–24, 158, 160, 161, **162–86**, 166. *See also* anti-bullfighting, animal welfare, animal rights

Anti-bullfighting, 1, 5, 7, 8, **17–24**, 43, 44, 45, 46, 47, **78–105**, 110, 111, **159–186**, 189, 190, 192, 197, 198, 210

Apes, 7, 17, **24–29**, 129, 149, 150, 152, 166

Araujo, Joaquín, xv, 38, 271

Arendt, Hannah, 16, 27–29, 218

Austerity, 214, 242, 265

Autopoesis, 250

Barcelona, xi, xvi, 5, 7, 22, 152, 153, 162, 166, 168, 169, 181, 189, 193, 195, 196, 197, **215–26**, 251, 254

Baroja, Pío, **138–40**, 142, 157

Beast/beastiality, 28, 46, 51, 71, 76, 100, 118, 132, 133, 135, 139, 140, 158, 194, 202

Bennett, Jane, 9, 219

Bergamín, José, 18, **39–40**, 41, 61

Berger, Pablo, 4, 52, **68–77**, 78

Biocapital, 236, 247

Bioeconomy, 7, 30, 220, 236, in Spain, **240–50**, 254, 256, **258–61**

Biomedicine/biomedical, 108, 236, 238, 241, 242, 246, 248

Biopolitics, 1–4, 6–10, 15–16, 129–30, 137, 189, 209, 213, 215, 216, 219, 232; alternative, xi, 6, 8, 10, 17, 32, 34, 41, 82, 104, 114, 128, 149, 158, 160, 178, 185, 251, 263, **268**, 270, 271; and Hispanism, 41; biopolitical turn, 8; and economy, 22–23; and animal rights, 24–29; and citizenship, 28; bullfighting biopolitics, **49–77**, 82, 160, 239, 240, 265, 266; in Noel, 92–94; in *Tiempo de silencio*, 106–110, 124, 127, 128, 148, 150; Nazi, 201; in *Lágrimas en la lluvia*, 236; in bioeconomy, 247

Bios, 3, 28, 187, 219, 270

Biotechnology, 7, 10, 12, 236, **237–50**, **258–61**

Biutiful, 6, 12, 212, **215–26**, 261, 269

Blancanieves, 52, **68–77**

Body, 6, 106–108, 111–12, 113–114, 122, 123, 126, 181–85, 216, 218, 219, 223, 226–27, 229, 230, 231, 232, 249, 252. *See also* Flesh

Bullfighting, 3, 7, 17, 50, 185, 252, 254, 265; bullfighting ban, 19, **166–67**, **168–69**; and multiculturalism, 20; and tradition, 20–21, 164, 165; and democracy, 21; and environment, 22; and economy, 22–23; and science, 30, 240; and Hispanism 38–47; in Lorca, 38–39, as animalization **50–63**; in *Matador*, 63–68; in *Blancanieves*, 68–77; comic, 69, 73; in Ortega, 79–80; bullfighting metaphors in *Tiempo de silencio*, **107–117**; in Forges, 160; and War on Terror, **187–210**

Buñuel, Luis, 4, **49–50**, 52, **53–55**, **61–63**, 64, 72, 107, 131, **140–43**, 164, 176, 266, 267

Caciquismo, 86

Cáncer, 30, 54, 55, 108, 109, 120, 122, 123, 124–26, 216, 218, 223, 225, 228, 229, 246, 269

Capeas, 58

Capitalism, 7, 30, 35, 37, 133, 134, 135, 137, 154, 163, 223, 231, 248, 254

Care, 17, 25, 71, 72, 119, 121, 125, 128, 129, 130, 136, 249, 251, 268

Casal, Paula, xvi, 20, 21, 25, 26, 27, 37–38, 149, 150, 159, 162, 164, 166, 200n8, 249–50, 256

Cases, Manel and Carmen, xi, 162, 195, 277

Castro, Américo, 40–41, 42, 238, 256, 266

Catalan: Parliament on bullfighting ban, 19, 21, 22, 24; separatism/autonomy, 45, 46, 80, 88, 167, 168, 186; nationalism, 166; culture, 167; capitalism, 223; knowledge economy 241

Catastrophe/catastrophic, 9, 32, 34, 35, 36, 93, 112, 113, 192, 213, 218, 224, 226–27, 229, 230, 237, 250, 261, 265, 269

"La caza" (Larra's), **96–97**

La Caza (Saura's), 70, 131, **143–48**, 157, 264

Cernuda, Luis, 8, 39, 41

Citizenship/ citizen, 4, 16, 18, 21, 27, 28, 43, 46, 50, 57, 58, 120, 136, 143, 152, 153, 155, 157, 168, 177, 182, 186, 194, 208, 209, 218, 238, 239, 243, 244, 267, 268, 270

City, 5, 13, 14, 18, 58, 142, 153, 154, 155, 161, 163, 183–84, 189, 193, 197; in *Biutiful*, 216, 218, 221, 222, 225; in *Tiempo de silencio* 107, 108, 109, 119, 122; in *Nocilla*, 226, 228, 229, 233; in *Lágrimas en la lluvia*, 235, 251, 267

Civilization, 4, 12, 16, 18, 28, 29, 32, 40, 42, 50, 51, 52, 54, 55, 63, 95, 132, 136, 140, 142, 151, 170, 186, 189, 200, 206, 209, 212, 213, 216, 224, 225, 227, 229, 234, 239, 248, 250, 257, 266, 267

Clark, Nigel, 9

Class, 18, 19, 48, 51, 53, 57, 58, 72, 79, 80, 81, 90, 105, 106, 107, 108, 111, 116, 118, 119, 120, 121, 122, 125, 126, 127, 133, 134n2, 135, 137, 140, 143, 146, 157, 160, 162, **179–81**, 234, 235, 254, 267

Climate change, 4, 32–34, 182, 235, 236, 262, 269, 270; and animal rights, 29

Coexistence, 6, 230, 231, 255

El coloquio de los perros, 151, 154

Complexity, 77, 133, 150, 229

Connections, xvii, 5, 6, 10, 88, 89, 103, 109, 185, 187, 192, 218, 229, 230, 241, 242, 252

Consumption/consumerism, xvi, 12, 16, 29, 30, 31, 32, 36, 37, 95, 98, 137, 169, 175, 181, 183, 225, 229, 231, 232, 233, 234, 246, 256, 271.

Copito de Nieve, 7, **151–54**, 202

Corcuera, José Luis, 19

"Corridas de toros" (Larra), **89–90**

Cortina, Adela, 24, 27, 150, 168

Costa, Joaquín, 86, 91, 98, 255

Crisis, 9, 12, 22, 26, 30, 32, 25, 36, 39, 40, 41, 42, 81, 104, 128, 149, 186, 187, 211, 212, 213, 214, 216, 226, 237, 263, 265, 268, 269, 270

Cruelty/anti-cruelty, 1, 5, 11, 14, 22, 24, 25n13, 46, 53, 60, 61, 62, 63, 69, 71, 71n13, 76, 88, 93, 96, 99, 101, 104, 118, 121, 122, 127, 144, 147, 156, 158, 160, 161, 162, 164, 165, 166, 171, 176, 177, 178, 179, 181, 186, 191, 193, 195, 196, 197, 198, 200, 209, 240, 266, 267

Dark ecology, 213, 216

Darwin, Charles/Darwinism 130, 132, 133, 134, 138, 148, 201

Dawkins, Richard, 234, 250

Death, 3, 7, 9, 13, 15, 17, 19, 21, 22, 23, 24, 25, 30, 34, 38, 40–43, 45, 46, 50, 51, 52, 55, 58–63, 238, 256, 265; in *Matador*, 63–68; in *Blancanieves*, 68–77; death in life in Larra, 84–85, 267; in Noel, 85, 94, 102; in *Tiempo de silencio*, 110–20, 124, 125, 137; in *La caza*, 143, 144; in animal rights' debate, 150, 159, 167, 168; in Mayorga, 153, 171–73, 204; in *Biutiful*, 213, 218, 223, 224, 225; in *Nocilla*, 227, 229, 230; in *Lágrimas en la lluvia*, 250, 251, 269. *See also* Killing

Degrowth, 32, 37

Democracy, 18, 27, 31, 53, 61, 68, 162, 168, 198, **202–10**, 214, 218, 232, 239, 242, 243, 251, 258, 260, 265, 270

Derrida, Jacques, 11, 16, 187, 188, 210

Desastres de la guerra (Goya's), **59–60**

Disability, 140, 145

Discrimination, xvii, 3, 4, 28, 61, 170, 209, 236

Disquiet/disquieting realism, 11, 212, 214, 216, 226, 227, 229, 252

Dog, 14n7, 20, 70, 117, 119- 122, 151, **154–57**, 192, 193, **202–10**, 232, 241, 250n2, 268

Don Quijote, 39, 127

Economy, 4, 7, 10, 13, 14, 18, 22, 29, 30, 32, 35, 36, 37, 43, 74, 78, 80, 85, 98, 107,110, 120, 121, 126, 129, 134, 135, 149, 181, 185, 186, 188, 213, 214, 215, 218, 219, 220, 222, 223, 225, 227, 230, 233, 234; bioeconomy, 235–61, 262, 263, 267, 270, 271

Education, 57, 74, 79, 103, 111, 163, 165, 169, 200, **204–5,** 208, 241, 255, 259, 267

Enlightenment, 7, 14, 53, 54, 84, 197, 238, 253, 254, 255, 256, 259

Environment, environmentalism, environmental studies, xii, xvii, 4, 5, 6, 7, 8, 9, 11, 12, 22, 29, **32–38**, 41, 46, 63, 119, 123, 124, 125, 131, 158, 160, 163, 171, 181–83, 185, 186, 212, 213, 214, 215, 218, 227, 228, 243, 244, 247, 248, 250, 252, 256, 267, 269, 270, 271

Epigenetics, 125

Esposito, Roberto, 3, 26, 28–29, 106, 107, 187, 210, 219, 262, 270

Ethics, xvii, 4, 8, 11, 12, 14, 19, 20, 21, 22, 23, 26, 28, 29, 37, 41, 44, 77, 81, 96, 100, 101, 102, 104, 105, 118, 119, 122, 127, 128, 130, 132, 135, 136, 137, 138, 139, 142, 149, 150, 157, 170, 183, 184, 187–210, 229, 231, 253

Exceptions, 20, 53, 61, 130, 189, 194, 198, 202, 206, 208, 266

Experiments on animals, 7, 27, 30, 31, 107, 109, 120, 121, 166, 171–3

Externalities, 37, 222, 223

Extinction, 12, 12n4, 15, 16, 22, 23, 171, 211, 212, 213, 214, 226, 269

Farm factories, 12–14, 23–25, 38, 171–73

Fascism, 143, 145, 154, 157, 257 *Se also* Nazism

Feminism, 10, 177, 186, 240

Ferdinand VII, 21, 49, 51–52, 55, 57, 197

Fernández Cubas, Cristina, 12, 168

Fernández Mallo, Agustín, 6, 12, 35, 212, **226–34**

Flamenquismo (in Noel), 85, 90, 95, 104

Flesh, 1, 46, 59, 62, 68, 72, 91, 93, 99, 101, 102, 107, 108, 111–12, 114, 118, 147, 181–85, 268

Forges, **159–60,** 161

Foucault, Michel, 9–10, 50, 59, 63, 76, 115, 130, 146, 157, 188, 219–20

Franco, Francisco/ Francoist 18, 19, 21, 39, 44, 45, 64, 80, 109, 125, 128, 144, 167, 175, 186, 193, 238, 239, 256, 257, 258, 259, 264, 265

Freedom, 12, 21, **50–51,** 53, 54, 55, 95, 115, 124, 151, 153–54, 168, 178, 214, 229, 233, 252, 267, 269

Freud, Sigmund, 131

"La ganadera" (Pardo Bazán's), **133–35,** 142

Garbage, 122, 171, 174, 175, 176, 179, 221–22

García Lorca, Federico (Lorca), 7, 13–14, 17, 18, **38–39**, 41, 42, 50, 51, 52, **60–62,** 74, 75, 78, 97, 104, 238, 266, 267

Garrido, Francisco, 21, 149, 166

Gender violence/domestic violence, 88, 96, 113, 123, 125, **175–79,** 195, 267

Genes, Genetic modification of organisms,

7, 10, **29-32,** 223, in *Tiempo de silencio,* 116, **120-26,** 234, **235-61**

González Iñárritu, Alejandro, **215-26**

Gópegui, Belén, 72, 248

Goya, Francisco, 49, 52, 54, 57, **59-60,** 62, 98, 109, 115-16

Goytisolo, Juan, 46, 87, 88, 95, 97, 98, 104, 119, 160, 192

Grandes, Almudena, 20, 21, 39, 168, 280

Green capitalism, 163, 248

Great Apes Project, xvi, 25, 149

Guantanamo, 152, 189, 191, 193, 194, 198, 199

Guerra, Alfonso, 19

Haeckel, Ernst, 130, 133, 134, 135, 142

Haraway, Donna, 6, 9, **10-11,** 16, 82, 106, 129, 162, 252

Health, 30, 31, 32, 37, 40, 42, 72, 85, 106-8, 123-26, 138, 244, 246, 247, 249, 256, 270, 271

Hispanism, 7, **38-47,** 86, 102, 104, 105, 169n9, 255

Human/humanity, xi, xii, xvii, xviii, 1, 3, 4, 5, 6, 7, 8, 9, 10, 11,12, 14, 15, 15, 17, 18, 19, 21, 24, 25, 26, 27, 28, 29, 30, 31, 32, 33, 37, 38, 40, 41, 42, 46n25, 49, 51, 53, 54, 55, 58, 59, 61, 62, 67, 69, 70, 71, 76, 77, 93, 94, 96, 99, 100, 101, 106, 107, 108, 110, **117-28, 129-58, 170-75, 181-85,** 189, **190-210,** 226, 227, 228, 229, 230, 231, 233, 234, 235, 236, 237, 245, 246, 247, 248, 249, 250, 251, 252, 253, 254, 255, 256, 261, 262, 264, 266, 267, 268, 269, 270

Humanities, xii, 8, 9, 33, 46, 122, 123, 125

Humanization, 129, 131, 132, 136, 142, 145, 193. *See also* Anthropomorphism

Holocaust, 13, 28, 130, 143, 146, 187, 280, 290, 187

Horta, Oscar, 21, 162, 170, 171

Human-animal divide, 4, 26, 107; in *Tiempo de silencio,* **118-20, 129-30;** in Mayorga, **151-158, 209**

Humanities, 8-9, 33, 124, 125, 213, 252

Hunting, 22, 70, 89, in Larra **96-97,** in Ortega, 104-05, in Pardo Bazán, 134, in Baroja, 139, in *La caza,* **143-48,** 157, by Spanish King Juan Carlos, **262-65**

Las Hurdes/ Tierra sin pan, 52, 53, **61-63,** 64, 141,164

Hyperobjects, 214, 224

Hypocrisy, 80, 128, 151, 198, 206, 208, 209, 263, 265, 266

ICREA, 240-41

Immigration, 19, 151,154, 158, 159, 218, 219, 220, 221, 222, 223, 225, 230, 293

Immunity, immune system, auto-immune, 4, 30, 108, 123, 187, 188, 210. *See also* cancer

Indignados, 36-37, 260

Intellectuals, 78, 81; (accommodated), 79-80; (marginal), **82-105**

Insects, 15, 131, 148, 212, 213, 228, 244

Jovellanos, Gaspar Melchor, 59, 95, 195, 197, 283

Juego y teoría del duende, **60-62**

Justice, xvii, 22, 32, 36, 42, 44, 71, 72, 73, 75, 103, 113, 116, 117, 119, 121, 134, 136, 137, 138, 139, 140, 205, 214, 222, 223, 232, 252, 253, 260, 269, 279

Killing, 3, 13, 15, 24, 29, 53, 57, 58, 60, 63, 64, 66, 68, 69, 70, 71, 72, 89, 96, 100, 109, 114, 115, 118, 121, 127,130, 135, 137, 157, 158, 161, 171-73, 189, 213, 226, 227, 251, 252, in *La caza,* 143-48. *See also* Death

King Juan Carlos, **262-65**

Knowledge economy, 237, 241, 257. *See also* bioeconomy

Laboratory, 30, 107, 109, 120, 123, 124, 243, 250, 253, 267. *See also* experiments on animals

Lafora, Antonio, 14, 23, 24

Lágimas en la lluvia, 6, 12, **235-237**, 261, 269

Larra, José Mariano, 4, 5, 50, 55, 78, **82-85**, 86, **89-90**, 93, **95-97**, **99-100**, 101, 266, 267

Latour, Bruno, 9, **10**, 120, 123

Law, xvi, 14, 17, 19, 21, 25, 26, 27n15, 35, 57, 134, 138, 139, 142, 144, 153, 162, 165, 166, 167, 169, 185, 187, 188, 189, 197, 199, 210, 214, 219, 236, 237, 244, 266

Life, **8-17**, 124, 126, 157, 185, 187, 188, 189, 212, 218, 219, 222, 223, 224, 226, 227, 236, 240, 247, 248, 249, 250, 251, 252, 253, 255, 256, 259, 261, 262, 268, 269, 270, 271

Limits, 11, 12, 37n20, 108, 119, 122, 204, 213, 213, 228, 247, 248, 249, 255, 256, 269

López de Uralde, Juan, 35, 166, 247

Lora, Pablo de, xvi, 21, 22-23, 165, 200n8

Love, xvii, 7, 65, 66, 69, 71, 72, 73, 74, 82, 83, 84, 85, 87, 88, 96, 97, 100, 110, 114, 115, 126, 153, 158, 162, 163, 175, 178, 205, 213, 231, 240, 251, 267, 271

Manolete, **64-66**, 74, 175, 193

Martínez Alíer, Joan, 32

Martín-Santos, Luis, 4, 5, 30, 46, 51, 66, 80, 88, 96, 104, **107-128**, 142, 200, 266

Masculinity, 7, 18, 21, 51, 53, 82, 90-91, 166, 177, 240; in Larra y Noel, **95-99**; in *Tiempo de silencio*, 114-15; in *La caza*, 144-148; (and the King's hunt), 265

Matador, 52, **63-68**, 74, 76, 78

Materialism, 183, 218

Matter, 9, 148, 216, 218, 223, 224, 227

Mayorga, Juan, 4, 6, 12, **150-57**, 158, 190, **202-10**

Mbembe, Achille, 17, 29, 41, 50, 160, 240, 265

Meat, 12-14, 24, 71

Media, 4, 7, 19, 20, 36, 135, 166, 169, 170, 174, 175, 182, 183,184, 187, 195, 209, 238, 244-46, 249, 253, 263, 269

Memory, 25, 71, 145, 150, 169, 215, 225, 237, 251, 252, 261

Metaphors, 5, 16, 17, 63, 104, **107-17**, 156, 176, 177, 203, 225, 251, 252, 267

Millás, Juan José, 21, 166, 189, **198-200**

Mitchell, Timothy, 18

Money, 6, 13n5, 22, 23, 25, 36, 57, 68, 69, 85, 93, 96, 98, 114, 146, 147, 163, 205, 215, 216, 219-23, 224-25, 230, 267

Montag, Warren, 22, 29, 134,135, 227

Montero, Rosa, 4, 6, 12, 21, 34, 166, 177, 189, 194, 197, **235-37**, 261, 263, 265, 268, 269

Morbid, 51, 57, 64, 66, 69, 74, 75, 76, 127, 216, 218

Morton, Timothy, 9, **11-12**, 34, 131, 213, 228

Mosterín, Jesús, 5-6, 19, 20, 21, 22-23, 25, 149, 165, 197, 248-49

Multiculturalism, 19, 20, 165

Multinational corporations, 7, 31, 216, 239, 240, 242, 247, 248, 256

Muñoz Molina, Antonio, 21, 166, 167, 197

Myxomatosis, **145-46**

Nagel, Thomas, **209-10**

National archive, 75, 160, 175, 176, 181-84, 185

National identity / national culture / nation building / nationalism, 7, **43**, 44, 47, 57–58, 59, 61, (taste) 78, 79, 80, 82, 88, 124, 166, 167, 168, 175, 182–83, 186, 254, 264, 265, 268

Nature-culture (Latour), natureculture (Haraway), 10, 28, 219, 228

Nazism/Nazi, 106, 130, 146, 188, 201. *See also* Fascism

Necro-economics, 22, 29, 43, 74, 78, 135, 227, 240

Necrophilia, 70, 75

Necropoetics, 38–39, 43

Necropolitics, 17, 29, 41, 50, 61, 74, 78, 160, 175, 189, 240, 265, 266; in Noel, 93. *See also* Thanatopolitics

Neoliberalism/neoliberal, **188, 242, 247, 256**

Nietzsche, Fredrick, 107, 130, 131, 137–38, 139, 148

Nocilla experience, 6, 35, 212, **226–34**, 261, 269

Noel, Eugenio, 4, 5, 55–56, 78, **82–83, 85–89, 90–95, 97–105**, 110, 113, 136, 266, 267

Nunca Mais, 35

Ortega y Gasset, José, **78–80**, 82, 85, 95, 104–5, 109, 116, 124, 126

Otherness, xvii, 4, 11, 44, 101, 131, 158, 164, 193, 260, 283

Pain, 14–15, 24 72, **99–104**, 114, 118, 119, 120, 122, 127, 148, 150, 166, 170, 177, 179, 183, 190, 191, 193, 201, 204, 268, 269

Palabra de perro, 151, **154–57**

Pardo Bazán, Emilia, 51, 131, **132–35**, 140, 141, 142, 157

La Paz Perpetua, 151, **202–10**

Pérez de Ayala, Ramón, 98n12

Le phântome de la liberté, **49–50,** 52, **53–55,** 56

Podemos, 271

Population, 23, 29, 32, 45, 50, 51, 57, 63, 108, 188

Posthumanism/posthuman, 142, 224, 226, 252, 269

Poverty/poor, 29, 30, 51–52, **62–63**, 181, **256**, 267; in *Blancanieves* 68–75; in *Las Hurdes*, 76; in *Tiempo de silencio*, 106–8; as deprivation, 116, 120–23, 125–28; in Pardo Bazán, 134–35; in *La caza*, 145, 188; in *Biutiful*, 223; in Unamuno, 256

Precautionary approach, 7, 34, 237, 244, **246–48**, 253–54

Prestige's disaster, 34–35, 226

Prison, 25, 54, 64, 91, 94, 95, 109, 114, 116, 117, 126, 136, 149, 152, 153, 154, 158, 166, 168, 187, 188, 191, 192, 193, 194, 195, 196, **202–10,** 230, 261, 266

Profit, 23, 32, 35, 213, 219, 220, 222n11, 223, 240, 242, 245, 260, 271

Progress, xvi, 19, 31, 44, 53, 54, 94, 105, 121, 135, 136, 212

Proyecto Gran Simio, 25–27

Public (in bullfighting) in Manolete and Matador, 63–68; in *Blancanieves*, 73; in *Tiempo de silencio*, 115, 126, 161; in *La Paz perpetua*, 203

Quadrillage, 108

Race/racism, xvii, 14, 27, 85, 93, 125, 143, 149, 154, 173, 188, 202, 203, 204, 209, 236, 267

Randomness, 229, 230, 231, 233

Rape, 69, 109, 116, 119, 159

Rebellion, 43, 49, 50, 54, 55, 57, 83, 84, 89, 119, 135, 153–54, 157

Regenerationism, 255, 256

Resilience, 108, 119, 125

Resistance, xvi, xvii, 84, 157, 208, 228, 233, 234, 253, 261

Resources, xvii, 13 n5, 30, 32, 36, 37, 95, 98, 182, 204, 213, 214, 223, 236, 242, 247, 256, 262, 267, 270

Riechmann, Jorge, xv, xvi, xvii, 21, 25, 26, 27, 30, 33, 35, 37, 101, 130, 162, 165, 185, 237, 239, 245, 247, 249, 255, 256, 260, 271

Rights, human and animal, xvi, xvii, xvii, 5, 7, 14n7, 15, 18, 29, **24–29,** 45, 57, 58, 81, 101, 103, 131, 148, 149, 150, 152, 155, **159- 86,** 189, 191, 194n4, 195, 197, 200, 200n8, 201, 205, 218, 219, 236, 240, 256, 268, 269, 271; and the environment, 26; and difference, 28, 271

Risk, 30, 31, 50, 76, 79, 91, 95, 105, 187, 205, 237, 239–40, 243, 244, 246, 247, 254, 259, 260, 266

Ruptural performance, 5, 161, **181–84**

Sacrifice, 4, 5, 20, 57, 60, 93, 94–95, 111–13, 119, 133, 135, 159, 164,175, 187, 189, 200, 206

Sadism/ sadomasochism, 74, 76, 188, 190, 191, 198, 200, 203

Sánchez Mejías, Ignacio, 58

Saura, Carlos, 70, **143–48,** 157, 264, 266

Savater, Fernando, 5–6, 15, 18, 22, 29, 168, 185–86, 190, 200–201, 209, 263, 267

Scape goat, 111–12, 115

Science, 6, 8, 9, 10, 13n5, 15n9, 16, **29–32,** 55; in *Tiempo de silencio,* 106, 107, 109, **120- 27,** 141; in *Biutiful,* 213, 223; in *Nocilla,* 227, 228, 229, **235–61**

Sentience, 170n11

Sex: in Foucault 63; in *Matador* 64–66; in *Blancanieves,* 69–71; in *Tiempo de silencio,* 107, 108, 109, 114–116, 188

Sexual dimorphism, 249

Shantytown, 109, 114, 115. *See also* slums and poverty

Singer, Peter, 99, 201

Siniestro total, 211–13, 215, 226, 232, 234

Situated knowledge, 10, 82, 162

Ska-p, 190, 198

Skin, 147–48, 181–82, 230

Slaughterhouses, 13–14, 23–24, 137, 189, 227, 228, 267

Sloterdijk, Peter, 195, 268

Slums, 30, 108, 110, 123, 125. *See also* shantytown

Smith, Adam, 22

"Spain is different," 88, 238, 239, 255, 256, 257, 259

Species, 212, 236, 262, 269

Speciesism, 99, 171–74, 326

Sustainability, 3, 21, 26, 169, 170, 227, 247, 250, 271

Synthetic biology, 4, 236, 237, 250, 251, 260. *See also* Genetic modification of organisms

Tauromaquia (Goya's), 59–60

Technology/technoscience, 6, 7, 10, 12, 15, 30, 32, 35, 40, 54, 59, 120, 136, 137, 188, 220, 223, 225, 228, 234, **235–61**

Terrorism, 20n11, 187, 189, 192, 195, 198, 199, 201, 202, 204, 205, 207

Thanatopolitics, 3. *See also* Necropolitics

Tiempo de silencio, 30, **107–28,** 148, 200n7, 247

Tierno, Galván, Enrique, 18, 61

Time, 25, 34, 72, 144, 148, 212, 215, 224, 225, 230, 233, 250, 271

Toledano, Ruth, 166, 189, 197

Toro embolao/ bou embolat, 164, 165, 166, 167

Torture, xvii, 4, 5, 7, 20n11, 23, 24, 25, 27, 29, 53, 71, 101, 122, 149, 150, 151, 152, 158, 159, 164, 168, **187–210**, 218, 252

Tradition, 19–20, 43–44, 58, 60, 159, 160, 164, 165, 167, 185, 190, 191; in *Blancanieves*, 68–77, 86

Transition, 162, 242, 258, 259

Transition Town Movement, 270–71

Las últimas palabras de Copito de Nieve, **151–54**

Unamuno, Miguel de, 39, 42–43, 60, 78, 85, 238, 239, 255, 256, 257, 266

Vegetarianism, xi, 14, 166, 170, 289

Vicent, Manuel, 19, 21, 22–23, 110–11, 177, 197

Violence, xvii, 5 , 17, 41, 43, 46, 50, 52, 53, 58, 60, 68, 69, 70, 72, 73, 75, 76, 88, 89, 91, 94, 96, 97, 98, 99, 101, 102, 111, 138, 141, 143, 145, 176, 177, 251, 265, 267; structural, 110, 116; symbolic, 110; gender, 113, 123, 125, 145; misdirected 113; slow, 218; state, 58, 88, 94, 142,185; institutionalized, 186; domestic, 195, 250

Vulnerability, xvii, 1, 11, 16, 33, 75, 76, 107, 144, 149, 193, 220, 256, 269

War, xvii, 4, 5, 6, 17, 20n11, 24, 29, 39, 40, 43, 46, 49, 51, 52, 53, 54, 57, 58n4, **59–60**, 61, 64, 70, 80, 81, 82n4, 88, 89, 90–92, 94–95, 97, 111, 130, 137, 138, 140, 141, 195, 263, 267, 268; antiwar, 186; in *La caza*, 143–48, 175; War on Terror, **187–210**; in *Lágrimas en la lluvia*, 236

Werewolf, 51, 117, 118

Wolfe, Cary, xvii, 16

Wolff, Francis, 21, 22, 24

Žižek, Slavoj, 187, 188, 206n10, 269

Zoé, 28

TRANSOCEANIC STUDIES
Ileana Rodríguez, Series Editor

The Transoceanic Studies series rests on the assumption of a one-world system. This system—simultaneously modern and colonial and now postmodern and postcolonial (global)—profoundly restructured the world, displaced the Mediterranean *mare nostrum* as a center of power and knowledge, and constructed dis-centered, transoceanic, waterways that reached across the world. The vast imaginary undergirding this system was Eurocentric in nature and intent. Europe was viewed as the sole culture-producing center. But Eurocentrism, theorized as the "coloniality of power" and "of knowledge," was contested from its inception, generating a rich, enormous, alternate corpus. In disputing Eurocentrism, books in this series will acknowledge above all the contributions coming from other areas of the world, colonial and postcolonial, without which neither the aspirations to universalism put forth by the Enlightenment nor those of globalization promoted by postmodernism will be fulfilled.

In Search of an Alternative Biopolitics: Anti-Bullfighting, Animality, and the Environment in Contemporary Spain
KATARZYNA OLGA BEILIN

Neoliberal Bonds: Undoing Memory in Chilean Art and Literature
FERNANDO A. BLANCO

Prophetic Visions of the Past: Pan-Caribbean Representations of the Haitian Revolution
VÍCTOR FIGUEROA

Transatlantic Correspondence: Modernity, Epistolarity, and Literature in Spain and Spanish America, 1898–1992
JOSÉ LUIS VENEGAS

Conflict Bodies: The Politics of Rape Representation in the Francophone Imaginary
RÉGINE MICHELLE JEAN-CHARLES

National Consciousness and Literary Cosmopolitics: Postcolonial Literature in a Global Moment
WEIHSIN GUI

Writing AIDS: (Re)Conceptualizing the Individual and Social Body in Spanish American Literature
JODIE PARYS

Learning to Unlearn: Decolonial Reflections from Eurasia and the Americas
MADINA V. TLOSTANOVA AND WALTER D. MIGNOLO

Oriental Shadows: The Presence of the East in Early American Literature
JIM EGAN

www.ingramcontent.com/pod-product-compliance
Lightning Source LLC
Chambersburg PA
CBHW021754230426
43669CB00006B/75